Fungal Plant Pathogens: Applied Techniques, 2nd Edition

FSC
www.fsc.org
MIX
Paper | Supporting
responsible forestry
FSC® C022174

Fungal Plant Pathogens:
Applied Techniques, 2nd Edition

_____

Edited by

Charles R. Lane

Fera Science Ltd, UK

Paul A. Beales

The Animal and Plant Health Agency, UK

Kelvin J.D. Hughes

The Animal and Plant Health Agency, UK

CABI is a trading name of CAB International

CABI	CABI
Nosworthy Way	200 Portland Street
Wallingford	Boston
Oxfordshire OX10 8DE	MA 02114
UK	USA

Tel: +44 (0)1491 832111
E-mail: info@cabi.org
Website: www.cabi.org

Tel: +1 (617)682-9015
E-mail: cabi-nao@cabi.org

A catalogue record for this book is available from the British Library, London, UK.

ISBN-13: 9781800620551 (paperback)
          9781800620568 (ePDF)
          9781800620575 (ePub)

DOI: 10.1079/9781800620575.0000

Commissioning Editor: Rebecca Stubbs
Editorial Assistant: Emma McCann
Production Editor: Shankari Wilford

Typeset by SPi, Pondicherry, India
Printed and bound in the UK by Severn, Gloucester

Contents

Contributors

Kinda Alraiss, Fera Science Ltd, Sand Hutton, York, YO41 1LZ, UK. E-mail: kinda.alraiss@fera.co.uk

Rachel Barker, Department for Environment, Food and Rural Affairs, Sand Hutton, York, YO41 1LZ, UK. E-mail: rachel.barker@defra.gov.uk

Victoria Barton, Fera Science Ltd, Sand Hutton, York, YO41 1LZ, UK. E-mail: victoria.barton@fera.co.uk

Paul A. Beales, The Animal and Plant Health Agency, Sand Hutton, York, YO41 1LZ, UK. E-mail: paul.beales@apha.gov.uk

Lucy Carson-Taylor, The Animal and Plant Health Agency, Sand Hutton, York, YO41 1LZ, UK. E-mail: lucy.carson-taylor@apha.gov.uk

David Cooke, The James Hutton Institute, Invergowrie, Dundee, DD2 5DA, Scotland, UK. E-mail: david.cooke@hutton.ac.uk

Ashleigh Elliott, Fera Science Ltd, Sand Hutton, York, YO41 1LZ, UK. E-mail: ashleigh.elliott@fera.co.uk

David Galsworthy, The Animal and Plant Health Agency, Sand Hutton, York, YO41 1LZ, UK. E-mail: david.galsworthy@apha.gov.uk

Steve Hendry, Forest Research, Northern Research Station, Roslin, Midlothian, EH25 9SY, UK. E-mail: steven.hendry@forestresearch.gov.uk

Kelvin J.D. Hughes, The Animal and Plant Health Agency, Sand Hutton, York, YO41 1LZ, UK. E-mail: kelvin.hughes@apha.gov.uk

Anthony Kermode, CABI Bioservices, Bakeham Lane, Egham, Surrey, TW20 9TY, UK. E-mail: a.kermode@cabi.org

Paul Kirk, Royal Botanic Gardens, Kew, Richmond, London, TW9 3AE, UK. E-mail: P.Kirk@kew.org

Charles R. Lane, Fera Science Ltd, Sand Hutton, York, YO41 1LZ, UK. E-mail: charles.lane@fera.co.uk

Michael Long, Fera Science Ltd, Sand Hutton, York, YO41 1LZ, UK. E-mail: michael.long@fera.co.uk

Aiga Ozolina, Fera Science Ltd, Sand Hutton, York, YO41 1LZ, UK. E-mail: aiga.ozolina@fera.co.uk

Ana Pérez-Sierra, Forest Research, Alice Holt Lodge, Farnham, Surrey, GU10 4LH, UK. E-mail: ana.perez-sierra@forestresearch.gov.uk

Belinda Phillipson, Department for Environment, Food and Rural Affairs, Horizon House, Deanery Road, Bristol, BS1 5AH, UK. E-mail: belinda.phillipson@defra.gov.uk

Matthew J. Ryan, CABI Bioservices, Bakeham Lane, Egham, Surrey, TW20 9TY, UK. E-mail: m.ryan@cabi.org

David Smith, Director Biological Resources, Julius Kühn-Institut (JKI), Bundesforschungsinstitut für Kulturpflanzen, Bundesallee 50, D-38116 Braunschweig, Germany. E-mail: d.smith@cabi.org

Christopher Thornton, Biosciences, Hatherly Building, University of Exeter, Prince of Wales Road, Exeter, EX4 4PS, UK. E-mail: C.R.Thornton@exeter.ac.uk

Jenny Tomlinson, Fera Science Ltd, Sand Hutton, York, YO41 1LZ, UK. E-mail: jenny.tomlinson@fera.co.uk

Joan Webber, Forest Research, Alice Holt Lodge, Farnham, Surrey, GU10 4LH, UK. E-mail: joan.webber@forestresearch.gov.uk

Foreword

I have had the pleasure of knowing the editors of this book, Charles, Paul and Kelvin, for over 20 years. Combined, they have possibly close to 100 years of knowledge in plant pathology, fungal diagnostics, and plant and seed health. This second edition of *Fungal Plant Pathogens* brings that wealth of expertise, along with the expertise of other contributors, to cover the very fundamental techniques involved in investigating fungal plant disease. The book includes useful new information on the highly topical subjects of sampling for tree pathogens, quality assurance, biosecurity and plant health engagement.

When I started my career as a plant pathologist back in the early 1990s, fungal detection methods were generally based on simple traditional methods, such as plating onto agar media, light microscopy and *in vitro* experimentation. At that time, fluorescent microscopy, the use of the green fluorescent protein and β-glucuronidase (GUS) transformation were becoming more popular methods to study fungal growth *in planta*. Alongside this, PCR was gaining real momentum for detecting plant pathogens and so began the rapid advancement of molecular plant pathology. Nearly 30 years on, I realise just how far we have come from the early 1990s. We are now able to understand complex interactions surrounding pathogenesis using a combination of powerful methods such as genomics, computational biology and bioinformatics.

Having worked in diagnostics and advisory for many years, I have seen the exponential growth of omics technologies (i.e. genomics, transcriptomics, metabolomics, proteomics) and bioinformatics to support this area of work. These are powerful and versatile tools that have revolutionized plant pathology and diagnostics and support many facets of these areas of work, including population and evolutionary studies, unravelling microbiomes and complex disease syndromes. In addition, they help us to understand the molecular mechanisms of disease resistance and susceptibility and allow us to assess how environmental conditions or cropping practices affect disease development and plant immunity. Outcomes from the knowledge generated from these technologies can hopefully lead to more environmentally sustainable plant products and practices in the future.

Alongside powerful omics and bioinformatics technology, applied plant pathology still plays an important role in providing biological context and practical support for the grower, farmer or forester. Symptom recognition, early pathogen detection systems, integrated pest management and disease sensing and forecasting are all essential in helping to respond to disease outbreaks and to understand environmental drivers for disease development and spread. Alongside this, early and accurate diagnosis of a disease problem can save a lot of time and money, and importantly can prevent taking a wrong corrective measure that could be costly and damaging. This book has a unique focus on some of the practical elements of plant pathology and the fundamental techniques applied for accurate diagnosis from symptom recognition, sample collection and examination of plant material through to pathogen isolation, identification and detection from a range of matrices. The diagnostic techniques covered range from traditional microbiological methods through to more complex serological and molecular methods. The mix of topics covered provides a solid foundation that anyone working in plant pathology or studying this subject area will find highly informative.

The inclusion of a chapter on quality assurance highlights the increasing awareness of the importance of quality management systems in plant health diagnostic laboratories and particularly in those that play a regulatory role in biosecurity. Internationally, there is an increasing demand for laboratories to have a quality management system to ensure that diagnostic results meet internationally accepted standards of quality, thereby providing assurance to their customers or clients of the performance, competency and quality of the test results produced. Setting up and maintaining a quality management system such as ISO17025 can be costly and complex. However, the overall outcome for a laboratory, once the system is established, is usually positive in providing that laboratory with a high level of credibility in relation to test results and in relation to biosecurity decisions that need to be made based on those test results.

The chapters on sampling for tree pathogens, biosecurity and plant health engagement are all highly topical. In the last 15–20 years, we have seen an upsurge in the numbers of invasive pests and pathogens moving globally, due largely to increased global trade in plants. The establishment of non-native pests and pathogens and the resulting damage is often exacerbated by climate change, which affects the behaviour of both the hosts and the pests or pathogens on them. Trees, woodlands and forests are being significantly harmed by several invasive pathogen species. In the UK, a Dutch elm disease (*Ophiostoma novo-ulmi*) epidemic in the 1970s resulted in the death of most mature English elms. More recently, ash dieback (*Hymenoscyphus fraxineus*) and various *Phytophthora* species have been causing harm in many new territories. Together with highly damaging bark beetles, such as the emerald ash borer, the Asian longhorn beetle and the bronze birch borer, this poses a serious threat to trees globally. As well as having a direct effect on trees, woodlands and forests, pests and pathogens can also have an indirect but broad impact on ecosystems and species that depend on the trees for their survival.

Tree health and tree pathology is complex due to the longevity and size of trees and the number of biotic and abiotic, including environmental, conditions that can affect them. The number of tree health specialists has been declining, partly due to lack of forest pathology training and higher education programmes. However, tree health specialists are often the first responders to pests and disease outbreaks and are usually the initial investigators of new and emerging pests and diseases. The need for these specialist skills is more apparent now than ever before. For this reason, the inclusion of a chapter on the fundamentals of sampling, processing and isolating tree pathogens is an important step in facilitating the exchange of specialist tree health knowledge. This may go some way to encourage others to develop the technical skills and capability needed to help support our trees, woodland and forests into the future.

Plant health regulations and quarantine are important aspects of plant health and diagnostic work for any organizations dealing with biosecurity pests and pathogens. It is essential that anyone working in this area has a good knowledge of plant health regulations governing the movement and handling of quarantine pests and pathogens. Government plant health regulations can be complex and often difficult to navigate. The inclusion of key aspects on the regulatory requirements for handling quarantine pathogens in this book provides access to this information in an easy to understand and digestible form.

The final chapter of this book provides useful insights and advice on developing the principles of an effective plant health engagement strategy. This chapter takes you step-by-step through the process of running a successful plant health campaign, from setting the purpose and objectives through to evaluating the success of the campaign. It is a welcome addition to the book as the need to broaden plant health messaging to plant industries and the wider public is becoming increasingly important. With the unprecedented pressures plant industries and the environment are facing from invasive pests and diseases, there needs to be collective responsibility to safeguard our plants, trees and environment. The positive benefits that plants and the environment have on our mental health have been realised during the COVID-19 pandemic. Plant pathologists and biosecurity specialists need to work closely with the government to contribute to the national conversation around plant health and biosecurity. Now is the time to leverage the current positive perception of nature to raise awareness of plant health issues as this is the first step in changing practitioner and public behaviour.

This book is relevant to plant pathology in all sectors – agriculture, horticulture and forestry. Its practical focus is a great resource for anyone interested in investigating fungal plant diseases and extending their knowledge and skills into the increasingly important topics of quality assurance, quarantine, plant health regulations and plant health engagement.

Dr Lisa Ward
Head of Pathology, Forest Research

Acknowledgements

———————————

The editors wish to thank the hard work of the chapter authors in completing their manuscripts to such a high standard and to deadline. We also wish to thank the Department of Environment Food and Rural Affairs (Defra), the Animal and Plant Health Agency (APHA) and Fera Science Ltd for helping us with time and resources to complete this book. This book would not have been possible without the support of our considerably talented colleagues in mycology and seed health at Fera for providing information about the latest techniques and sharing their knowledge generously. We also wish to thank former colleagues, in particular, Dr Roger Cook, for imparting their wealth of knowledge and passion about plant pathology to us all. We would also like to thank David Crossley for his excellent photographic work and provision of images.

1 Introduction to Fungal Plant Pathogens

Charles R. Lane[1]* and Paul Kirk[2]

[1]*Fera Science Ltd, York, UK;* [2]*Royal Botanic Gardens, Kew, London, UK*

1.1 Introduction

Fungal plant pathogens play an important role in all our lives as they can threaten food security, economic prosperity and natural environments. As far back as biblical times, references have been made to the impact of such pathogens on animal and human health, as in the case of ergot poisoning (due to sclerotia of *Claviceps purpurea*), which causes a condition known as St Anthony's fire that was recorded in mediaeval times (Cooke, 1977). However, it was not until progress made in the 18th and 19th centuries that diseases such as potato late blight (caused by the fungus-like organism *Phytophthora infestans*) started to be understood (Kamoun, 2009). Currently, it is estimated that plant diseases result in an average of about 5–10% loss in food production in developing countries (but up to 40% in some countries) (Strange and Scott, 2005; Kirk *et al.*, 2008). In more recent times, we have seen increasing popular awareness and concern about the ability of fungal plant pathogens to damage natural environments, as exemplified by diseases such as ash dieback. This fungal disease, caused by *Hymenoscyphus fraxineus*, was first identified as causing dieback of *Fraxinus* species in Poland (Kowalski, 2006), although later was discovered in Asia where it has co-evolved to be a relatively harmless disease (Nielsen *et al.*, 2017).

The disease has spread unchecked across Europe. Since arriving in the UK in 2012 it has spread widely, leading to extensive damage to the natural environment (Hill *et al.*, 2019).

In this book, for ease of reference and to save repetition for the reader, organisms from kingdom Fungi (true fungi) and fungal-like organisms from kingdoms Chromista and Protozoa (see Section 1.2) are collectively referred to as 'fungi'.

Looking forward, there are several challenges that will alter our perceptions about fungal plant pathogens and their impacts.

Advances in molecular biology

The use of molecular techniques is not only challenging fungal taxonomic concepts but also enhancing our ability to detect and identify these pathogens. The evolution of faster, cheaper and simpler methods, such as sequencing and real-time PCR (polymerase chain reaction), have had a huge impact in the past few decades on the diagnosis of plant pathogens (Henson and French, 1993). New molecular techniques like next generation sequencing are now becoming increasingly used and relied on by mycologists in providing even more information on molecular diversity and gene function (Quirino *et al.*, 2010;

* E-mail: charles.lane@fera.co.uk

DOI: 10.1079/9781800620575.0001

Jain *et al.*, 2021). The challenge now is to harness this knowledge to manage fungal diseases more effectively.

Globalization of trade

The United Nations Environment Programme (UNEP), the Food and Agriculture Organization of the United Nations (FAO) and the General Agreement on Tariffs and Trade (GATT) under the framework of the World Trade Organization (WTO) largely govern international movement of plants and phytosanitary activities. The FAO's International Plant Protection Convention (IPPC) sets out rules and policies to limit the spread of pathogens in international trade. Implementation of these controls is being placed under increasing pressure owing to greater movement of goods by air transport and as temperature-controlled containerized sea freight. The increase in global trade in plants and plant-based goods has been fuelled by consumer demand for out-of-season produce such as cut flowers, vegetables, and fruit and exotic plants, all of which could harbour harmful fungal pathogens. For example, consumer demand for the all-year-round availability of stone fruits such as cherries, plums and peaches has increased the threat from introduction of alien species of *Monilinia*, the cause of fruit brown rot (van Leeuwen, 2001). In Europe, *Monilinia laxa* and *M. fructigena* are now being displaced by the predominately North American species *M. fructicola* (Abate *et al.*, 2018). In addition to well-established introduced plant health threats due to trade globalization, such as potato famine (*Phytophthora infestans*) and Dutch elm disease (*Ophiostoma novo-ulmi*), plant material sourced from new areas of production continues to result in new threats. For example, new and emerging fungal diseases from ash dieback (*Hymenoscyphus fraxineus*) and sudden oak death or ramorum blight (*Phytophthora ramorum*) have all been introduced to new territories and caused significant harm.

Changes in agricultural practices

Agriculture is continually diversifying. Originally, it involved introduced plants or crops such as potatoes, coffee, tea and rice, but more recently it has included oil palms, soybeans, maize and vines grown far from their centres of origin. Endeavours to improve plant varieties that produce greater yields and better quality have led to the use of biotechnology to generate genetically modified organisms (GMOs). However, the introduction of beneficial traits – such as disease or herbicide resistance conferred by transformation using 'alien genes' from other organisms – has not been without some controversy, as seen with GM crops (Levidow and Boschert, 2008; Fedoroff *et al.*, 2010; Taheri *et al.*, 2017), but gene editing offers a potential alternative. The diversity and geographical origin of hosts is leading to new host–pathogen combinations, the evolution of new pathogenic fungal strains and, on some occasions, has occurred in the absence of natural biological control agents; these are all contributing to increased damage to the crops concerned. On a more local level, changes in agricultural practices – such as greater use of protection from the elements (e.g. in polytunnels) and reduction in the availability of irrigation – are also altering how crops are grown, the diseases that are encountered and the pathogens that are observed.

Climate change

Environmental factors play a key role in disease development by affecting fungal growth and sporulation, infection processes and pathogen survival. Predictions of global warming and changes in weather patterns are likely to affect diseases directly and indirectly (Coakley *et al.*, 1999; Willis, 2018; Fones *et al.*, 2020; Jeger, 2022). Expectations of increasingly warmer weather, milder wetter winters, hotter drier summers and more extreme weather events such as unusually high summer temperatures and more winter storms will all be of significance (Hansen *et al.*, 2005). Pressure to mitigate climate change factors may assist agriculture's endeavours to become more sustainable as energy inputs and irrigation are reduced. These practices, in turn, can affect the health status of crops and environmental conditions that will influence diseases. Additionally, consumers are becoming increasingly aware of 'food miles' and the geographical origin of plants and food, and some are also adopting a more 'plant-based' diet in response to climate change. This awareness may affect where plants and plant products are sourced and vary seasonal demand, and ultimately may affect the geographical distribution of some fungal plant pathogens.

Concerns about climate change have also led to changes in land usage and the type of crops being grown. Until recently, the emphasis for many countries was entirely about food production and food security. Although this continues to be a priority for many countries, there has been a move in recent decades towards thinking about impacts on biodiversity and the natural environment (Cunningham *et al.*, 2015). Climate mitigation by tree planting has become internationally adopted as a mitigation strategy, but this is being challenged by the numerous incursions of pests and pathogens threatening tree health. Recent examples of fungal plant pathogens include those that cause diseases such as ash dieback (*Hymenoscyphus fraxineus*), plane wilt (*Ceratocystis platani*), sweet chestnut blight (*Cryphonectria parasitica*) and Eucalyptus rust (*Puccina psidii*), to highlight just a few from the UK plant health risk register (https://planthealthportal. defra.gov.uk/pests-and-diseases/uk-plant-health-risk-register/index.cfm; accessed 1 March 2023).

Pesticides

In order to maintain production levels and ensure food safety, fungal diseases must be controlled effectively. Although synthetic organic chemicals have been widely available since the mid-1940s, there has been a move to reduce the availability and use of pesticides in response to consumer concerns and political demands (Ragsdale and Sisler, 1994; Hahn, 2014). Although most growers use some aspects of integrated pest management, such as crop varietal selection or disease monitoring, reduction in the availability of pesticides will put even greater emphasis on the need to understand the biology and management of fungal plant pathogens.

1.2 Taxonomy – Categorizing Diversity

Taxonomy is the science of the classification and assignment of organisms to defined categories in a ranked system. The basic ranking for most fungal species, rising to kingdom at the highest level, is detailed in Table 1.1.

The phrase 'fungus', however, encompasses a diverse group of organisms that have been classified in different ways and now occur across three different kingdoms. In the late 1960s, a new kingdom Fungi was established to differentiate fungi from other eukaryotes. In the early 1980s kingdom Chromista led to further separation of oomycetes (water moulds) and kingdom Protozoa for the myxomycetes (slime moulds) from the 'fungi'. There is no doubt, with advances in ultrastructural, biochemical and, especially, molecular biological studies, the revision of the structure of kingdoms and their contents will continue.

The term fungus is still widely used to encompass all these organisms. However, the term fungus-like organism (FLO) is used in some circumstances, such as plant health legislation, to describe the 'lower' or 'simpler' organisms in kingdoms Chromista and Protozoa. In this book, the term fungus is used in its broadest sense for organisms from all three kingdoms.

As their taxonomy is continuously under review, a commonly accepted view based on three kingdoms (Fungi, Protozoa and Chromista) and significant phyla is presented in Table 1.2. Common examples of plant pathogens are included.

There has increasingly been a move away from taxonomy based entirely on phenotypic characters such as colony characteristics (e.g. growth rate, colony colour) and morphological features (e.g. fruiting body type, spore size and colour) to greater molecular characterization. The use of gene typing has revolutionized fungal taxonomy and challenged existing species concepts, leading to both amalgamation and division.

Table 1.1. Major ranks in fungal nomenclature using a powdery mildew (*Erysiphe pisi*) as an example.

Kingdom	Fungi
Phylum	Ascomycota
Class	Leotiomycetes
Order	Erysiphales
Family	Erysiphaceae
Genus	*Erysiphe*
Species	*Erysiphe pisi*
Subspecies	*Erysiphe pisi* var. *pisi*

Table 1.2. Overview of kingdoms (based on Kirk *et al.*, 2008).

Kingdom	Phylum	Common examples
Chromista	Oomycota	Downy mildews, *Phytophthora*, *Pythium*, *Albugo*, etc.
Fungi	Ascomycota	*Taphrina*, yeasts, powdery mildews, *Ceratocystis*, *Nectria*, *Glomerella*, *Diaporthe*, *Eutypa*, *Pleospora*, *Venturia*
	Basidiomycota	Rusts, smuts, mushrooms and toadstools
	Chytridiomycota	*Olpidium*, *Synchytrium*
	Zygomycota	*Mucor*, *Rhizopus*
Protozoa	Amoebozoa	Slime moulds

In general, phenotypic characters are qualitative and open to interpretation on whether a character is present or not; for example, the type of spore-bearing structure (e.g. acervulus or pycnidium) has been used to categorize fungi. However, due to the diversity of fungi and the influence of factors such as the substratum the fungus is growing on and the environmental conditions, there is a continuum of variation and no simple either/or situation. Nucleotide sequence data initially pose less difficulty as they are not likely to be affected by these external factors, although in practice they can be just as hard to interpret because the number of genes and the length of DNA that can be sequenced are increasing steadily, and the reliability of molecular data in the public domain varies. However, when new boundaries have been established by molecular methods, phenotypic behaviours, when re-examined, may parallel the taxonomic changes that have been made, as demonstrated by the revision of the powdery mildews (Braun *et al.*, 2002; Braun and Cook, 2012).

Recent estimates suggest there are 2.2 to 3.8 million species of fungi on Earth (Hawksworth and Lücking, 2017). It is estimated that there are only 120,000 accepted species, with many new fungi yet to be described or studied (Hawksworth and Lücking, 2017). Fungal diseases on commonly grown cultivated plants are usually well documented, with good taxonomic information, so it is relatively uncommon to encounter new taxa on these plants. Yet, with greater interest in the damage to previously less well-studied plants, such as those found in the natural environment, and with greater globalization of trade and crop diversity, in addition to climate change, there are ever-increasing opportunities to encounter new species (Jones and Baker, 2007). For example, the identification in 2001 of *Phytophthora ramorum* as the causative agent of the disease commonly known as sudden oak death has led to intensive monitoring of woodlands and the discovery of new species such as *Phytophthora kernoviae* (Brasier *et al.*, 2005) and other woodland-inhabiting species (Hansen, 2015).

1.3 Nomenclature – Assigning Names

The ability to identify and assign a name to an organism is fundamental to any scientific investigation. Nomenclature has an internationally agreed mechanism for assigning names (in the case of fungi, provided by the International Code of Nomenclature for algae, fungi and plants [ICN]) and is based on the principles first introduced by Linnaeus. The Code provides a framework to ensure that names are unambiguous and to prevent confusion. In determining the correct name for an organism, a number of steps must be completed. These include effective publication, indication of the type specimen or culture, and ensuring that the name is legitimate and meets other the requirements of the Code (Kirk *et al.*, 2008).

The status of names can be checked for many fungi by reference to online databases, principally those provided by the Fungal Databases of the Systematic Mycology and Microbiology Laboratory of the US Department of Agriculture's Agricultural Research Service (Farr and Rossman, 2022) and the Index Fungorum (available at http://www.indexfungorum.org/Names/Names.asp; accessed 1 March 2023), or by reference to published literature.

While searching for information about an organism, it is important to be aware of previous names and other fungal growth states (see Section 1.4). For example, while searching for information about *Glomerella acutata*, it is important to be aware that until 2001 (when the teleomorph was described) this was known as *Colletotrichum acutatum* (Guerber and Correll, 2001). This name, although taxonomically correct until 2001, is still in common usage. It is also important to consider the most appropriate name. For example, most growers are familiar with *Botrytis cinerea* as the cause of grey mould, but would not recognize the name of *Botryotinia fuckeliana*, which is taxonomically correct.

In 2012, the ICN introduced a new proposal commonly referred to as 'one fungus one name'. This proposal supported the abolition of dual nomenclature (teleomorph names and anamorph names), and in most cases the sexual name takes precedence. However, depending on committee agreement, sometimes the asexual state that is widely known may take precedence. For names published prior to 1 January 2013, either the anamorph or the teleomorph name may be used, depending on a rule of 'first named takes precedence'. After this date, only a single name for a new fungus is valid. Therefore, it is always prudent to check the current taxonomy of an organism based on the above resources cited.

1.4 Pleomorphism – Accounting for Variable States

Fungi may be as simple as a budding yeast, or as complex as a rust fungus with up to five life-cycle stages and spore types present on two separate hosts. Many fungi have simple life cycles based on asexual reproduction ('Imperfect fungi'); these are referred to as anamorphs. There are also fungi that have a sexual stage in their life cycles ('Perfect fungi'); the sexual stages are referred to as teleomorphs. The presence of different states in a life cycle is referred to as pleomorphism – the term encompasses both teleomorphs and anamorphs. Reference to the teleomorph includes all anamorphs (some fungi may have different anamorphs, which are referred to as synanamorphs), and thus applies to the whole fungus (referred to as the holomorph).

1.5 Ecological Groups

Fungi are heterotrophic organisms that are dependent on external organic carbon sources, which are broken down and absorbed through the production of enzymes. They may be categorized into three groups:

- Parasitic – living on or in and obtaining nutrients from another organism (the host), while not providing any benefit to the host plant; these fungi may be described as biotrophic (the host is colonized while alive) or nectrotrophic (the host is first killed and then colonized).
- Saprobic – obtaining nutrients from dead organic material.
- Symbiotic – in a mutual relationship that benefits both the fungus and its partner.

Fungal plant pathogens are organisms that cause plant disease (a condition where normal functions are disrupted or harmed, and usually expressed as symptoms, e.g. leaf blight, root rot, wilt). The pathogens may employ different modes of nutrition during their life cycles. Fungi may live on the exterior of hosts (epiphytes) or within host tissue (endophytes). Although the term endophyte strictly just describes the location of the fungus within the host tissue, it has become common to think of endophytes as causing symptomless infections. The period from infection to development of symptoms is described as the latent period and may be variable in length, from several days to many months or even years. Pathogenic fungi capable of symptomless colonization may also be referred to as causing a quiescent infection, in which, once again, although the organism may be present within the host, no symptoms are observed.

As this brief introduction illustrates, fungi (encompassing fungus-like organisms) are extremely diverse; only some are plant pathogens, but these may be extremely damaging to natural environments and crops. Virtually all plants, at some point during their lifetime, are affected by one fungus or another, resulting in a range of symptoms and damage. This book draws on many years of expertise of plant pathologists involved in the diagnosis of key fungal diseases to help guide readers through the complex process of diagnosing and studying these fascinating organisms.

References

Abate, D., Pastore, C., Gerin, D., De Miccolis Angelini, R.M., Rotolo, C. et al. (2018) Characterization of *Monilinia* spp. populations on stone fruit in South Italy. *Plant Disease* 102, 1708–1717.

Brasier, C.M., Beales, P.A., Kirk, S.A., Denman, S. and Rose, J. (2005) *Phytophthora kernoviae* sp. nov., an invasive pathogen causing bleeding stem lesions on forest trees and foliar necrosis of ornamentals in Britain. *Mycological Research* 109, 853–859.

Braun, U. and Cook, R.T.A. (2012) *Taxonomic Manual of the Erysiphales (Powdery Mildews)*. Fungal Biodiversity Centre (CBS Biodiversity Series No. 11), Utrecht, the Netherlands.

Braun, U., Cook, R.T.A., Inman, A.J. and Shin, H.-D. (2002) The taxonomy of the powdery mildew fungi. In: Belanger, R.R., Bushnell, W.R., Dik, A.J. and Carver, T.L.W. (eds) *The Powdery Mildews: A Comprehensive Treatise*. APS Press, St Paul, Minnesota, pp. 13–55.

Coakley, S.M., Scherm, H. and Chakraborty, S. (1999) Climate change and plant disease management. *Annual Review of Phytopathology* 37, 399–426.

Cooke, R. (1977) *Fungi, Man, and His Environment*. Longman Group, London.

Cunningham, S., MacNally, R., Baker, P., Cavagnaro, T.R., Beringer, J. et al. (2015) Balancing the environmental benefits of reforestation in agricultural regions. *Perspectives in Plant Ecology Evolution Systems* 17, 301–317.

Farr, D.F. and Rossman, A.Y. (2022) Fungal Databases. U.S. National Fungus Collections, ARS, USDA, Washington, DC. Available at: https://nt.ars-grin.gov/fungaldatabases/ (accessed 3 June 2022).

Fedoroff, N.V., Battisti, D.S., Beachy, R.N., Cooper, P.J.M., Fischhoff, D.A. et al. (2010) Radically rethinking agriculture for the 21st century. *Science* 327, 833–834.

Fones, H.N., Bebber, D.P., Chaloner, T.M., Kat, W.T., Steinberg, G. and Gurr, S.J. (2020) Threats to global food security from emerging fungal and oomycete crop pathogens. *Nature Food* 1, 332–342.

Guerber, J.C. and Correll, J.C. (2001) Morphological description of *Glomerella acutata*, the teleomorph of *Colletotrichum acutatum*. *Mycologia* 93, 216–229.

Hahn M. (2014) The rising threat of fungicide resistance in plant pathogenic fungi: *Botrytis* as a case study. *Journal of Chemical Biology* 7, 133–141.

Hansen, E.M. (2015) *Phytophthora* species emerging as pathogens of forest trees. *Current Forestry Reports* 1, 16–24.

Hansen, J., Nazarenko, L., Ruedy, R., Sato, M., Willis, J. et al. (2005) Earth's energy imbalance: confirmation and implications. *Science* 308, 1431–1435.

Hawksworth, D. and Lücking, R. (2017) Fungal diversity revisited: 2.2 to 3.8 million species. *Microbiology Spectrum* 5, 10.

Henson, J.M. and French, R. (1993) The polymerase chain reaction and plant disease diagnosis. *Annual Review of Phytopathology* 31, 81–109.

Hill, L., Jones, G., Atkinson, N., Hector, A., Hemery, G. and Brown, N. (2019) The £15 billion cost of ash dieback in Britain. *Current Biology* 29, 315–316.

Jain, A., Singh, H.B. and Das, S. (2021) Deciphering plant–microbe crosstalk through proteomics studies. *Microbiological Research* 242, 126590.

Jeger, M.J. (2022) The impact of climate change on disease in wild plant populations and communities. *Plant Pathology* 71, 111–130.

Jones, D.R. and Baker, R.H.A. (2007) Introductions of non-native plant pathogens into Great Britain: 1970–2004. *Plant Pathology* 56, 891–910.

Kamoun, S. (2009) Plant pathogens: oomycetes (water mold). In: Schaechter, M. (ed.) *Encyclopedia of Microbiology*, 3rd edn. Elsevier, Amsterdam, pp. 689–695.

Kirk, P.M., Cannon, P.F., Minter, D.W. and Stalpers, J.A. (eds) (2008) *Ainsworth & Bisby's Dictionary of the Fungi*, 10th edn. CAB International, Wallingford, UK.

Kowalski, T. (2006) *Chalara fraxinea* sp. nov. associated with dieback of ash (*Fraxinus excelsior*) in Poland. *Forest Pathology* 36, 264.

Levidow, L. and Boschert, K. (2008) Coexistence or contradiction? GM crops versus alternative agricultures in Europe. *Geoforum* 39, 174–190.

Nielsen, L.R., McKinney, L.V., Hietala, A.M. and Kjær, E.D. (2017) The susceptibility of Asian, European and North American *Fraxinus* species to the ash dieback pathogen *Hymenoscyphus fraxineus* reflects their phylogenetic history. *European Journal of Forest Research* 136, 59–73.

Quirino, B.F., Candido, E.S., Campos, P.F., Franco, O.L. and Krüger, R.H. (2010) Proteomic approaches to study plant–pathogen interactions. *Phytochemistry* 71, 351–362.

Ragsdale, N.N. and Sisler, H.D. (1994) Social and political implications of managing plant diseases with the increased availability of fungicides in the United States. *Annual Review of Phytopathology* 32, 545–557.

Strange, R.N. and Scott, P.R. (2005) Plant disease: a threat to global food security. *Annual Review of Phytopathology* 43, 83–116.

Taheri, F., Azadi, H. and D'Haese, M. (2017) A world without hunger: organic or GM crops? *Sustainability* 9, 580.

van Leeuwen, G.C.M., Baayen, R.P. and Jeger, M.J. (2001) Pest risk assessment for the countries of the European Union (as PRA area) on *Monilinia fructicola*. *EPPO Bulletin* 31, 481–487.

Willis, K.J. (ed.) (2018) *State of the World's Fungi 2018*. Royal Botanic Gardens, Kew, London.

2 Examination of Plant Material

Charles R. Lane*

Fera Science Ltd, Sand Hutton, York, UK

2.1 Introduction

Examination of plant material is an essential part of any diagnosis, but without gathering supporting information about the problem an accurate solution is unlikely to be achieved. The information may be in a written format and include: origin of plants, previous cropping history, cultural practices, agrochemical treatments, weather, etc.; it may involve soliciting information from the grower, such as timing, distribution and spread; or it may require collecting samples of symptomatic and healthy plants, propagating material, seeds, soil or water for testing. The process is improved by listening to growers' and advisers' opinions on causes and then deciding on the need for a field visit. Before beginning an investigation, it is essential to understand what the customer wants from the diagnosis in order to plan the work and to provide advice on the feasibility of the testing that is required, the likelihood of success, costs and timescales. There is no benefit to the customer in carrying out a detailed and expensive investigation which takes several months to complete and then reporting back when what the grower needs is a quick and presumptive diagnosis in order to make a rapid decision on what to do next. Diagnosticians must be

prepared to revise the customer's expectation of the outcomes as the diagnosis evolves and be willing to request more information and samples with time. A successful diagnosis will involve dialogue and collaboration among laboratory-based staff, field consultants and growers that draws on their specific knowledge about the crop and the problem.

It is important to understand that diagnosis is a two-part process involving detection followed by identification, although these terms are sometimes used rather interchangeably. Typically, detection refers to establishing the presence of an organism, identification refers to determining which organism it is and diagnosis refers to describing the entire process. It is quite common to consider the sensitivity, specificity and level of confidence desired for any diagnosis. On some occasions, it may be necessary to identify accurately an organism to a known species or even subspecies; this would be essential in certain circumstances, such as statutory plant health action. However, where an organism is less well understood or even new to science, the level and confidence of the identification must be tempered. From a grower's perspective, the identification level required in the first instance may be as simple as whether decay is due to a

* E-mail: charles.lane@fera.co.uk

DOI: 10.1079/9781800620575.0002

fungus or a bacterium, so that consideration as to pesticide application can be made quickly. As the grower begins to consider the implications of the initial diagnosis and to manage the disease more effectively, a more accurate identification and quantification of inoculum levels may lead to a more detailed diagnosis. On certain occasions though, for example in the case of obligate parasites such as mildews or rusts, an accurate identification early on can predict potential crops at risk because of the frequently host-specific nature of these fungi.

A further area of confusion over terminology can be reference to diseases and disorders (Federation of British Plant Pathologists, 1973). Disorders are non-infectious and caused by environmental (or abiotic) factors such as temperature, rainfall, nutrient status, toxic chemicals and genetic anomalies, etc., while diseases are infectious and caused by pathogens such as fungi, bacteria and viruses (and some also by nematodes) that bring about a harmful deviation from the normal functioning of physiological processes. However, some diagnosticians do not refer to diseases and disorders but rather to infectious and non-infectious diseases – caused by biotic and abiotic factors, respectively.

2.2 Symptom Recognition

The accurate description of symptoms is essential to prevent confusion when growers, advisers and laboratory staff are working together in order to ensure that the problem observed in the field is the one investigated by the laboratory (see Protocol 2.1). This becomes particularly important if samples arrive in a poor condition as it may not be obvious at first what the grower is concerned about. A good example of this is with amenity turf diagnosis: etiolation of the grass commonly occurs in transit and is usually a result of light deprivation during that process.

The determination of the cause(s) of symptoms is aided by examination of plant material in a methodical order, both in the field and in the laboratory. In this way, important underlying problems will not be overlooked and conclusions will be scientifically justified. Understandably, it is very tempting to be drawn into only examining and investigating the causes of dramatic symptoms, although these may have only occurred owing to a more subtle or underlying problem elsewhere. A good example of this is the dieback of containerized hardy ornamental nursery stock. This is frequently due to root disease (such as that caused by *Phytophthora*) or poor cultural conditions (such as waterlogging), but to the casual observer opportunistic pathogens such as *Botrytis* (grey mould) may be more obvious on the dying aerial parts. Actions to control the grey mould may help to alleviate the problem temporarily but will not deal with the underlying causes. When trying to tease apart a complex problem, it is as helpful to know which parts of the plant appear healthy as which appear unhealthy; this helps to focus efforts in other (non-obvious) areas. A suggested 'Symptom record' pro forma is presented in Protocol 2.2. A useful technique used in tree health citizen science, such as Observatree (https://www.observatree.org.uk/; accessed 1 March 2023), is to ask surveyors to take three pictures: an overall picture of the tree in its setting, a picture illustrating where the main symptom is found and a close-up of the symptoms of concern. This encourages surveyors to work in a structured and methodical way, gathering information that will help laboratory staff make a more informed diagnosis. The information is then submitted online through Tree Alert (https://www.forestresearch.gov.uk/tools-and-resources/fthr/tree-alert/; accessed 1 March 2023) that again demands a structured approach for gathering and reporting information.

2.3 Sampling

Collecting samples provides a useful 'snapshot' of the problem but only at that time. Gathering samples with a range of symptoms and levels of intensity (including material considered to be healthy) is recommended as it will increase the likelihood of an accurate diagnosis. It is important to remember that aerial symptoms may be the result of problems lower down, so wherever possible whole plants or at least specimens from all parts of the plant should be examined in the field and, where appropriate, sampled and submitted for laboratory testing. It is essential to include the leading edge (the interface between healthy and diseased tissue) wherever this is present as there is little or no benefit in submitting dead or severely rotten plants.

Good packaging and rapid transit are required to limit sample deterioration. The principal concerns are decay (accelerated by overheating) and contamination with growing media. In general, whole plants should be placed in a large, strong plastic bag and the root ball tied off securely at the collar; the bag should be loosely sealed at the top to prevent spillage. Samples should be placed in a strong, secure cardboard box packed with absorbent paper (such as crumpled-up newspaper) with the relevant paperwork in a separate sealed plastic bag (Fig. 2.1).

Once samples have been collected, these should be packaged and shipped to the diagnostic laboratory without delay and, if possible, arrive the next day. In warmer climates, if possible, samples should be refrigerated (but not frozen) until shipped. Sending samples over the weekend or public holidays should be avoided wherever possible. Special quarantine precautions may be required before dispatch, as discussed in Chapter 12. If smaller samples such as leaves or stems need to be sent, in general these should be wrapped in dry, absorbent paper and placed preferably in a paper bag or, if unavailable, a plastic bag (Fig. 2.2).

Plant material prone to rapid decay, such as fruits, flowers or tubers, should be wrapped individually in dry paper (Fig. 2.3).

Any material in an advanced state of decay should not be sent. If specimens are expected to arrive quickly (in less than 24 h), bags may be sealed to prevent spillage; for longer periods of transit, bags should be folded over or tied loosely to prevent spillage, but should not be sealed unless there are quarantine concerns. Soil samples should be double bagged in heavy-gauge plastic bags tied securely at the top (Fig. 2.4).

Water samples should be placed in robust plastic bottles, securely sealed, and placed in a heavy-gauge plastic bag sealed at the top to prevent spillage (Fig. 2.5).

Cultures and water samples should be shipped to arrive the next day. Consideration should be given to the use of freezer packs and/or insulated packaging materials or cool boxes to prevent overheating.

2.4   Recording Symptoms, Testing Procedures and Results

To ensure a thorough diagnostic process it is essential to record sample details, symptoms, symptom intensity, testing procedures and results in a clear and orderly fashion so that it is easy to see what has been done and what stage the diagnosis has reached.

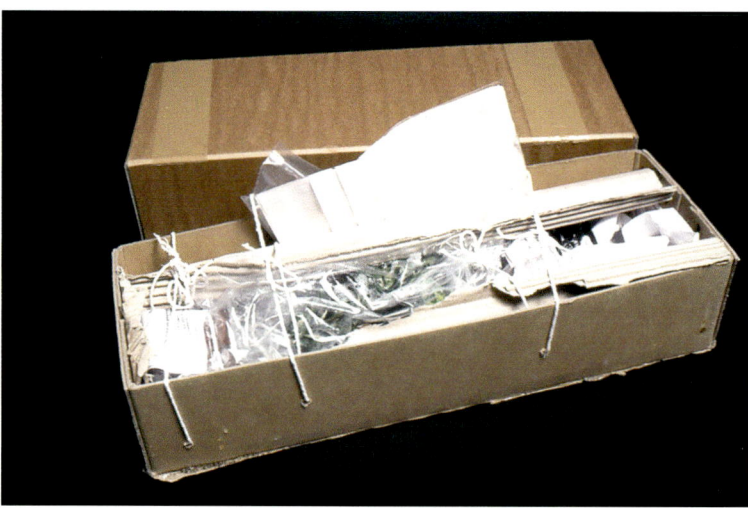

Fig. 2.1. Packaging of a whole plant. The sample should be placed in a strong, secure cardboard box packed with absorbent paper (such as crumpled-up newspaper) with the relevant paperwork in a separate sealed plastic bag. (© UK Crown copyright – courtesy of Fera.)

Fig. 2.2. Packaging of small samples. Smaller samples, such as leaves or stems, should be wrapped in dry, absorbent paper and placed preferably in a paper bag or, if unavailable, a plastic bag. (© UK Crown copyright – courtesy of Fera.)

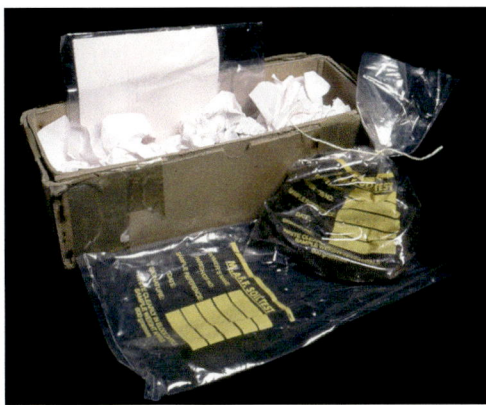

Fig. 2.4. Packaging of soil samples. Samples should be double bagged in heavy-gauge plastic bags tied securely at the top. (© UK Crown copyright – courtesy of Fera.)

Fig. 2.3. Packaging of rotting potato tubers. Plant material prone to rapid decay, such as fruit, flowers or tubers, should be wrapped individually in dry paper. (© UK Crown copyright – courtesy of Fera.)

Fig. 2.5. Packaging of water samples. Samples should be placed in a robust plastic bottle in a heavy-gauge plastic bag sealed at the top to prevent spillage. (© UK Crown copyright – courtesy of Fera.)

Structured recording of symptoms at the time of sampling is an important first step. Useful guidance is provided for submitting samples to Tree Alert (https://www.forestresearch.gov.uk/tools-and-resources/fthr/tree-alert/what-do-you-need-to-make-your-report/; accessed 1 March 2023).

You will be asked to provide:

- The date of your observation.
- The type of location in which your observation was made.
- The location of the problem you are reporting, including country, and either a grid reference (a 10-figure grid reference (GR) is preferable, but a 6- or 8-figure GR is acceptable) or a point on a map.

- The number of trees affected and their approximate size measured as trunk/stem diameter.
- The type of tree affected (conifer or broadleaf), its common name and species.
- Information about the problem you have observed, including location on the affected tree (crown, stem or base) and the nature of the symptom(s).
- Photographs of your observation. Ideally, three good-quality photographs showing (1) the affected tree(s) in context, (2) the problem/symptom in context and (3) the details of the symptom.

Once the sample has been received in the laboratory, it is important to maintain the same systematic approach to recording information. An example of a pro forma for a mycology laboratory diagnosis 'Examination record' is given in Protocol 2.3. This allows information to be recorded at the time of examination and can be added to during the diagnosis. It can also be used to review the entire process when reporting results to customers and dealing with follow-up enquiries. Using this method, it is clear what the symptoms were on receipt. This is because when examining incubated material, possibly several weeks later, it can be quite easy to lose sight of the problem that the customer was initially concerned about. The method also helps to collate results in the case of a multidisciplinary testing process and to highlight shortcomings in sampling. This is particularly important in a large diagnostic laboratory where one person may not see a sample through the entire process. There are also a number of key quality assurance elements that must be included with respect to sample receipt, condition and type, and also with respect to the nature and duration of the tests, testing by other disciplines and sample storage – in addition to the involvement of diagnosticians, test results, conclusions and contact with customers. Not only are all of these the basis for all scientifically sound diagnoses, they are also essential for any quality assurance scheme (see Chapter 13). Furthermore, the complexity of the problem and the diagnosis will influence the level of detail recorded and the evidence required.

As mentioned above, along with the 'Examination record' (Protocol 2.3) there should also be a 'Laboratory test results' record. The latter provides a simple and clear way of recording the most common tests used, such as incubation, isolation and floating, and of detailing set-up and check dates and which diagnostician is involved. Further sections on storage and/or subsampling may also be included. Comment boxes should be used to record presence/absence of fungal structures. As some fungi may need weeks or even several months of incubation before they produce features that permit identification, it is helpful to indicate how long this could take so that samples are not disposed of prematurely.

Central to any diagnostic laboratory is a system to identify and track samples by using a unique reference number or identifier for each sample. The most common system is a two-part reference number that uses the year (e.g. 2022) as the first part and a sequentially increasing numbering system during the year as the second part (e.g. 20228976). As this number is used not only on paperwork but also on all Petri dishes, incubation boxes and small items such as microcentrifuge tubes, the provision of pre-printed labels can be helpful.

2.5 Recording Severity/Disease Assessment Scales

Disease assessment scales are used to record the relative proportions of healthy to affected tissues and are frequently scaled from 0 to 10 or from 1 to 4. They may be used to help predict yield losses or as part of a plant health certification scheme where certain tolerances are acceptable (Ebbels, 2003). For some crops and diseases there may be published assessment scales (sometimes referred to as disease assessment keys), such as the disease assessment key for common potato skin diseases (James, 1971; Anon, 1985), but a simple key can be devised as necessary (Fig. 2.6).

Assessing disease levels relies frequently on experience in assigning levels to determined categories, but other methods have been developed that use image capture and analysis software for two-dimensional objects such as leaves. However, simpler techniques for image capture, such as tracing outlines, photographing or photocopying, and analysis using squared paper or weighing outlines traced on paper can be just as effective. Three-dimensional objects such as potato tubers or fruits are much more difficult to analyse and tend to rely more on visual examination using the human eye rather than on image capture or processing.

2.6 Visual Examination

The simplest way to examine specimens is by direct observation with the naked eye; in the field and also to a lesser extent in the laboratory, this can be aided by the use of a hand lens or magnifying glass. These typically offer a 5–20 times magnification and can be helpful in discriminating

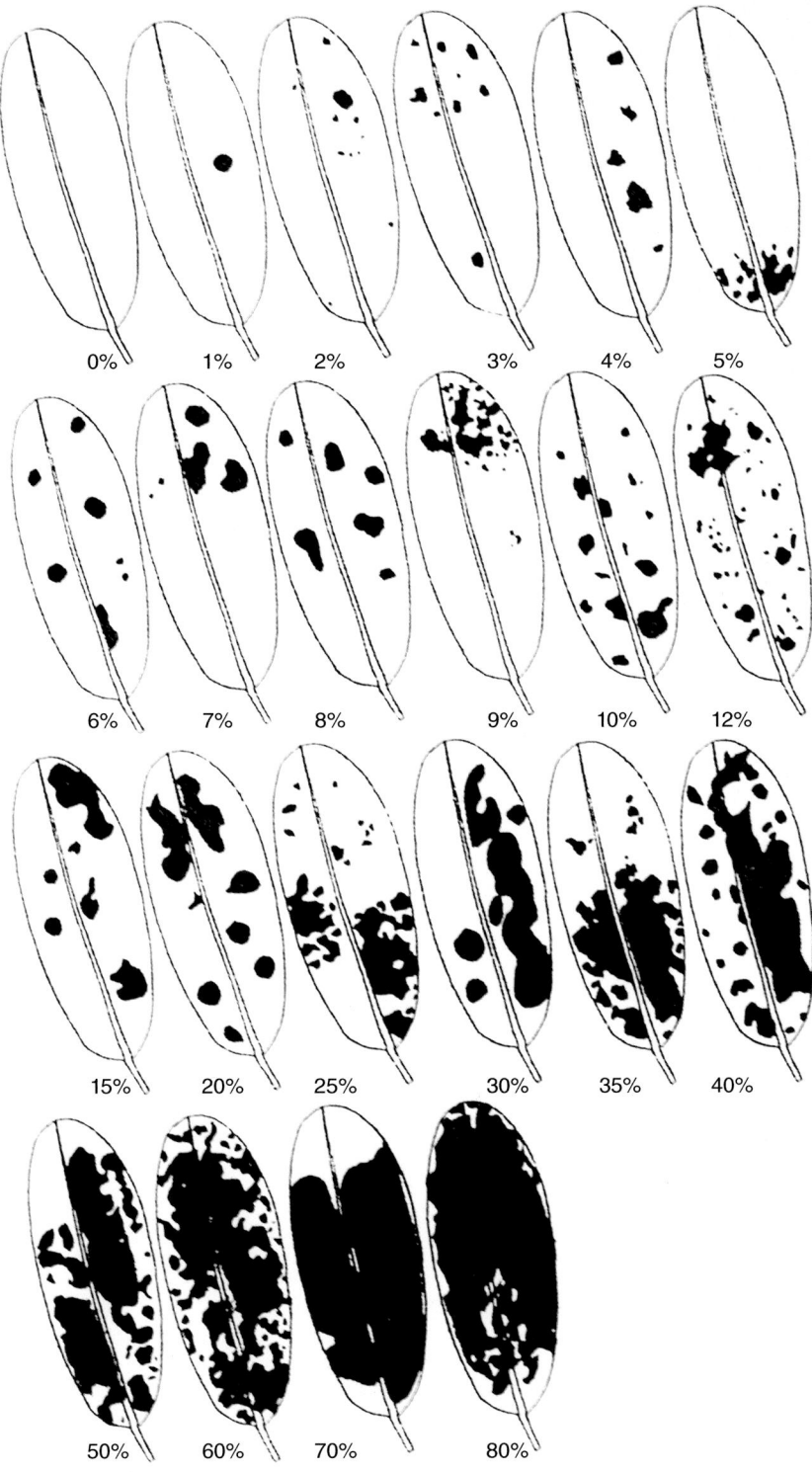

Fig. 2.6. Disease assessment key to determine levels of damage for a foliar pathogen. (© Paul Beales.)

different types of macroscopic fungal structures. In the laboratory, a higher magnification using a dissecting microscope (typically 6–50 times magnification, but up to 80 times) is commonly used to examine specimens. Ideally, these microscopes should be mounted on a height-adjustable stand, preferably with a 'swing arm', to permit the examination of a wide size range of objects (Fig. 2.7).

A high-powered LED or tungsten light source and fibre optic illumination from above the specimen, preferably as a ring light source mounted around the objective lens rather than one or more point light sources, is the best approach. Illumination from beneath the specimen can sometimes be helpful but is not commonly used. A compound microscope is essential for the examination of fungal structures mounted on glass slides at a much higher magnification (Chapter 6). Typically, a series of objective lenses (×10, ×20, ×40) and a ×100 oil immersion lens should be available, with a usual eyepiece magnification of ×10. A powerful LED or tungsten light source (at least equivalent of 20 W but preferably 100 W) should be used, and additional features such as phase contrast or differential interference contrast are helpful. An eyepiece graticule and micrometre for calibration are essential, and a series of neutral density filters may also be useful. If any immunofluorescence work is planned, then additional filters of the correct wavelength and a UV light source will be required. Wherever possible, ergonomically designed microscopes should be used to prevent discomfort for long-term users.

Greater magnification can be achieved by using either a scanning electron microscope (SEM) (Fig. 2.8) or a transmission electron microscope (TEM) (Fig. 2.9), although the latter is more commonly used for ultrastructural studies rather than for routine diagnosis.

SEM magnification typically varies between 1000 and 10,000 times and permits the examination of surface features such as the ornamentation of spores (Cook *et al.*, 1997). Different sample preparation steps are required, depending on the type of SEM available; these range from the simplest, which uses ambient conditions, to the gold coating of frozen samples.

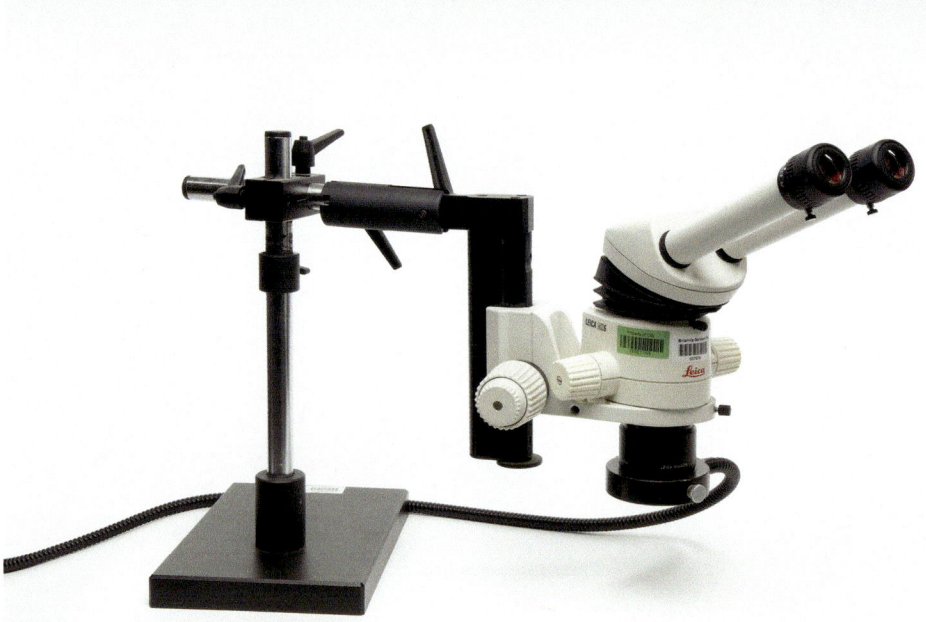

Fig. 2.7. Dissecting microscope mounted on a height-adjustable, swing-arm stand. (© UK Crown copyright – courtesy of Fera.)

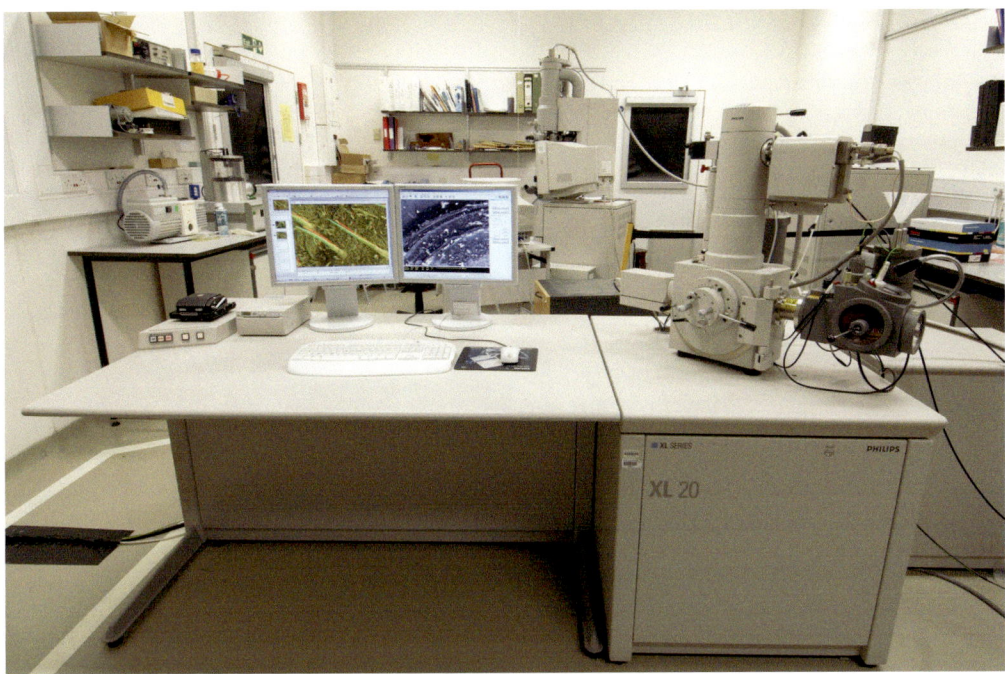

Fig. 2.8. Scanning electron microscope. (© UK Crown copyright – courtesy of Fera.)

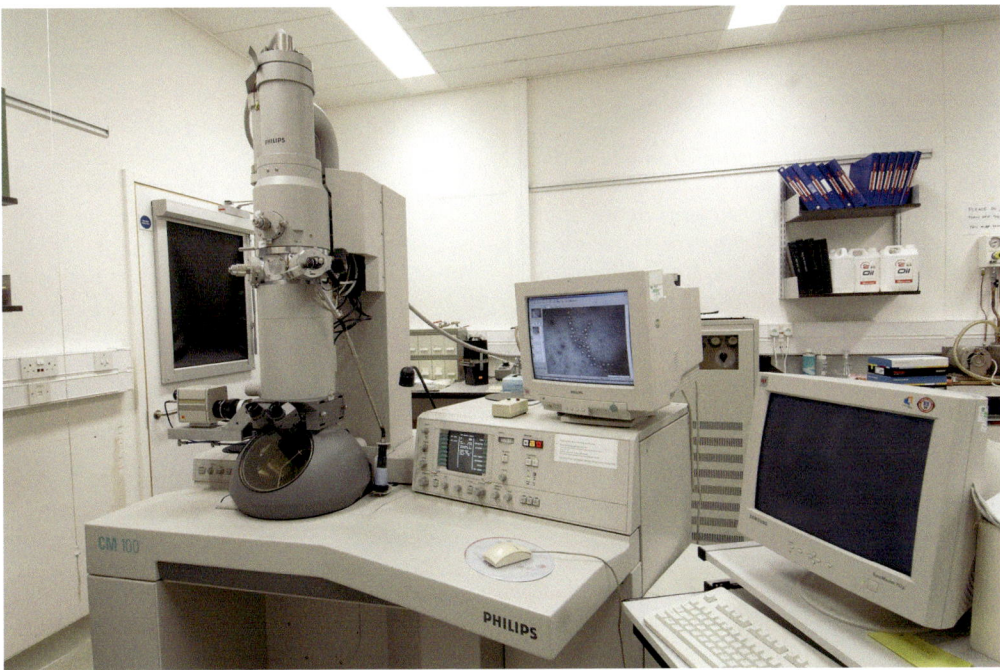

Fig. 2.9. Transmission electron microscope. (© UK Crown copyright – courtesy of Fera.)

2.7 Image Capture and Analysis

Recording symptoms and signs has become increasingly easy and cost-effective owing to the availability of relatively cheap high-resolution digital cameras, although a good written description with drawings is still very effective. However, digital cameras offer a simple way of recording symptoms, both in the field and in the laboratory, for use by growers, advisers and diagnosticians alike. In the laboratory, an LED or tungsten lighting set-up is preferable to the use of a built-in flash or natural light if more professional results are required. Many dissecting and compound microscopes have attachments that are compatible with digital and film cameras. Digital cameras have the advantage of providing instant results, which can be useful when trying to obtain sharply focused pictures that are notoriously difficult to achieve when taken through a microscope. Additionally, digital images are easy to document, store and share with interested parties.

More recently, image capture and analysis software has been available that offers further opportunities to measure and analyse the morphological features of fungi. These measurements can be simple parameters such as length and breadth, but the software permits many objects to be measured rapidly and the data exported into a spreadsheet such as Excel ready for statistical analysis. More advanced features, such as area, can be calculated automatically. There have been a number of attempts to quantify qualitative features such as spore shape (Gottwald *et al.*, 2001), analyse the development of mycelial systems (Donnelly *et al.*, 1999), make identification based on spore shape (Ordynets *et al.*, 2021) or quantify fungal sporulation (Muskat *et al.*, 2021). However, the difficulty in the application of this technology is in the accurate differentiation of the object from the background, referred to as 'thresholding'. This can be very difficult to achieve, as many fungal structures are hyaline (colourless) and thus present contrast difficulties. The small size of fungal structures and the difficulty of obtaining non-touching 'objects' poses further problems, resulting in laborious data gathering. None the less, this technology has great potential and has been used successfully to separate closely related taxa such as *Fusarium* based on the shape of the macroconidia (Gottwald *et al.*, 2001), although as yet it cannot be used to identify fungi based on spore outlines alone. However, advances in biomedical sciences have demonstrated what can be achieved with image analysis. Petra (2018) has described the development of an automated image acquisition, sample handling and image-interpretation system for identifying airborne fungi in order to monitor fungal contaminants in bio-aerosols.

2.8 Legal Requirement to Report All New and Unusual Findings Immediately

An essential component of good plant biosecurity is to ensure the early reporting of new and unusual harmful organisms. This is governed internationally by the World Trade Organization's General Agreement on Tariffs and Trades (GATT), which includes agreement on the application of sanitary (humans and animals) and phytosanitary (plants) measures (WTO-SPS) (https://www.wto.org/english/tratop_e/sps_e/sps_e.htm; accessed 1 March 2023). The International Plant Protection Convention (IPPC) (https://www.ippc.int/en/; accessed 1 March 2023) requires all countries to establish an official national plant protection organization (NPPO) that is responsible for plant health. As a result of any diagnostic or investigational work, you are legally required to report immediately any potential or suspicious findings of a new or unusual fungal plant pathogen. The IPPC provides a list of all NPPOs and contact details worldwide (https://www.ippc.int/en/countries/nppos/list-countries/; accessed 1 March 2023). Reporting such suspicions may be through online services such as Tree Alert in Great Britain or 'hotlines' such as Australia's Exotic Plant Health Hotline (https://www.planthealthaustralia.com.au/wp-content/uploads/2019/11/Discovered-something-new-researcher.pdf; accessed 1 March 2023) or South Africa's Emergency Plant Pest Hotline (https://pir.sa.gov.au/biosecurity/plant_health/emergency_and_significant_plant_pests; accessed 1 March 2023). Larger NPPOs may require state reporting, such as the

United States Department of Agriculture (USDA) (https://www.aphis.usda.gov/aphis/ourfocus/planthealth/ppq-program-overview/sphd; accessed 1 March 2023). Following the report, further identification work may be required to establish the identity of any harmful organism and its origin. The NPPO can then decide if the organism poses a significant plant health risk and whether any statutory action is required.

Protocol 2.1

Checklist of principal steps and considerations when examining plant samples

This protocol should be used in conjunction with the more detailed and specific protocols described in Chapter 6.

1. Before examining the samples, check:
   - the growers' comments and determine urgency;
   - that host details and sample numbers tally;
   - that sample is fit for examination (determine transit time);
   - that the sample sent is relevant to the problem reported;
   - the pesticide application schedule;
   - for health and safety or quarantine issues;
   - for any flying insects before opening bags; and
   - whether samples can be examined fully on the day of receipt, and if not store appropriately to prevent further decay.
2. Clean and decontaminate laboratory surfaces and equipment.
3. Examine samples.
   - 3.1. Roots/tubers/bulbs/corms.
     Symptoms – check for malformation, physical damage, galls/cysts, root congestion, rotting, etc.
     If rotting, ascertain if dry or wet, deep or superficial, major or minor, whether extending into crown.
     Visual examination – initially examine the sample without disturbing any attached soil, then gently remove soil using a blunt instrument or by shaking and re-examine; finally, wash off any soil and check again. At each step examine surfaces under a dissecting microscope and look for mycelium, fruiting bodies, sclerotia, runner hyphae, bootlaces, etc. Cut transverse and longitudinal sections through larger pieces of material (e.g. tubers) and look for any staining, discoloration or decay.
     Microscopic examination – make slide preparations and look for fungal structures (see Chapter 6).
     Tests – isolate from the leading edge, float, bait, incubate, check soil pH/electrical conductivity (EC).
     Enquire about cultural problems (e.g. waterlogging), storage issues (e.g. frost, overheating, oxygen starvation), ground compaction/hard pan, groundwater pollution (e.g. sewerage leaks), planting or potting date, onset of symptoms, host range affected, distribution (patchy or widespread), previous cropping history, nutritional analyses, application of chemicals (e.g. pesticides or growth regulators such as sprout suppressants), etc.
   - 3.2. Stem base.
     Symptoms – check for lesions, cankers, galls, adventitious roots, vascular staining, evidence of invertebrate pest damage (e.g. frass, webbing, slug/snail slime), etc.
     Visual examination – look for mycelium, fruiting bodies (e.g. sclerotia, pycnidia), cut transverse and longitudinal sections, look for vascular discoloration, graft incompatibility (e.g. poor growth, weak union, root suckers), etc.
     Microscopic examination – make slide preparations and look for fungal structures, vascular staining.
     Tests – isolate from the leading edge, incubate and/or float in Petri dishes with Petri's mineral solution.
     Enquire about vertebrate pest damage (e.g. rabbits, deer), physical damage (e.g. grass strimming), weather damage (e.g. wind rock, frost, drought, rock salt application in winter), soil improvers and mulches, etc.

3.3. Stems/trunks.

Symptoms – check for physical damage/wounds, lesions, cankers, galls, vascular staining, invertebrate pest damage (e.g. insect bore holes, webbing, frass), etc.

Visual examination – look for mycelium, fruiting bodies, toadstools, bracket fungi, cut transverse and longitudinal sections, etc.

Microscopic examination – make slide preparations and look for fungal structures, vascular or heartwood staining.

Tests – isolate from the leading edge, incubate, float.

Enquire about vertebrate pest damage (e.g. squirrels, deer), invertebrate pest damage (e.g. wood-boring insects), whether any mushrooms/toadstools have been seen, physical damage (e.g. lightning damage).

3.4. Leaves/flowers/fruits.

Symptoms – examine upper and lower surfaces and for larger specimens cut transverse and longitudinal sections to determine extent of decay. Check for any discoloration or spots and ascertain distribution (e.g. old/new growth, interveinal (restricted by veins)/veinal (crossing veins), marginal, apical/distal, single/numerous/coalesced, etc.), malformation/distortion (e.g. atypical shapes and colour, prolific hair production, epinasty). Record coloration of spots and surrounding tissue, whether spots are necrotic, chlorotic or water soaked. Record if spots are flat, sunken or raised. Check for physical damage/ wounds, lesions, cankers, galls, invertebrate pest damage (e.g. insect bore holes, webbing, frass).

Visual examination – look for mycelium, fruiting bodies, mildews, rust pustules, oedema (raised corky tissue), slime moulds, etc.

Microscopic examination – make slide preparations and look for fungal structures.

Tests – isolate from the leading edge, incubate, float.

Enquire about pesticide applications, weather damage (e.g. hail, frost, sun/ wind scorch, windblown soil/sand).

Protocol 2.2

Symptom record pro forma

Mycology - Diagnosis Form : Sample 201

INITIAL EXAMINATION　　　　　　　　**Diagnostician**: .... **Date**:..................

| Sample condition: | good | adequate | poor | unfit for testing | further samples requested |

Comments:

| Sample details on receipt: | whole plant | roots | stem base / crown | stems |
| | leaves | flowers | fruits | tubers / bulbs / corms |

Size or quantity of sample ....

SYMPTOMS: (circle descriptions as appropriate)

Roots　　　　　healthy　　　　rotting *　　　　　　　　damage (distorted / congested)
　　　　　　* is rotting　major / minor　　tap root / lateral roots / fine roots　　extending into crown
Comments

Stem base / Crown　healthy　　　　rotting　　　　lesion　　　　vascular staining　　　　damage
Comments

Aerial parts　　healthy　　lesions / cankers　　vascular staining　　wilt　　dieback　　damage
Comments

Leaves / flowers / fruits　healthy　　spots *　　rots　　pustules　　premature fall　damage
　　* is spotting　discrete / diffuse　　necrotic / chlorotic / water soaked　scorch / oedema　rust / P mildew / D mildew
　　Distribution　single / numerous / coalesced　new growth / old growth　all over / limited　upper surface / lower surface
Comments

Tubers / bulbs / corms　healthy　　rots　　　　lesions　　　　damage
　　　　　　　　　dry / wet　　superficial / deep-seated　apical / basal / lateral
Comments

Comments:

Photograph? Y / N
PHOT no.................. Digital Image no...................... 　Put on Image Library Y / N 　as reference image? Y / N

Protocol 2.3

Examination record pro forma

Mycology - Diagnosis Form : Sample 201

INCUBATION

damp box / dry box / other ....

Result

By: .... date set up:...............

Checked by: .... date set up:...............

Re-checked by: .... date set up:...............

ISOLATION

bleach / runing water / none

PDA / PDA(S) / ¼ PDA / ¼ PDA(S)

TWA / PARP / PARP-H

Result

By: .... date set up:...............

Checked by: .... date set up:...............

Re-checked by: .... date set up:...............

FLOAT

in Petri's solution / other ....

Result

By: .... date set up:...............

Checked by: .... date set up:...............

Re-checked by: .... date set up:...............

OTHER TESTS

soil pH / Soil EC ....

Result

By: .... date set up:...............

Checked by: .... date set up:...............

Re-checked by: .... date set up:...............

TaqMan Y / N   No. of tests ....

PCR Y / N   No. of tests ....

Further Action:

Diagnostician: .... **Date**:...............

Pass (whole/part) of sample to: Bac   Vir   Ent   Nem   Other (specify) ....

Keep sample in lab: Keep sample in cold store   Discard

Is this sample finished with?   Y / N   Diagnostician .... Date ....

References

Anon (1985) *Disease Assessment Manual for Crop Variety Trials*. National Institute of Agricultural Botany, Cambridge, UK.

Cook, R.T.A., Inman, A.J. and Billings, C. (1997) Identification and classification of powdery mildew anamorphs using light and scanning electron microscopy and host range data. *Mycological Research* 101, 975–1002.

Donnelly, D.P., Boddy, L. and Wilkins, M.F. (1999) Image analysis – a valuable tool for recording and analysising [analysing] development of mycelial systems. *Mycologist* 13, 120–125.

Ebbels, D.L. (2003) *Principles of Plant Health and Quarantine*. CAB International, Wallingford, UK.

Federation of British Plant Pathologists (1973) *A Guide to the Use of Terms in Plant Pathology*. Phytopathological Papers No. 17, FBPP [now British Society for Plant Pathology (BSPP)], Reading, UK.

Gottwald, S., Germeier, C.U. and Ruhmann, W. (2001) Computerized image analysis in *Fusarium* taxonomy. *Mycological Research* 105, 206–214.

James, W.C. (1971) An illustrated series of assessment keys for plant diseases, their preparation and usage. *Canadian Plant Disease Survey* 51, 39–65.

Muskat, L.C., Kerkhoff, Y., Humbert, P., Nattkemper, T.W., Eilenberg, J. and Patel, A.V. (2021) Image analysis-based quantification of fungal sporulation by automatic conidia counting and gray value correlation. *MethodsX* 8, 101218.

Ordynets, A., Keßler, S. and Langer, E. (2021) Geometric morphometric analysis of spore shapes improves identification of fungi. *PLoS ONE* 16, e0250477.

Petra, P. (2018) Identifying fungi spores, yeast, bacteria by opto-electronic imaging and image processing and identification for protecting human health. *Current Trends Biomedical Engineering & Biosciences* 11: 555806.

References of general interest

Bolton, M.D. and Thomma, B.H.J. (2012) *Plant Fungal Pathogens – Methods and Protocols*. Humana Press, Totowa, New Jersey.

Burns, R. (2009) *Plant Pathology – Techniques and Protocols*. Humana Press, Springer Science and Business Media, New York.

Carlile, M.J., Watkinson, S.C. and Gooday, G.W. (2001) *The Fungi*, 2nd edn. Elsevier Academic Press, Amsterdam.

Dugan, F.M. (2006) *The Identification of Fungi: An Illustrated Introduction with Keys, Glossary, and a Guide to Literature.* APS Press, St Paul, Minnesota.

Kavanagh, K. (2005) *Fungi: Biology and Applications*. John Wiley and Sons, Chichester, UK.

Lucas, J.A. (2020) *Plant Pathology and Plant Pathogens*. Wiley-Blackwell, Hoboken, New Jersey.

Mueller, G.M., Bills, G.F. and Foster, M.S. (2004) *Biodiversity of Fungi: Inventory and Monitoring Methods*. Elsevier, Amsterdam.

Spooner, B. and Roberts, P. (2005) *Fungi*. The New Naturalist Library No. 96. Harper Collins, London.

Waller, J.M., Richie, B.J. and Holderness, M. (1998) *Plant Clinic Handbook*. CAB International, Wallingford, UK.

Waller, J.M., Lenné, J.M. and Waller, S.J. (2001) *Plant Pathologists' Pocketbook*, 3rd edn. CAB International, Wallingford, UK.

3 Detection of Fungal Plant Pathogens from Plants, Soil, Water and Air

Michael Long[1]*, Aiga Ozolina[1] and Paul A. Beales[2]
[1]*Fera Science Ltd, Sand Hutton, York, YO41 1LZ, UK;* [2]*The Animal and Plant Health Agency, Sand Hutton, York, YO41 1LZ, UK*

3.1 Introduction

Approaches to detecting fungal plant pathogens vary, depending on the medium and requirements of detection survey. When examining plant material, direct visualization of symptoms on the host looking for fungal structures (signs) maybe sufficient as often there is a clear relationship between symptoms and signs, e.g. leaf spotting and the presence of rust pustules (Fig. 3.1) or white powdery spores of a mildew. However, further lab techniques may be required to induce fungal growth and sporulation of cryptic infections (Fig. 3.2) and from complex substrates. Immature fungal structures may be present on the host material, so conclusive identification cannot be made. Alternatively, fungal structures may be present but cannot be observed – for example, because of the overgrowth of saprophytic/opportunistic microorganisms or being 'deep seated' within the plant tissue. Some fungal structures (e.g. sclerotia) may be detected directly from soil, but more often specific techniques, including the use of bait material, are needed. Detection from other media such as water and air always requires specialist methods, described later in this chapter.

The evolution of serological and molecular methods (Chapters 8–10) has made it possible to detect fungi in these difficult matrices. Yet, owing to the resources required to develop and validate these methods, they are generally more suited to testing for individual pathogens in specific situations. However, traditional detection methods such as incubation, isolation and baiting are still commonly used because they are simple, sensitive, cost-effective and permit screening for a wide range of viable pathogens.

The desired level of confidence in the diagnosis will dictate the range of methods employed and experience will help to refine the processes and methods deployed. It is preferable to isolate the fungus from the host because this will permit a more reliable identification and, if cultures are available, further studies. However, this is not always possible, for example when dealing with an obligate parasite that cannot be grown in pure culture, or if isolates cannot be purified from the initial isolation plates or their preparation is economically impractical.

3.2 Incubation to Encourage Sporulation

Conditions that stimulate growth and sporulation should be guided by the possible pathogens expected based on the symptoms observed. For

---

* E-mail: michael.long@fera.co.uk

DOI: 10.1079/9781800620575.0003

Fig. 3.1. *Puccinia striiformis* affecting wheat. Clear relationship between leaf chlorosis, spotting symptoms and signs of rust fungus. (From Shutterstock.)

Fig. 3.2. Basal stem infection along with root necrosis symptoms, but no immediately obvious signs of fungal infection. (From Shutterstock.)

some studies, it may be possible to define quite precise conditions of temperature, humidity and light based on previous sample experience or information gleaned from the published literature. However, for problems with unknown causes, generic conditions relevant to the host's growing condition should be used.

Most incubation chambers typically consist of a large, clear-sided plastic box with a lid that seals tightly (Fig. 3.3), although alternatives such as inflated plastic bags or Petri dishes may also be used (Figs 3.4–3.5).

Plant material is either placed directly on moist (but not saturated) tissue or filter paper, or

Fig. 3.3. Clear-sided, disposable box with moist paper used as a damp chamber. (© UK Crown Copyright – courtesy of Fera.)

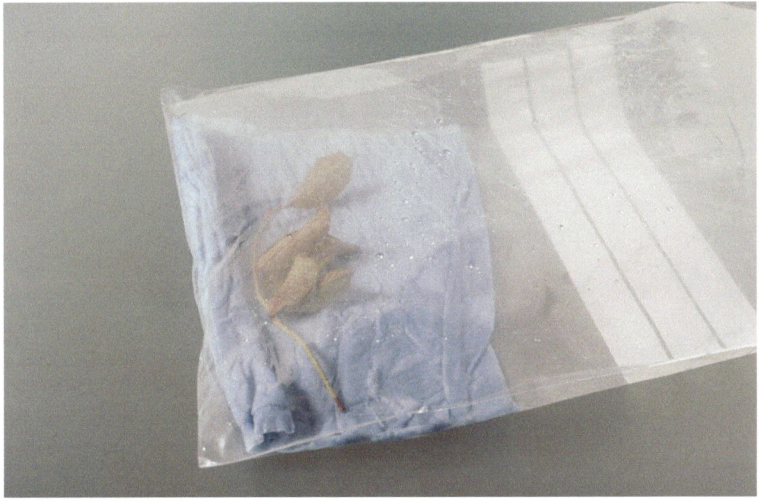

Fig. 3.4. Inflated plastic bag with moist paper used as a damp chamber. (© UK Crown Copyright – courtesy of Fera.)

on a platform above the water or, in the case of some fruits and tubers, specimens should be placed on dry paper. Full details are provided in Protocol 3.1. Chambers are frequently incubated at 18°C to 22°C in natural daylight, but an incubator may be used to provide controlled conditions. Chambers should be checked frequently, depending on host/pathogen (this is usually every 2–3 days, although some pathogens sporulate within 1 day (e.g. potato tuber pieces infected with blight (*Phytophthora infestans*)). Ensure the chamber does not dry out for the duration of the incubation. An atomizer can be used to maintain humidity by spraying the inside of the chamber. Chambers should be checked frequently at the beginning to prevent rapid overgrowth by common saprophytic fungi or bacteria, or infestation by invertebrates such as mites or fly larvae. Typically, mycelial growth will occur after several days, sporulation after 3–5 days and fruiting

Fig. 3.5. Petri dish with moist filter paper used as a damp chamber. (© UK Crown Copyright – courtesy of Fera.)

bodies may form after 5–7 days, but it may take many weeks and even months for the culture to mature and spores to form. Commonly, there is an initial flush of fast-growing fungi (frequently saprophytes, such as *Rhizopus*, *Mucor* or *Cladosporium*). Systemic pathogens and fungal fruiting bodies such as sclerotia will take longer to develop. Some fungi, such as *Botrytis cinerea* (the cause of grey mould), that may not sporulate in culture often do so more readily on incubated plant material.

Water moulds, such as *Phytophthora* or *Pythium*, can be induced to sporulate by floating infected plant material in Petri's mineral solution (see Fig. 3.6 and Appendix 1), filtered unsterile/ sterile pond water or sedge peat.

3.3 Isolation from Plant Tissue onto Culture Medium

To identify consistent morphological features of plant pathogenic fungi that grow readily in culture, isolate the organism by culturing it on to a simple or semi-selective culture medium under controlled conditions. Methods are described in Protocol 3.2.

Decontamination and sterilization of samples are important steps of any isolation protocol. Most samples will require some form of decontamination to remove common saprophytes because these will rapidly outcompete potential pathogens of interest. Remove by rinsing in flowing water any growing medium such as soil as it may both harbour pests and diseases not of direct interest and can also reduce the effectiveness of sterilizing agents. Good packaging along with rapid transport to the laboratory (Chapter 2) will help to reduce contamination and decay. A wide range of sterilants may be used, but commonly either 70% ethanol or 10% sodium hypochlorite (domestic bleach) are used (see Appendix 3). Older texts may refer to the use of mercuric chloride ($HgCl_2$), but this is now not recommended due to health and safety implications. The duration of treatment must be determined by experimentation but typically can last anywhere from 30 s to 5 min. Sterilants may be removed before plating onto culture medium by

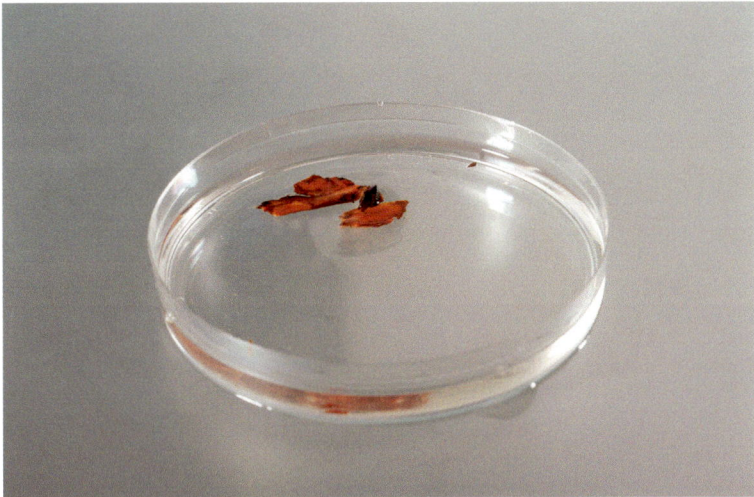

Fig. 3.6. Sections of strawberry crown material floating in Petri's mineral solution to encourage sporulation of *Phytophthora* or *Pythium*. (© UK Crown Copyright – courtesy of Fera.)

blotting onto sterile filter paper or washing in sterile distilled water and blotting dry. Small pieces of tissue should be transferred aseptically onto a range of basic media such as tap water agar, potato dextrose agar or cherry decoction agar, unless specific nutrient requirements are known (see Appendix 1). An example of plating out from symptomatic host plant tissue is shown in Figs 3.7 and 3.8.

There is a vast range of semi-selective nutrient agar media for isolation of specific organisms to encourage sporulation, such as SNA for *Fusarium* and PARP-H for *Phytophthora*. A comprehensive list is provided by Shurtleff and Averre (1997). In general, media with low levels of carbon and nitrogen and those containing plant extracts (such as decoction agars) encourage fungal sporulation. Using a slightly acidic medium (pH 6–6.5) or the addition of antibiotics such as streptomycin or penicillin will help reduce bacterial growth. Culture plates should usually be labelled on the base (not on the lid) with a sample reference number and date of isolation, invert so the lid is on the bottom and incubate. This prevents condensation dripping back on to developing colonies and affecting their growth. Seal plates with electrical tape, Parafilm® or clear sticky tape. Porous tapes are available (such as those used to secure bandages, e.g. Micropore) and these have the benefit of preventing plates from drying out too quickly, while still preventing both the invasion or escape of other organisms (especially mites) and contamination if plates are accidentally knocked over. Following initial isolation, plates should be checked every 2–3 days and sub-cultured from the colony margin/single spores if pure isolates are required.

3.4 Isolation from Leaf Spots and Blights

Various fungal pathogens can cause spots to develop on leaves. These pathogens often spread their spores by wind or splash dispersal between host plants. Fungal leaf spots are generally visible as necrosis on the leaves, in discrete spots on the upper and lower surfaces. These spots are often pale brown to black, and roughly circular, almost always with discrete margins. When further developed, they may coalesce to form large irregular shaped blotches, or blight the entire leaf.

While some fungi cause quite characteristic necrosis on leaves, many have similar symptoms, but can be distinguished by closer examination using a dissecting microscope due to the presence or absence of certain fungal structures. Hyphomycete fungi can produce white or brown mycelium over the necrotic spot. These include commonly encountered leaf-spot pathogens of

Fig. 3.7. Leaf dieback of lavender showing the leading edge between healthy and necrotic tissue. (© UK Crown Copyright – courtesy of Fera.)

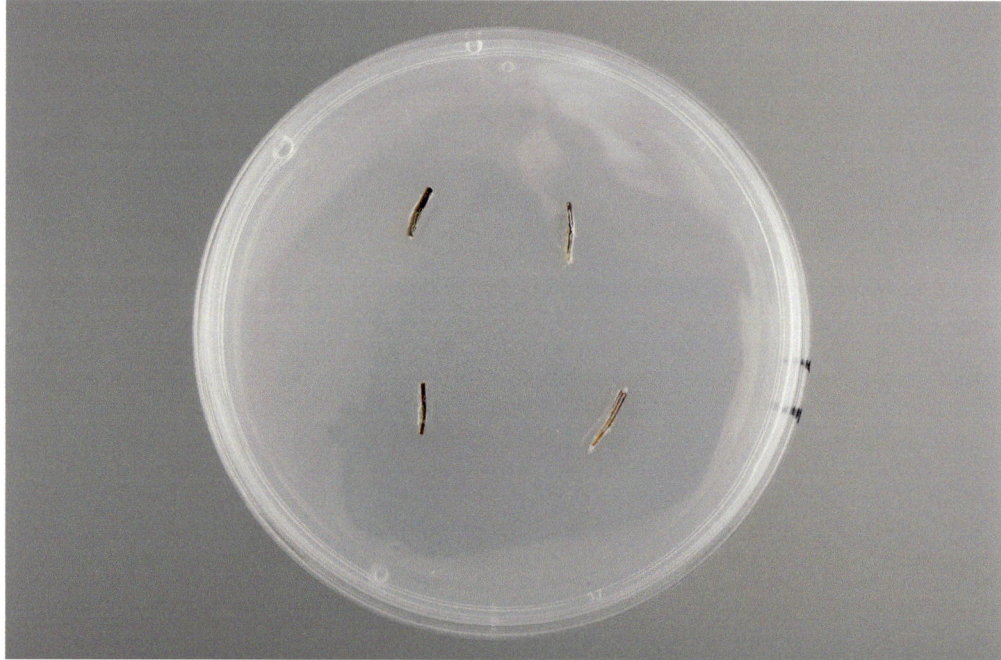

Fig. 3.8. Plant material including the leading edge between healthy and necrotic tissue aseptically transferred to growing culture medium. (© UK Crown Copyright – courtesy of Fera.)

the fungal genera *Ramularia* and *Cercospora*. Coelomycetes, on the other hand, such as *Colletotrichum* spp., can produce fruiting bodies containing their spores in cups on the leaf surface. However, the rusts (order Pucciniales) produce pustules containing a mass of spores, often on the underside of the leaf, whereas smuts (mostly in the order Ustilaginales) produce characteristic powdery black spores. Lastly, a variety of leaf spots and blights are caused by oomycetes such as *Phytophthora ramorum* and *P. kernoviae*, downy mildews and Albugo 'white rusts' (Fig. 3.9).

Most leaf-spot diseases are caused by fungi, but a few are caused by bacteria, foliar nematodes or other pathogens. Insect pests can also produce leaf-spotting symptoms, particularly Eriophyid mites and gall wasps, as well as the abiotic disorder called oedema. These three characteristically cause leaf tissue distortion and/or swelling so can be easily confused, but they can be distinguished by microscopic examination. Eriophyid mites trigger excessive hair growth on the leaf underside, gall wasp larvae develop in uniform, hollow domes and oedemas are generally not hollow but irregular swellings due to excess water (Fig. 3.10).

Protocol 3.3 explains the procedure to follow to perform microscopic visual examination of a sample of leaf material with symptoms of leaf spotting, to determine which causal agent may be present. If the causal agent cannot be ascertained on initial examination, guidance is given on isolating and incubating the pathogen. More detailed instructions on these steps can be found in Protocol 3.2 (Isolation from leaves, stems, flowers and fruits) and Protocol 3.1 (Incubating samples in damp chambers).

3.5   Baiting for Fungal Pathogens

Baiting (using bait plants or parts/pieces thereof) is a particularly useful tool for detecting fungal or fungus-like pathogens from what can often be difficult substrates (e.g. soil and water). This is described in Protocol 3.4. The technique enables large amounts of the specific medium to be tested, detects viable propagules (useful when determining whether a disease is pathogenic) and can be used as both a field and a laboratory diagnostic aid. In the field, a bait can be used to detect pathogens across numerous environments, from areas with limited access over long periods of time. For example, bait testing a watercourse or flowing stream will attract *Phytophthora* zoospores from many thousands of litres of water. Baiting can also be used in the laboratory as a precursor to other tests, e.g. the detection of *Plasmodiophora brassicae* (the causal agent of clubroot disease) from soil using bait plants (Protocol 3.3(b)). The bait plants can then be subjected to molecular testing or direct visualization to detect the pathogen (see Chapters 8 and 9). Figure 3.11 shows the constituents of a baiting device (bag of baits, or 'Bobs'), also described in Protocol 3.4.

Fig. 3.9. Downy mildew on *Limonium* caused by *Peronospora statices*. (© UK Crown Copyright – courtesy of Fera.)

Fig. 3.10. (a) Eucalyptus gall wasp and (b) oedema on *Eucalyptus*. (© UK Crown Copyright – courtesy of Fera.)

A wide range of bait plant materials can be used, including leaves, stems, roots, seeds, seedlings, cotyledons, roots and tubers. Selection of material depends on the pathogen and the environment of interest. Baits may be enclosed in bags or cages to aid recovery from the test medium. They must be disease free, and wherever possible not have been recently treated with fungicides (preferably not within 6 weeks); they may require surface sterilization before use. In some circumstances, a range of plant bait materials may be enclosed in bags or cages if the aim of the study is to obtain a broad range of fungal pathogens or the pathogen has a wide

Fig. 3.11. A baiting device (bag of baits, or 'Bob') consisting of muslin or cheesecloth square, small pebbles and healthy leaf squares (single species or mix of different leaves susceptible to a range of *Phytophthora* spp.). (© UK Crown Copyright – courtesy of Fera.)

host range. The age, cultivar and type of plant most suited to baiting should be determined by experiment, and this may also help to determine the specificity, sensitivity, recovery rate and optimum baiting time. Baiting times may vary from a few days to even several weeks or months, depending on the biology of the pathogen. However, if left for longer periods, baited material itself can decay and may harbour a wide range of saprophytic organisms, making recovery of the target organism challenging. It might be possible to detect the pathogen through direct observation (hand lens/microscope), incubation, isolation and serological or molecular methods, as appropriate.

3.6   Isolation of Soil Fungi

There are several techniques for determining the total number of microorganisms in soil. With careful manipulation of the culture media and conditions, cultural methods can be used to detect the fungal pathogens of interest. Numerous methods have been described for isolating fungi from soil, including suspension plating (dilution plating), Warcup's soil plates, particle filtration (soil washing) and various adaptations of these, which are comprehensively reviewed in Dhingra and Sinclair (1985). For plant pathogens, suspension plating (Protocol 3.5) is perhaps the most frequently used owing to its simplicity and the fact that it can be tailored to different pathogens by using semi-selective media. In general, suspension methods favour fungi that sporulate profusely or exist primarily as spores, while soil-plate methods are better for fungi that exist as mycelium. Molecular techniques are increasingly being developed to test for pathogens by real-time PCR or sequencing (Chapters 9 and 10).

3.7   Spore Trapping

Many fungal pathogens, such as rusts or mildews, produce airborne spores, and a wide range of air/rain-sampling devices have been developed to detect them. There are three commonly used approaches: impactor devices, suction traps or sticky surface traps. Popular traps include Burkhard's volumetric spore trap (Fig. 3.12) and the Rotorod sampler. Trap selection depends on the end use required – determination of either spore load (concentration of spores for a given volume of air) or the number of spores deposited on a surface. Protocol 3.6 describes simple methods for trapping spores.

Fig. 3.12. Burkhard spore trap for airborne spore collection. (© UK Crown Copyright – courtesy of Fera.)

The development of serological (Chapter 8) and molecular techniques (Chapters 9 and 10) has generated new opportunities for detecting and enumerating fungal propagules from trapping devices that have previously relied on direct observation or cultural methods. For example, Rogers *et al.* (2009) have quantified airborne inoculum of *Sclerotinia sclerotiorum* – a damaging pathogen of over 400 plant species worldwide that causes diseases such as a serious stem blight of oilseed rape (canola) – using quantitative PCR. DNA can be extracted from wax-coated plastic tapes (such as those used in the Burkhard (Hirst-type) spore trap and rotating arm traps). A linear relationship between ascospore numbers and *S. sclerotiorum* was found which permitted highly sensitive and specific spore enumeration, thereby negating the need for laborious visual examination.

3.8 Size-selective Sieving for Fungal Propagules

Soil-borne fungi that produce relatively large, thick-walled propagules, such as sclerotia, microsclerotia or sporangia, may be detected according to their physical size. A variety of wet or dry sieving methods have been developed for quantifying propagules of these pathogens, e.g.

the sclerotia of *Sclerotinia cepivorum* (the causative agent of onion white rot), the microsclerotia of *Verticillium dahliae* (which causes wilt on a wide range of crops), the sporangia of *Synchytrium endobioticum* (the causal agent of potato wart disease) or the bunt spores of *Tilletia* spp. from grain samples. A specific method used for quantifying the microsclerotia of *Verticillium dahliae* in soil is described in Protocol 3.7. The sieving methods used for *Tilletia* spores and for the soil-borne sporangia of *Synchytrium endobioticum* are shown in Figs 3.13 and 3.14, respectively.

When sieving soil, a representative sample is first air dried, while for grain a suitable sample is soaked in water. Sub-samples are passed through a series of metal sieves typically ranging from 2 mm to 20 μm in diameter, aided by water or shaking. Propagules can then be removed from specific sieves depending on their known size range. Enumeration may occur by direct observation aided by a variety of techniques, such as chloroform extraction or the use of sucrose gradients to remove debris. Other techniques include culturing on semi-selective media to determine the number of colony forming units (CFU) or testing using serological or molecular techniques. These methods are frequently very time consuming and they require considerable skill to obtain consistent and reliable results.

Fig. 3.13. Size-selective sieves for *Tilletia* spores. (© UK Crown Copyright – courtesy of Fera.)

Fig. 3.14. Vibrating, wet-sieving apparatus for soil-borne pathogens such as *Synchytrium endobioticum*. (© UK Crown Copyright – courtesy of Fera.)

3.9 Establishing Pathogenicity (Koch's Postulates)

Classically, pathogenicity is confirmed by completing a series of procedures that establish four criteria, referred to as Koch's postulates. These were formulated in the 1880s by the medical bacteriologist Robert Koch while he was investigating anthrax (caused by *Bacillus anthracis*), but have become accepted for other pathogens, including certain fungi, viruses and viroids, as well as nematodes and parasitic plants. The postulates require that:

1. The suspected causal agent must be consistently associated with the disease symptoms.

2. It must be isolated from the diseased plant and grown in pure culture.

3. When a pure culture of the suspected causal agent is inoculated into a healthy susceptible plant, the host must reproduce the same disease symptoms as those that were observed originally.

4. The same causal agent must be isolated from the freshly diseased tissue and it must have the same characteristics as the organism that was first isolated.

An example of the artificial inoculation of a potted plant of *Chamaecyparis* with *Phytophthora* to prove the pathogenicity of the organism (to part fulfil Koch's postulates) is shown in Fig. 3.15.

For many plant pathogens, establishing Koch's postulates is possible, although at times it can be difficult. However, for obligate parasites such as rusts and mildews that cannot be grown in pure culture (criteria 2, above), these cannot be followed strictly. In these cases, it may be possible to inoculate a healthy susceptible plant directly with spores from the initial diseased plant (criteria 3), thus avoiding the need for criteria 2. The characteristics of the organism should then be the same as those observed on the initial sample (criteria 4).

In practice, the establishment of Koch's postulates is mainly undertaken when a disease is discovered on a new host plant, and it is important to prove that the organism observed is the primary cause of the problem. Protocol 3.8 describes the practical procedures that can be followed to establish Koch's postulates.

Fig. 3.15. *Chamaecyparis* artificially inoculated with *Phytophthora* to prove pathogenicity as part of Koch's postulates. Inoculation points are wrapped in a sterile damp cotton wool taped with waterproof tape such as Parafilm®. (© UK Crown Copyright – courtesy of Fera.)

Protocol 3.1

Incubating samples in damp chambers

Materials

- Absorbent paper
- Clear-sided boxes
- Plastic trays
- Plastic bags
- Petri dishes

Method

1. Choose a suitable box, Petri dish or plastic bag with adequate space to fit the sample comfortably.

2. Place a dry piece of clean absorbent paper within the incubation chamber.

3. Moisten the paper (do not saturate) and place the plant sample within the chamber, seal and incubate under controlled conditions (18–25°C, 12–16 h light). A windowsill with natural lighting is often sufficient, although if using this method, be careful not to expose samples to extreme temperatures and light fluctuations such as are frequently encountered on some windowsills. Fruits and tubers should be placed on dry not damp paper to reduce bacterial soft rotting.

4. For woody material, stems or crowns it can be helpful to cut transverse or longitudinal sections to reveal the vascular tissue.

5. Check every 2–3 days, remove any test material heavily contaminated with fast-growing saprophytes such as *Rhizopus* or *Penicillium*.

6. Incubate for at least 7–10 days and up to 4–8 weeks as appropriate. However, for some fungi the incubation period may need to be longer. Check moisture levels and re-wet/replace paper as necessary.

Protocol 3.2

Isolation from leaves, stems, flowers and fruits

If the tissue is totally degraded and no leading edge can be found, isolation of the primary organism may be difficult owing to numerous secondary/opportunistic organisms overrunning the primary fungus of interest. More extensive surface decontamination may be required if this is the case or, alternatively, a bait test (see Protocol 3.4) may aid the isolation of the primary organism. Some fungi produce structures (e.g. sclerotia) that can easily be removed from plant tissue or spores that can be picked off directly using a sterile needle.

Materials

- Scalpel, sharp scissors or single-edged razor blades
- Mounted needles or disposable hypodermic needles
- 70% ethanol solution
- 10% bleach (sodium hypochlorite)
- Petri dishes containing appropriate nutrient agar medium
- Empty Petri dishes for bleaching
- Sterile distilled water
- Forceps

Method

1. Flame sterilize a needle, scalpel or scissors, or use a new single-edged razor or needle. The implement needed depends on the host and how much fungus material there is.

2. Select tissue showing symptoms of interest. Choose an area with a leading edge (i.e. the area between healthy and affected tissue – generally, where the organism is most actively growing and the most reliable place for successful isolations) and cut out with the implement selected.

3. Surface decontaminate to remove secondary and opportunistic fungi and bacteria that could cause contamination on the isolation plates. If looking for *Phytophthora* or *Pythium* species, surface decontaminate by rinsing thoroughly in sterile water. If not looking for *Phytophthora*, place the selected part of the sample in 10% bleach for 2–5 min, depending on the type of plant material you're isolating from:

Leaves: Those with a thick cuticle (e.g. *Ilex* spp.) can be surfaced decontaminated by wiping with a 70% ethanol solution. Sections of thinner leaves should be placed in a solution of 10% bleach in a Petri dish for 30 s to 5 min relative to their thickness. Rinse with three changes of sterile distilled water.

Stems: Green shoots can be surface decontaminated in 10% bleach as described for leaves above. Older more woody stems can also be surface decontaminated in 10% bleach (for 3–5 min) or immersed in 70% ethanol for 3 min. Alternatively, the affected woody tissue can be held with forceps and scorched gently in a flame.

Flowers: Surface decontaminate in 10% bleach for up to 1 min. Rinse with three changes of sterile distilled water.

Fungal structures: It is advisable to surface decontaminate structures such as sclerotia by immersing in 10% bleach for 3–5 min.

4. Use the selected sterile implement to place on or immerse into the agar. Typically place four pieces per plate. Note: The vast majority of commonly occurring plant pathogens and spoilage fungi will grow very well on potato dextrose agar (PDA). This can be prepared from a commercial formulation. For the first isolation the media may be supplemented with antibiotics to counteract any bacteria, e.g. penicillin and streptomycin. Incubation temperature may be selected based on what is known about the fungus; most commonly occurring plant pathogens grow quite well at between 17°C and 22°C.

5. Label the Petri dish with a unique reference number and date. If there is more than one plate, then differentiate with a further label such as 'Leaves' and 'Stems' to avoid confusion when reading the plates.

6. Incubate for between 5 and 7 days, observing every couple of days to follow culture progress and check if pure. At the early stage of incubation, the organism can be sub-cultured on to fresh medium if the culture is not pure.

7. Subculture from the margin of the colony, as appropriate, to obtain pure cultures if required.

8. If no growth occurs after 7 days, consider re-isolating but reduce stringency of the sterilization process.

Protocol 3.3

Identifying the causal agent of leaf spots and blights

Fungal leaf spots are often visible as brown to black necrosis on the leaves, in discrete, roughly circular spots on the upper and lower surfaces. As they develop, they may coalesce to form large irregular blotches or blight the entire leaf.

> **Materials**
>
> - Dissecting microscope
> - Cotton/trypan blue (0.1%) in lactoglycerol
> - Glass microscope slides
> - Flame heating device (e.g. gas or spirit burner)
> - Fine mounted needle or sterile hypodermic needle
> - Sharp single-edged razor blade or scalpel
> - Coverslips
> - Compound microscope
> - Clear adhesive tape
> - Scissors
> - 10% bleach (sodium hypochlorite)
> - Forceps
> - Potato dextrose agar (PDA)
> - Clear-sided plastic boxes
> - Extra absorbent paper towel
> - Small plastic bags
> - P_5ARP-H (PARP-H) agar
> - Petri dishes
> - Petri's mineral solution

Method

1. Place leaf material under a dissecting microscope and select the lowest magnification ($\times$10). Examine the upper and lower surfaces of a range of different spots and/or blights, then increase magnification up to $\times$60.

2. Record in Table 3.1:
- The colour of the margin and centre of spots (Fig. 3.16).
- If found on the upper/lower surface of leaves.
- The extent and distribution of symptoms, if from tip, edges of leaves, base or veins.

- If lesions are limited by the leaf veins or traverse them.
- Presence of any mycelium, conidiophores and/or setae (dark thorn-like structures) and if they are associated with the leaf spots.
- If pustules are present (raised spore masses that rupture epidermal leaf tissue (Agrios, 2005)) on upper and/or lower surface, and what colour they are (Fig. 3.17).
- If suspected fruiting bodies (small black bumps) are present.
- Excessive hair growth on leaf underside.
- Irregular or regular swelling of leaf tissue on upper and/or lower surface.

3. *Suspected fruiting bodies and pustules.*

3.1. Place a drop of lactoglycerol onto the surface of a microscope slide.

3.2. Under the dissecting microscope, carefully pick off several suspect fungal fruiting bodies with the tip of an alcohol-flamed mounted needle or sterile hypodermic needle and place them into the lactoglycerol drop. If it is not clear how the spores are attached, cut a thin section out of the fruiting body with a razor blade or scalpel.

3.3. Examine the slide under the dissecting microscope to confirm that structures were successfully removed and are correctly orientated for examination. Excessive amounts of host material can be teased away and discarded.

3.4. Gently lower a coverslip onto the slide then press down lightly to expel any air bubbles. If necessary, holding the slide at one end, gently warm it over a spirit burner flame until any air pockets start to expand, then remove the slide from the flame and place on the bench to cool.

3.5. Transfer the slide to a compound microscope and proceed to section 6.

4. *Aerial mycelium, chemical deposits or unknown structures.*

4.1. Cut off a small piece of clear adhesive tape (up to 30 mm long) and gently press adhesive side down onto the leaf material.

Fig. 3.16. Leaf spot (caused by *Mycocentrospora acerina*) on *Prunus avium*. (© UK Crown Copyright – courtesy of Fera.)

Fig. 3.17. *Puccinia modiolae* rust pustules on the underside of a *Lavatera* leaf. (© UK Crown Copyright – courtesy of Fera.)

4.2. Lift the tape off and place it adhesive side down onto a small drop of lactoglycerol on a microscope slide.

4.3. Press down gently to remove air bubbles, then warm very carefully to avoid boiling which may melt the tape. Allow slide to cool and transfer it to a compound microscope and proceed to section 6.

5. *Suspect oedema (irregular swelling of leaf tissue).*

5.1. Under the dissecting microscope and using a razor blade, take a 1-mm thick cross-section slice at the sites of the leaf swelling. Steady the blade by pushing it down along a fingernail and include some healthy tissue either side of the bump to compare the cell morphologies.

5.2. Add the cross-section slice to lactoglycerol on a slide and examine under the compound microscope.

5.3. The cells of leaves with oedema increase in size towards the centre of the bump.

6. *Examine the mounted material with a compound microscope.*

6.1. Select a low power objective (e.g. ×100) and locate the suspect fungal structures. Record morphological features in Table 3.1.

6.2. Expel contents of any fruiting bodies by very carefully and gently pressing down on the coverslip. Alternatively, the slide can be removed from the microscope and gently tapped with a pencil.

6.3. Re-examine spores and spore-producing structures under ×400 magnification.

7. *Isolations and incubations.*

To encourage growth and/or sporulation of the suspected primary pathogen if it cannot be confidently determined under the microscope:

7.1. With flame-sterilized scissors, cut out and place four sections of the leading edge of leaf spots of different leaves in 10% bleach for 2 min.

7.2. Plate out onto PDA with forceps, and incubate for 5–7 days, observing every couple of days, and subculture onto new agar plates during early growth stages if any contaminants are growing with it.

7.3. Place some remaining leaves with representative symptoms in a plastic box on a damp piece of paper towel, as described in Protocol 3.1.

8. *Aerial Phytophthora species.*

For brown or black leaf spots/blighting (Fig. 3.18). See EPPO (2006) Protocol for *Phytophthora ramorum* for more detail:

8.1. With flame-sterilized scissors, cut out four representative lesions to include the leading edge. Place these in a small plastic bag.

8.2. Add enough tap water to the bag to cover the sample, and soak for at least 15 min.

8.3. Gently agitate the bag and pour off the wash water twice.

8.4. Plate onto PARP-H selective agar with sterile forceps and incubate at room temperature for 5–7 days.

8.5. If the mycelium typical of *Phytophthora* is present but no sporangia are observed, float pieces of agar containing the mycelium in a clean Petri dish, containing Petri's mineral solution.

Fig. 3.18. Brown leaf blighting caused by *Phytophthora ramorum* on *Arbutus*. (© UK Crown Copyright – courtesy of Fera.)

Table 3.1 Leaf examination checklist. Fill in as fully as possible.

General details	
Location	
Date of collection	
Reference number	
Host species	

Continued

Table 3.1 Continued.

Leaf spot characteristics	
Leaf spot	Colour in centre: Colour at margin: Upper surface ❑ Lower surface ❑ Both surfaces ❑ On leaf tip ❑ edges ❑ base ❑ veins ❑ Limited by leaf veins ❑ Crosses veins ❑
Mycelium	Present ❑ Absent ❑ Associated with leaf spots ❑ Not associated with leaf spots ❑
Conidiophores	Present ❑ Absent ❑
Setae	Present ❑ Absent ❑
Pustules	Present ❑ Absent ❑ Upper surface ❑ Lower surface ❑ Both surfaces ❑ Colour of pustule: Colour of contents:
Fruiting bodies	Present ❑ Absent ❑
Excessive hair growth on underside	Present ❑ Absent ❑
Leaf swelling	Present ❑ Absent ❑ Upper surface ❑ Lower surface ❑ Both surfaces ❑ Regular ❑ Irregular ❑

Protocol 3.4

Baiting

(a) Baiting from water – testing for *Phytophthora* spp. and *Pythium* spp.

Materials

- Muslin cloth (cheesecloth) 15 × 15 cm square
- 6 g of stones (autoclaved) per Bob
- 1 × rhododendron (*Rhododendron catawbiense*) leaf cut into roughly 2-cm lengths using surface-sterilized scissors/scalpel, 1 × camellia leaf, 1 × sprig or small branch of hebe, 2 pine needles, 4 blades of pre-autoclaved grass (for *Pythium* testing only)
- 1 polystyrene float (packing chips are suitable) per Bob
- 2.5 m string
- Polythene bags
- P$_5$ARP-H (PARP-H) or PARP semi-selective agar medium
- Bamboo cane/stick to attach bait.

Method

1. Place the autoclaved stones, float and four of each 2 cm leaf piece/leaf into the centre of the muslin square and gather the corners together so that there are no gaps.

2. Tie together tightly with the length of string, securing with a knot. Wrap the string around the fastening again and secure with another knot. Repeat this again so the bag has been fastened by three knots in total.

3. Wet the bags of baits (referred to as 'Bobs') by dipping them into tap water, then place in a plastic bag and seal. This will prevent the leaves drying out and help to keep them fresh. Keep the Bobs refrigerated until required (up to 1 month) or freeze (up to 6 months). Use within 2 days if kept at room temperature.

4. Select watercourse to test. Baits work most efficiently in flowing water. Use one Bob for every 0.5 km of stream/river.

5. Place Bobs in test water (they should float just below the surface of the water). Secure string to bank using a cane/stick, label and leave for 3–5 days.

6. Collect Bobs and place in polythene bag prior to testing. (Note that leaves should be tested as soon as possible following removal from water.)

7. Testing procedure. Remove string, place leaves into a small polythene bag. Place grass blades (if present) into a separate bag.

8. Add 50 ml tap water to bag and wash leaves by rubbing from the outside of the bag. Repeat three times, or until the wash water becomes clear.

9. Using surface-sterilized forceps place four pieces of baited leaf material onto PARP-H medium (use PARP medium for grass blades).

10. Incubate plates at 18–25°C for 6 days (12 h light/dark cycle).

11. Examine plates for typical morphological features of *Phytophthora* spp. (PARP-H medium) or *Pythium* spp. (grass blades on PARP medium). (See Protocol 6.7.)

(b) Baiting from soil

Materials

For an indirect test:

- Large sandwich boxes (to accommodate 1 kg of soil)
- Petri's mineral solution
- Bait leaves
- Muslin cloth

For a direct test:

- Bobs as described in Protocol 3.4(a)
- Bamboo or equivalent cane

Method

1. *Indirect test*

Remove 1 kg soil made up from 10 g cores from 100 points from a 1 ha test area.

Place test soil into sandwich boxes and fill with Petri's mineral solution until the solution is 1 cm above the top of the soil.

Float detached bait leaves (rhododendron, camellia, hebe, pine and grass blades (*Pythium* only) as described in Protocol 3.4(a) for the preparation of Bobs) on the surface of Petri's mineral solution.

The bait leaves can be separated from the soil material by wrapping in a piece of tied autoclaved muslin cloth, to make locating the pieces easier if required.

It is advisable to run a negative control alongside test material using sterile soil.

Replace lid and incubate at 18–25°C for 7 days.

Identify the presence or absence of *Phytoph-thora* spp. or *Pythium* spp. (see Protocol 6.7).

2. Direct test

Bobs produced in Protocol 3.4(a) for testing water can be used directly in soil. However, it is essential that the soil tested always remains wet (this can be achieved by regular watering by hand if the ground is dry).

Dig a small hole into test soil *c*.10 cm deep.

Place bait bag and cover with soil. Ensure string remains above ground level and secure/mark with a bamboo cane or equivalent. Leave for 3–5 days.

Identify presence or absence of *Phytophthora* spp. or *Pythium* spp. (see Protocol 6.7).

Protocol 3.5

Quantitative assessment of fungal colonies from soil

Materials

- Trays (sterile)
- Beaker (sterile)
- Sterile distilled water
- Stirrer
- Pipette (sterile)
- Potato dextrose agar (PDA) amended with streptomycin and penicillin antibiotics
- Spreader (sterile)
- 70% ethanol solution

Method

1. Air dry 250 g (wet weight) test soil. Spread evenly over a disinfected (sprayed and wiped with 70% ethanol) plastic tray.

2. Prepare a 1×10^{-1} soil dilution by weighing 1 g of air-dried soil into a small sterile beaker and add 10 ml sterile distilled water. Shake well or mix with a magnetic stirrer to attain homogeneity.

3. Using a sterile pipette, transfer 1 ml of the suspension to 9 ml sterile distilled water, label: 1×10^{-2}.

4. Repeat step 3 until a dilution series up to 1×10^{-6} has been achieved (ensure thorough mixing before pipetting at each step).

5. Pipette 100 µl of 1×10^{-2}, 1×10^{-4} and 1×10^{-6} dilutions onto PDA medium amended with streptomycin and penicillin (other non-selective media can be used, although amending with antibiotics is generally necessary to inhibit bacteria within the preparation). Ensure plates are suitably labelled.

6. Spread diluent over surface of agar using a disposable or sterile bent plastic or glass rod ('hockey stick'). Use a new plastic rod, or re-sterilize the glass rod, between dilutions.

7. Incubate under suitable conditions for 2–7 days. In general, total fungal counts from sources such as field soil in temperate locations can be incubated at around 18–20°C with a 12 h light/dark cycle.

8. Calculate the number of colonies per gram of soil (i.e. number of colonies × dilution factor). For example, if you counted 50 colonies on 1×10^{-6} dilution, you would have $50 \times 1,000,000 = 5 \times 10^7 \, g^{-1}$ soil.

Protocol 3.6

Spore trapping

(a) Trapping on deposition slide traps

Materials

- Tap water agar (TWA)
- Microscopic slides
- Large coverslips
- Bamboo cane
- Mounting fluid, e.g. cotton blue in lactoglycerol/water

Method

1. Pour molten, sterile TWA on to microscope slides to a depth of a few millimetres and allow to set.

2. Depending on the nature of the work to be undertaken, the slide can be set up in the location of interest either horizontally or attached to a small bamboo cane or rod and placed vertically (an angle of 42° has proved useful when trapping rust spores; Gregory and Stedman, 1953). It is advisable to protect the slide from rain or strong winds where possible.

3. Remove slide after 1 day, add a few drops of mounting fluid and carefully place large coverslip on slide.

4. View under a compound microscope and record number/range of spores.

(b) Rain trapping – e.g. this method is good for trapping aerial spores of *Phytophthora* spp.

This is a simple but effective method for trapping aerially dispersed spores of *Phytophthora* spp.

Materials

- 500 ml to 5 l screw-lid plastic bottle (new or sterilized)
- Electrical tape
- Bamboo cane
- Plastic funnel
- Measuring cylinder
- Rhododendron leaves (preferably *R. catawbiense* var. 'Cunningham's White')

Methods

1. Fix plastic bottle to bamboo cane at base so that the bottle faces upwards (use appropriately sized bottle depending on length of time the rain trap will be left in its location).

2. Place in location of interest. To test for rain splash of *Phytophthora* from the ground, place at ground level, or to test for deposition in rain (or being washed out from overhanging shrubs or trees), place 1–2 m above the ground.

3. Unscrew bottle lid and replace with funnel.

4. Leave in test location for desired time, then remove trap and record the amount of water captured in a measuring cylinder. Assess the presence of *Phytophthora* by one of the two methods.

Bait testing (qualitative and spore viability test)

5. Float 4 × 1 cm square pieces of rhododendron (wipe leaves with 70% ethanol and allow to evaporate before cutting) on the surface of collected water. Incubate for 3 days, wash and then culture any *Phytophthora* spp. on semi-selective medium (see Protocol 3.4), or if available carry out appropriate molecular test on leaves.

Filter (quantitative test)

6. Filter collected water through an appropriately sized filter paper (depending on the size of spore). The filter paper can then be tested by quantitative real-time PCR.

Protocol 3.7

Quantification of microsclerotia of *Verticillium dahliae* in soil (after Harris *et al.*, 1993)

Materials

- Metal sieves (2 mm, 20 µm and 160 µm)
- Balance
- Screw-top bottle
- Distilled water
- Shaker
- Pipette
- Incubator
- Dissecting microscope
- Semi-selective media

Method

1. Collect a representative soil sample (see Protocol 3.4(b)) as the pathogen can be distributed unevenly across a field.

2. Keep the sample cool (<20°C) and dispatch to a laboratory for testing as soon as possible after sampling.

3. Air dry soil in the laboratory until sieving through a 2 mm mesh is possible (*c.*10% moisture content).

4. Sieve soil and take a 25 g subsample. Place in a screw-top bottle and add 100 ml of distilled water.

5. Shake vigorously for 1 h, preferably on a reciprocating shaker operating at 270 oscillations min^{-1}.

6. Wash soil suspension through nested 160 µm and 20 µm sieves with tap water.

7. Back flush through the 20 µm sieve, collect the sievate and add 100 ml distilled water.

8. Shake the suspension thoroughly, then remove a 2 ml aliquot with a wide orifice pipette and disperse over the surface of the semi-selective media. Repeat until ten plates have been aliquoted, shaking the suspension between each plate.

9. Incubate at 22°C for 2 weeks.

10. Wash plates thoroughly with distilled water (remove the agar from the plates while washing to remove any soil particles that are trapped between the agar and the base of the dish).

11. Drain the plates, blot dry carefully and incubate for a further 2–3 weeks.

12. Examine plates with a dissecting microscope for colonies of *Verticillium dahliae* typified by the production of microsclerotia.

13. Calculate the number of colony forming units (CFU) g^{-1} soil.

Protocol 3.8

Establishing Koch's postulates

> **Materials**
>
> - Healthy test plant(s)
> - Fungal culture for testing
> - Scalpel
> - 70% ethanol
> - Large plastic bag
> - Small brush (e.g. camel hair paintbrush)
> - Plant labels
> - Cotton wool
> - Sterile distilled water

Method

1. Choose a suitable part of the plant (e.g. stem, leaf, flower), depending on where the initial disease symptoms were observed, and label as inoculated or control.

2. Either prepare a spore suspension or excise a small agar plug/block ($c.0.5-1$ cm) and place carefully onto an unwounded surface and a wounded surface. Wounding may be achieved by abrading/scoring the epidermis or cutting a slit into the plant tissue. For obligate parasites such as rusts and mildews, transfer spores using a sterile fine brush on to the plant.

3. For the negative control, if appropriate, wound another, comparable, part of the plant in the same way but do not inoculate.

4. Wherever possible use whole plants, but if this is impractical detached plant parts can be used.

5. It may be advisable to place a small piece of damp cotton wool onto the infected area for 24–48 h or to cover the infection site with a small plastic bag for a similar period to maintain humidity at the beginning of the test. Do not leave on longer as this will attract opportunistic fungal growth.

6. Place the plant in a large damp chamber (surfaces may be sprayed with water to raise humidity further).

7. Incubate and check for disease symptoms. Conditions may also induce sporulation of opportunistic pathogens, which may interfere with assessment.

8. Once disease symptoms are present, record appearance and isolate from the leading edge. For obligate parasites, where isolation is not possible, check the identity of the organism associated with the symptoms.

9. Check for consistent isolation of the pathogen and purify typical isolates to compare identity of the causal agent with the original finding.

10. If the same symptoms and pathogen are observed as on the original infected plant then Koch's postulates have been satisfied.

References

Agrios, G.N. (2005) *Plant Pathology*, 5th edn. Academic Press, San Diego, California.

Dhingra, O.K. and Sinclair, J.B. (1985) *Basic Plant Pathology Methods*. CRC Press, Boca Raton, Florida.

EPPO (2006) *Phytophthora ramorum*. Bulletin OEPP/EPPO Bulletin 36, 145–155.

Gregory, P.H. and Stedman, O.J. (1953) Deposition of airborne *Lycopodium* spores on plane surfaces. *Annals of Applied Biology* 40, 651–674.

Harris, D.C., Yang, J.R. and Ridout, M.S. (1993) The detection and estimation of *Verticillium dahliae* in naturally infested soil. *Plant Pathology* 42, 238–250.

Rogers, S.L., Atkins, A.D. and West, J.S. (2009) Detection and quantification of the airborne inoculum of *Sclerotinia sclerotiorum* using concentrated PCR. *Plant Pathology* 58, 324–331.

Shurtleff, M.C. and Averre, C.W. (1997) *The Plant Disease Clinic and Field Diagnosis of Abiotic Diseases*. APS Press, St Paul, Minnesota.

4 Detection of Fungal Plant Pathogens in Seeds

Kelvin J.D. Hughes[1] and Victoria Barton[2]*

[1]*The Animal and Plant Health Agency, Sand Hutton, York, UK;* [2]*Fera Science Ltd, Sand Hutton, York, UK*

4.1 Introduction

Most arable crops are grown from the annual planting of seeds, which have the potential to carry plant diseases caused by bacteria, fungi, viruses and nematodes (Maude, 1996). Fungal pathogens can be either within the seed, such as the embryo pathogen *Ustilago nuda* of wheat, or on the seed, such as for many *Penicillium* species. When present, fungal pathogens can cause direct effects that include seedling collapse – commonly called 'damping off' and caused by *Pythium* and *Phytophthora* species, through to loss of crop vigour and yield caused by fungi including *Fusarium*, *Aspergillus*, *Penicillium* and *Claviceps* species. Many of these fungi can also produce mycotoxins (including ochratoxin, ergot alkaloids and patulin) in crops and in their stored products and these can affect animals and humans (Frisvad and Thrane, 2004).

Bunts and smut fungi such as *Tilletia* spp. can also taint food products such as flour. Additionally, seed-borne pathogens can raise production costs through the need for increased disease surveillance, greater pesticide application and, if a quarantine disease is introduced, statutory regulation which may lead to the loss of market opportunities for many years. For example, it has been estimated that if the quarantine pathogen *Tilletia indica*, commonly known as Karnal bunt, was introduced into cereal production in the UK in an area the size of Cambridgeshire (50,000 ha), this would cost at least €13.47 million in year one alone, owing to loss of yield, downgrading of product and loss of market (Sansford *et al.*, 2006).

To monitor seed health, seed lots are normally screened visually for macroscopic contaminants including sclerotia and ergots caused by *Sclerotinia*, *Botrytis* and *Claviceps* species, as well as soil and plant debris or trash which can carry the spores and mycelium of many fungi (see Fig. 4.1). Seeds may also be tested for seed-borne infection (Richardson, 1990) using methods developed by organizations such as the International Seed Testing Association (ISTA; https://www.seedtest.org; accessed 1 March 2023) and the International Seed Federation (ISF; https://www.worldseed.org; accessed 1 March 2023). Before harvest, crop inspections, which are commonly referred to as growing session inspections, may be used to help certify the health of seed crops.

This chapter describes the basic principles used in seed health monitoring, including sampling

---

* E-mail: victoria.barton@fera.co.uk

DOI: 10.1079/9781800620575.0004

Fig. 4.1. Macro-contaminants that can be found in seed lots (in this case, wheat, *Triticum aestivum*) include (left to right) ergots of *Claviceps* sp., soil, other seeds, broken seeds, nematode-infected seed, insects and true seeds. (© UK Crown Copyright – courtesy of Fera.)

and testing methods that can be adapted to suit most requirements. These should be used in conjunction with the identification methods described in Chapters 2–8. In general, there is increasing interest in using molecular techniques (Chapters 9 and 10) which can simultaneously perform both pathogen detection and identification. Finally, the chapter highlights various web-based resources for diagnosing fungal plant pathogens in seeds.

4.2 Sampling

Seed testing is usually performed on subsamples taken from larger seed lots, especially when dealing with large-volume crops such as cereals, grasses or brassicas. When a subsample is taken, it should be as representative of the whole seed lot as possible. Many books exist on seed sampling; these include the handbooks of Hall (2022) and of the Association of American Seed Control Officials (AASCO, 2016) and the procedures that are described in them should be followed. Key

points include mixing the sample well before sampling to ensure homogeneity and ensuring that the sample taken is of an appropriate size to detect the target pathogen at the level required. Consideration should also be given as to whether any seed treatment could affect sampling or testing; for example, are there enough seeds for testing, how long will testing take, do the seeds need to be returned after testing (e.g. for unique breeding seeds) and has a safety assessment been carried out for chemically treated seed? If sampling and testing criteria cannot be met in full, this should be borne in mind when interpreting the test results.

Generally, a sample of 2500 seeds or more is examined when looking for macro-contaminants including sclerotia, ergots, soil and plant trash (ISTA, 2022b), while typically for microscopic seed-borne fungi, 400 seeds are tested to give a 95% confidence of detecting at least a 1% level of infection. Commonly, this testing is destructive, which can cause problems if the seeds are small or highly valuable, although some non-destructive methods have been developed

(Abdullahi *et al.*, 2001; Roberts *et al.*, 2002; Agarwal, 2006; Gleason, 2009). Table 4.1 presents data on suggested seed sample weights for purity testing for some common crop species.

4.3 Visual Examination

Most seed is visually inspected using a low-power dissecting microscope or a magnifying lens, as described in Protocol 4.1, before other forms of testing. This approach is non-destructive, easy and quick to carry out, and generally gives a good assessment of seed quality, for example, of the level of physical damage – which can reduce germination. Macro-contaminants such as sclerotia can also be observed, as shown in Fig. 4.1, as can many fungal fruiting bodies such as pycnidia, acervuli, cleistothecia or sori, as shown in Fig. 4.2 and Table 4.2.

Protocol 4.1 contains details of the inspection of seed samples for macroscopic and microscopic contamination, and Table 4.2 presents examples of fungal fruiting bodies which may be seen on seed.

4.4 Agar Plating

Agar can be used to amplify infection on or from within seeds by acting as a water source to rehydrate test seeds and a nutrient source to stimulate the growth of any fungi present. Potato dextrose agar (Protocol 4.2) is commonly used for this purpose as this allows the growth of all fungi. The addition of antibiotics, or adjusting the pH, can affect fungal and bacterial growth and help to distinguish different species and counteract the effects of saprophytic organisms, which may mask slower growing seed-borne pathogens. Antibiotics including penicillin (5 units ml^{-1}) and streptomycin (10 units ml^{-1} final concentration) can be used to restrict bacterial growth. Hymexazol (30% active ingredient used at 75 µl l^{-1}) similarly stops the growth of *Pythium* species. Details of other amendments and media can be found in the *Plant Pathologist's Pocketbook* (Waller *et al.*, 2001).

In general, seeds are pushed into the agar and incubated, preferably in controlled environment facilities, for 7–14 days at 20–25°C under alternating 12 h cycles of light and dark. Near UV light (black light) is frequently used during the light phase as this can stimulate spore production, and spore morphology is commonly used to identify fungi (Chapter 6). Illumination can also have negative effects on fungal growth and sporulation (Yusef and Allam, 1967). Scoring the agar with a sterile needle or the addition of hard surfaces to agar, such as sterile pieces of filter paper, may also encourage sporulation at the point where the growing mycelium comes into contact with the scored area or paper. Allowing seeds to germinate can also help to draw out any internal infection on to the agar or permit fungal growth on the seedlings that develop (see Fig. 4.3).

Test seeds may be surface sterilized to differentiate between the presence of internal and external fungal infection. This is done by placing the seeds in 10% bleach (with a final concentration

Table 4.1. 1000 seed weights for a selection of common crop species and suggested weight for purity testing.

Seed species	1000 seed weight (g)[a]	Recommended sample weight for purity testing (g)[b]
Barley (*Hordeum vulgare*)	41.9	120
Onion (*Allium cepa*)	3.32	8
Brown bent grass (*Agrostis canina*)	0.06	0.25
Oilseed rape (*Brassica napus*)	3.3	10
Poppies (*Papaver* sp.)	0.06	5
Peas (*Pisum sativum*)	139.9	900
Spinach (*Spinacia oleracea*)	6.8	25
Sunflowers (*Helianthus annuus*)	38.8	200
Wheat (*Triticum aestivum*)	37.07	120

[a]1000 seed weight from Kew Seed Information Database (2022).
[b]Information from ISTA (2022b).

Fig. 4.2. Fruiting bodies (pycnidia) produced by *Septoria apiicola* on celery seeds (*Apium graveolens*). (© UK Crown Copyright – courtesy of Fera.)

Table 4.2. Fruiting bodies of surface-borne fungi that may be seen on seeds.

Host plant	Pathogen(s)	Structure seen
Beetroot and sugarbeet (*Beta* spp.)	*Uromyces betae*	Sori
Spinach (*Spinacia oleracea*)	*Neocamarosporium betae*	Pycnidia
	Colletotrichum dematium	Acervuli
Parsnip (*Pastinaca sativa*)	*Erysiphe heraclei*	Cleistothecia
Corn salad (*Valerianella locusta*)	*Subplenodomus valerianae*	Pycnidia
Parsley (*Petroselinum crispum*)	*Septoria petroselini*	Pycnidia
Celery (*Apium graveolens*)	*Septoria apiicola*	Pycnidia

of 0.5% level of active ingredient for 1–10 min, as described in Protocol 4.3), washing in continuous flowing water for 30 min or more, or by a short immersion in 70% alcohol. Excessive sterilization should be avoided, as this can kill any internal infection and prevent seed germination. However, the removal of surface-borne organisms, which often tend to be saprophytes, allows internal infection to be revealed which would otherwise be masked.

4.5 'Blotter' Testing

Blotter chambers can be used to amplify infection present on or within seeds in a similar way to the use of agar plates, although for large seed numbers blotters are simpler and cheaper. As with agar plating, surface sterilization of seeds can be used to help discriminate internal from external infection, and to reduce masking effects from saprophytes that are present.

It is difficult to add selective agents into blotter systems, which makes it harder to select the growth of target organisms than with agar plating. However, the method can be adapted to promote the development of certain pathogens. For example, freezing blotter trays at −20°C for 24 h makes the separation of *Pyrenophora graminea* and *Pyrenophora teres Drechsler* on barley (*Hordeum vulgare*) seeds easier (Mathur and Kongsdal, 2003). Freezing can also encourage fungal sporulation and stop or inhibit germination; for example, it is recommended for the detection of *Alternaria dauci* on carrot (*Daucus carota*) seeds

(ISTA, 2022a). Protocol 4.4 describes how to set up a simple blotter test using Petri dishes and damp filter paper which generally will allow the testing of 10–20 small seeds. For large seeds, or when greater capacity is required, a larger lidded chamber (such as shown in Fig. 4.4) can be

Fig. 4.3. Agar plate test showing fungal infection coming from germinated wheat seed (*Triticum aestivum*). (© UK Crown Copyright – courtesy of Fera.)

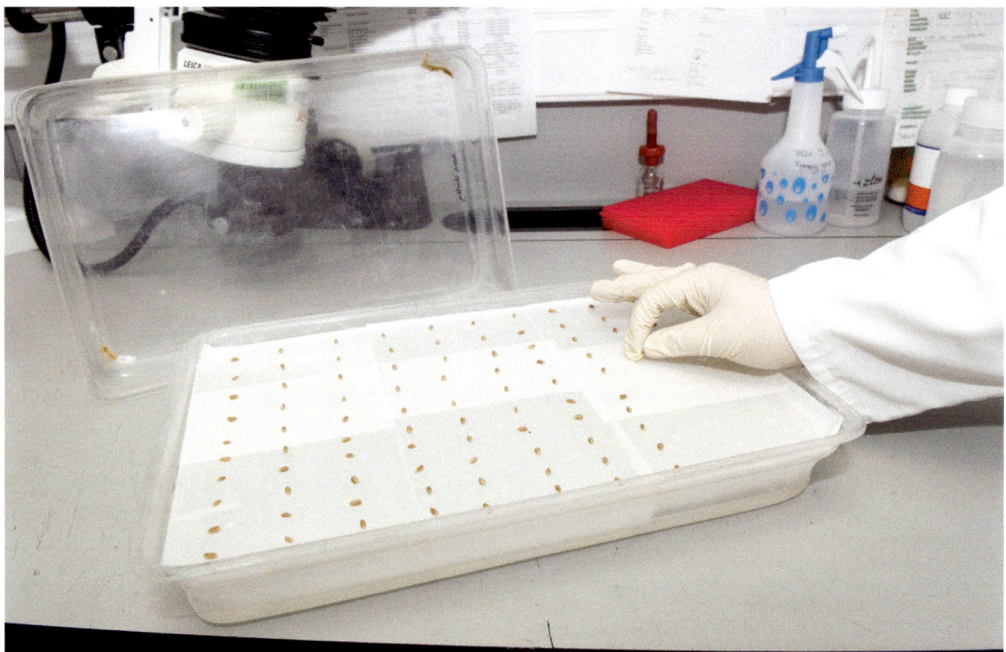

Fig. 4.4. Seeds being arranged on a germination tray before incubation as part of a fungal blotter test. (© UK Crown Copyright – courtesy of Fera.)

used. Seeds are placed on damp paper on a platform above a small reservoir of water within the chamber. Water is then drawn into the paper via a wicking cord placed under the paper, or by placing the edges of the paper into the water. The whole chamber is then put into a plastic bag to reduce evaporation. As with the agar method, testing should, if possible, be performed in a controlled environment to deliver reproducible results between tests.

4.6 Wash Testing

Many fungal pathogens can be carried on the surface of seeds, such as the teliospores of smuts and bunts, including *T. indica*, the oospores of downy mildews and various spore stages of rust fungi. Generally, these can be dislodged by washing

with a mild aqueous detergent solution, followed by visual examination under a compound microscope. As washing generally uses volumes that are too large to examine in their entirety, the washings can either be subsampled or concentrated by filtration or centrifugation (Protocols 4.5 and 4.6). This process should be gentle so as not to destroy or deform any fungal structures which may be needed for identification.

4.7 Web-based Resources

Many web-based resources are available that can help diagnose seed-borne pathogens. The sites listed in Table 4.3 have details of peer-reviewed methods or information that has been used by the authors over many years and have been found to contain accurate information.

Table 4.3. Useful websites for the diagnosis of seed-borne disease.

Website	Content	URL[a]
European and Mediterranean Plant Protection Organization (EPPO)	Testing protocols and pathogen information	https://www.eppo.int
International Seed Federation (ISF)	Seed testing protocols	https://www.worldseed.org/our-work/seed-health/ishi-methods/
International Seed Testing Association (ISTA)	Protocols and links	https://www.seedtest.org
Royal Botanic Gardens, Kew, UK	Seed data	https://ser-sid.org/
USDA[b] Germplasm Resources Information Network	Plant names, seed images	https://npgsweb.ars-grin.gov/gringlobal/search
USDA/Iowa State University	Seed testing protocols	https://seedhealth.org

[a]All websites accessed 1 March 2023.
[b]US Department of Agriculture.

Protocol 4.1

Inspection of seed samples for macroscopic and microscopic contaminants

Method

Detection of contaminants and superficial fungal structures

1. Test samples should be representative of the seed lot on which results are to be reported. Samples should be mixed and subsampled as described in Section 4.2.

2. Weigh the sample on a balance to at least two decimal places so that the percentage of any contaminants found can be calculated.

3. Spread out the sample on one side of the tray. Gradually work through the sample under the magnifying lens with a blunt knife or flat spatula, moving the sample across the tray.

4. Place contaminants into separate piles on the tray away from the remainder of the true seeds. Common contaminants include ergots of *Claviceps* sp. and other fungal sclerotia, soil, other seeds, broken seeds, and nematode- and insect-infected seed. Seeds showing signs of pathogen infection, including the presence of pycnidia, mycelial growth (sometimes seen as pink coloration when caused by *Fusarium* sp.) or unusual staining should also be separated and further examined using a light microscope or other means.

Identification of fungal contamination using light microscopy

5. Pick off superficial fungal structures using a sharp instrument such as a sterile syringe needle. Softening the seed by soaking in water for 5 min may help.

6. Place fungal material on a glass slide with a small drop of mounting fluid, e.g. water or lactoglycerol cotton blue, and place a coverslip on top.

7. Examine under a compound microscope and describe structures.

Reporting results

8. Weigh the separate components of the seed lot and calculate their presence as a percentage of the entire sample examined.

9. Record the presence of any fungi seen based on using macroscopic and microscopic examination as described already.

Protocol 4.2

General agar plating method to test seeds for seed-borne pathogens

> **Materials**
>
> - Potato dextrose agar with antibiotics if required
> - 9 cm Petri dishes
> - Incubator at *c.*20°C under 12 h light or near-UV light and 12 h dark if available

Method

Preparation of potato dextrose agar

1. Suspend 39 g of powdered agar (Oxoid, CM0139, agar 15 g l^{-1}, dextrose 20 g l^{-1}, potato extract 4 g l^{-1}) in 1 l of distilled water.
2. Bring to the boil to dissolve agar completely.
3. Sterilise by autoclaving agar at 121°C for 15 min.
4. Allow agar to cool to *c.*50°C.
5. Add antibiotics if needed to suppress bacterial growth. 0.6% penicillin sulphate (Sigma Aldrich P-8431) +2% streptomycin sulphate (Sigma Aldrich S6501-50G). To prepare the antibiotics, dissolve 0.6 g of penicillin sulphate and 2.0 g streptomycin sulphate in 100 ml of distilled water, filter sterilize through a 0.22 µl filter and dispense into a sterile container and store in the fridge. Add 10 ml of stock solution to each 1 l of media.
6. Pour agar under aseptic conditions into Petri dishes.
7. Allow agar to set before use.

Preparation and incubation of sample

8. Surface sterilize seeds if required (see Protocol 4.3 for details).

9. Label agar plates as required with details of the sample under test and the date on which plating was performed.
10. Aseptically submerge seeds (ten per plate) by pushing into agar so that the seeds are evenly distributed. For large seeds, such as beans, the number of seeds per plate may need to be reduced to five.
11. Stack the plates in a plastic bag and seal tightly.
12. Incubate plates at *c.*20°C under 12 h near-UV light and 12 h dark for 5–14 days.

Fungal identification

13. Examine plates for the amount and type of fungal infection present. Initially, classify the range of colonies based on morphological characteristics. Typically, *Alternaria* spp. are dark green/brown to black, *Fusarium* spp. are yellow to deep red and *Cochliobolus* spp. are grey to black.
14. Make microscope preparations of fungi. Use a needle or scalpel to lift small samples of fungal colonies of interest, place in a small drop of mounting medium on a glass slide and place a coverslip on top before examining under a compound microscope. Clear adhesive tape such as Sellotape™ can be used to sample the colonies; the tape is then stuck over a small amount of mounting fluid on the slide. This technique can also be useful in that it removes the need for coverslips.
15. A magnification of between ×200 and ×400 is usually required to observe any diagnostic features present.

Protocol 4.3

Simple method for surface sterilization of seed

Materials

- Beaker
- Domestic thin bleach (<5% sodium hypochlorite)
- Sterile water
- Sieve/tea strainer

Method

1. Count out seeds from the subsample of seed under test.

2. Place the seeds in a beaker containing 10% aqueous bleach, using enough sterilant to immerse them.

3. Soak small seeds such as those of viola in the bleach solution for 1 min, and larger seeds, including peas and beans, for 10 min. Add an extra minute of contact time if seeds are waxy, as for *Brassica* spp., or have a large surface area, e.g. nasturtium.

4. Pour off the bleach and dispose of safely.

5. Rinse excess bleach from the seeds by adding excess sterile water to the seeds in the beaker, then pour off instantly using a small sieve or tea strainer, again disposing of the wash water safely.

Protocol 4.4

Simple blotter method to test for seed-borne pathogens

Materials

- 9 cm Petri dishes
- Sterile water
- 9 cm circular filter paper
- Microscope slides, mounting fluid and coverslips
- Compound microscope

Method

Preparation of blotters

1. Label Petri dishes as required with details of the sample under test.
2. Place two circles of filter paper on top of each other in each dish and moisten with approx. 2 ml water; do not drench.

Plating out and incubation of seed

3. Surface sterilize seed if required (see Protocol 4.3).

4. Aseptically place seed on the damp filter paper within each Petri dish. Place up to ten seeds per plate so that seeds are evenly distributed. If large seeds such as beans are being tested, then it may be necessary to reduce the number of seeds per plate to five.
5. Place a lid over each plate, carefully stack them and place in a sealed bag for incubation.
6. Incubate plates at $c.20°C$ under 12 h near-UV light, 12 h dark for up to 14 days, adding further sterile water to each plate if they appear to be drying out.

Fungal identification

7. Examine plates under a dissecting microscope for the amount and type of fungal growth. Infection is usually obvious, as mycelium often glistens under illumination.
8. Identify the fungi present, making microscope preparations if required.

Protocol 4.5

Simple wash test for the detection of *Peronospora farinosa* f. sp. *spinaciae* (causative pathogen of downy mildew of spinach) spores

Method

Washing of sample

1. Soak approximately 1000 seeds (about 6.8 g) in 50 ml of sterile water and stir for about 5 min.

Separation of spores

2. Filter the water through the muslin cloth to separate the wash water from the seed.

3. Centrifuge the water suspension at about $1600 × g$[1] for 5 min.

Fungal identification

4. Resuspend the precipitate in 5 ml of water and examine 10 drops of 10 µl each on microscope slides under a compound microscope to look for characteristic oospores (Francis *et al.*, 1983; Inaba *et al.*, 1983).

Note

[1] Equation for calculating RCF (× g) from RPM: RCF = $1.12r_{max}$ × (RPM/1000)^2, where r_{max} = radius (mm) from the centre of rotation to the bottom of the centrifuge tube. RCF = relative centrifugal force, RPM = revolutions min^{-1}.

Protocol 4.6

Wash, filtration and centrifugation test to assess the presence of *Tilletia* teliospores in cereal or grass seeds

Materials

- Conical flask
- 0.01% Tween-20 aqueous wash water
- Flask shaker
- 20 and 53 µm sieves
- Pasteur pipettes

Method

1. Weigh out approximately 4000 seeds of cereal seeds (wheat, barley or oats, approx. 150 g) or grass seeds (approx. 0.32 g) and place in a conical flask with about 100 ml 0.01% Tween-20 aqueous solution.

2. Place flask on a shaker at approx. 250 oscillations min^{-1} for 3 min to release any teliospores.

3. Pour wash water through a 53 µm sieve which will hold back grain and other loose material but will let through wash water and teliospores.

4. Pour sieved water through a finer filter (20 µm; EPPO, 2018) which will retain *Tilletia* teliospores, including *T. indica*, *T. walkerii*, *T. horrida* and *T. caries*.

5. Using a Pasteur pipette wash trapped teliospores to one edge of a 20 µm sieve, then remove the spores by drawing into the pipette.

6. Dispense contents of pipette into a centrifuge tube and precipitate spores by spinning tube at $1000 \times g^2$ for 3 min.

7. Using a new pipette, draw off excess water and replace with 50–100 µl of water for viewing on slides under a compound microscope. Alternatively, Shear's solution may be used as this evaporates less from slides and for some teliospores, such as *T. indica*, gives more comparable measurements to published sizes (Mathur and Cunfer, 1993).

Identification

8. Compare spore size, colour and morphology to published descriptions as referenced in the EPPO protocol PM 7/29 (3) *Tilletia indica* (EPPO, 2018).

Note

[2] Equation for calculating RCF ($\times g$) from RPM: RCF = $1.12 r_{max} \times (RPM/1000)^2$, where r_{max} = radius (mm) from the centre of rotation to the bottom of the centrifuge tube. RCF = relative centrifugal force, RPM = revolutions min^{-1}.

References

AASCO (2016) *Handbook on Seed Sampling*. Association of American Seed Control Officials, Ithaca, New York.

Abdullahi, I., Ikotun, I., Winter, S., Thottapphilly, G. and Atriri, G.I. (2001) Investigation on seed transmission of cucumber mosaic virus in cowpeas. *African Crop Science Journal* 9, 677–684.

Agarwal, V.K. (2006) *Seed Health*. International Book Distributing Co., Charbagh, Lucknow, India.

EPPO (2018) PM 7/29 (3) *Tilletia indica*. *EPPO Bulletin* 48, 24.

Francis, S.M., Williams, R.J., Byford, W.J. and Berrie, A.M. (1983) *Peronospora farinosa. CMI Descriptions of Pathogenic Fungi and Bacteria*. Set 77, No. 765. Commonwealth Mycological Institute, Kew, UK.

Frisvad, J.C. and Thrane, U. (2004) Mycotoxin production by common filamentous fungi. In: Samson, R.A., Hoekstra, E.S. and Frisvad, J.C. (eds) *Introduction to Food and Airborne Fungi*, 7th edn. Centraal Bureau voor Schimmelcultures (CBS), Utrecht, The Netherlands, pp. 321–331.

Gleason, M.L. (2009) Assays for seed-borne bacteria: what they can and can't do. In: XI International Symposium on the Processing [of] Tomato. *Acta Horticulturae* 823, 231–234.

Hall, G. (2022) *ISTA Handbook on Seed Sampling*. International Seed Testing Association (ISTA), Bassersdorf, Switzerland.

Inaba, T., Takahashi, K. and Morinaka, T. (1983) Seed transmission of spinach downy mildew. *Plant Disease* 67, 1139–1141.

ISTA (2022a) *International Rules for Seed Testing. Annexe to Chapter 7: Seed Health Testing Methods. Method 7-001a: Detection of Alternaria dauci in Daucus carota (carrot) seed by blotter method*. International Seed Testing Association, Bassersdorf, Switzerland. Available at: https://www.seedtest.org/en/international-rules-for-seed-testing/seed-health-methods-product-1054.html (accessed 1 March 2023).

ISTA (2022b) Minimum size of working sample. In: *International Rules for Seed Testing*. International Seed Testing Association, Bassersdorf, Switzerland, Chapter 2, p. 6, Section 2.5.2.1. Available at: https://www.seedtest.org/api/rm/2T5MWR67H75U7NA/ista-rules-2022-02-sampling.pdf (accessed 1 March 2023).

Kew Seed Information Database (2022) *Seed Weights*. Available at: https://ser-sid.org/ (accessed 1 March 2023)

Mathur, S.B. and Cunfer, B.M. (1993) Chapter 3.2: Karnal bunt. In: *Seed-Borne Diseases and Seed Health Testing of Wheat*. Danish. Government Institute of Seed Pathology for Developing Countries, Copenhagen, pp. 31–43.

Mathur, S.B. and Kongsdal, O. (2003) *Common Laboratory Seed Health Testing Methods for Detecting Fungi*. International Seed Testing Association (ISTA), Bassersdorf, Switzerland.

Maude, R.B. (1996) *Seedborne Diseases and their Control: Principles and Practice*. CAB International, Wallingford, UK.

Richardson, M.J. (1990) *An Annotated List of Seed-borne Diseases*, 4th edn. International Seed Testing Association, Bassersdorf, Switzerland.

Roberts, S.J., Brough, J. and Chakrambarty, S. (2002) Non-destructive seed testing for bacterial pathogens in germplasm material. *Seed Science and Technology* 30, 69–85.

Sansford, C., Baker, R., Brennan, J., Ewert, F., Gioli, B. *et al.* (2006) *EC Fifth Framework Project QLK5-1999-01554: Risks Associated with Tilletia indica, the Newly-listed EU Quarantine Pathogen, the Cause of Karnal Bunt of Wheat. Deliverable Report DL 6.1: Report on the Risk of Entry, Establishment and Socio-economic Loss for Tilletia indica in the European Union; Deliverable Report 6.5: Determination and Report on the Most Appropriate Risk Management Scheme for Tilletia indica in the EU in Relation to the Assessed Level of Risk*. Available at: https://pra.eppo.int/pra/6de1b57b-07ec-4362-97a8-566e37679831 (accessed 1 March 2023).

Waller, J.M., Lenné, J.M. and Waller, S.J. (2001) *Plant Pathologists' Pocketbook*, 3rd edn. CAB International, Wallingford, UK.

Yusef, H.M. and Allam, M.E. (1967) The effect of light on the growth and sporulation on certain fungi. *Mycopathologia* 33, 81–89.

5 Collecting and Processing Samples for Isolation of Tree Pathogens

Ana Pérez-Sierra[1]*, Steve Hendry[2] and Joan Webber[1]

[1]*Forest Research, Alice Holt Lodge, Farnham, Surrey, UK;* [2]*Forest Research, Northern Research Station, Roslin, Midlothian, UK*

5.1 Introduction

Collecting samples for the purposes of isolating and identifying pathogens is often a key step in the process of diagnosing the cause(s) of ill health in trees or when undertaking research into tree diseases. When collecting material for research to investigate a specific disease, the nature and quantity of samples required is frequently known in advance; however, for diagnosis, sample collection must be directed by careful observation of the symptoms. It must be done mindful that symptoms may be the result of damage by an agent other than a pathogen (e.g. insects, abiotic factors or a nutrient deficiency) which will require different approaches to collection and processing of samples from those described here.

Where the presence of a pathogen is suspected, samples must be obtained from tissues where the damaging organism is likely to be located and preferably actively growing (i.e. in the process of infecting healthy material) if attempts to isolate or extract it are to succeed. It is important to remember that pathogens are not necessarily present in all parts of the tree displaying symptoms (Gregory and Redfern, 1998). For example, discoloured leaves or needles can be due to direct infection by a foliar pathogen, but the foliage of trees with diseased shoots, branches, stems or roots may show similar symptoms. Paying close attention to the nature, and in particular the distribution, of the tree's symptoms should allow the seat of damage (the position on the branching structure where the actual damage which is eliciting symptoms is located) to be observed or deduced (Strouts and Winter, 2000). Sampling from dead, dying, discoloured or deformed tissue at the seat of damage using the techniques described in this chapter provides the best prospect of successfully isolating any pathogen which is present.

5.2 Sampling

Recommended equipment

In many cases, adequate samples can be obtained from trees using a few well-chosen tools and accessories. As a minimum, these should be:

- **Waterproof notebook and pencil** for recording details of the symptoms displayed by the sampled tree, their distribution and other relevant information such as local soil type, soil wetness/compaction, degree of exposure, etc. Details of the individual samples taken, including the identifying numbers allocated to them (matching those

* E-mail: ana.perez-sierra@forestresearch.gov.uk

©CAB International 2023. *Fungal Plant Pathogens: Applied Techniques, 2nd Edition*
(eds C.R. Lane, P.A. Beales and K.J.D. Hughes)
DOI: 10.1079/9781800620575.0005

recorded on the appropriate sample bags) and the numbers of corresponding photographs must also be noted. If required, sheets from the notebook can be cut to create labels for insertion into the sample bags, as described below.

- A **pocket-knife** (Fig. 5.1) for cutting bark to locate and excise lesion material. A knife with a folding blade of 10 cm in length (shorter blades provide insufficient clearance for the user's hand when cutting into larger branches and stems) and preferably with a blade-locking mechanism is recommended, along with a **knife sharpener**. A sharp knife is much safer to use than one with a dull edge and frequent use of the sharpener is usually required since cutting woody tissues quickly blunts even the best blades; a knife with a locking blade is also safer than a standard penknife with a spring-retained blade, since the latter type can inadvertently close during use and cause injury. However, it should be noted that in some countries, such as the UK, it is illegal to carry locking folding knives, fixed-blade knives and knives longer than 3 inches (7.62 cm) in a public place without a further defence. Therefore, a knife of the type recommended should not be carried unless it is being used for sampling or an activity for which its possession can reasonably be justified.
- A **pair of anvil secateurs** (Fig. 5.2) for removing foliar, twig and branch samples with the capacity to cut material of at least 25 mm diameter. Although potentially less clean cutting than bypass action secateurs, anvil secateurs are more durable (particularly when used for the removal of thicker branches) and are easier to sharpen.
- **Tree marking (flagging) tape** (Fig. 5.3) for indicating trees from which samples have, or should subsequently be, taken so that they can be easily identified later. Also useful for creating labels, holding together bundles of small samples and securing the necks of larger sample bags. Only label trees if you have permission from the owner.
- **Sample bags and permanent marker pen** for protecting and identifying collected samples. As well as protecting sampled material from contamination following collection, appropriate bags serve to prevent desiccation and should be air-tight when sealed. Clear polythene food bags with integral grip-seals (not slider seals) are ideal for smaller samples (up to about 40 cm in length) and will be all that is required in many cases. However, a limited supply of heavy-duty clear polythene bags in larger sizes up to 2 m × 1 m is useful for accommodating long branches or sections through stems – after sample collection these should be securely sealed using flagging tape, as noted above. Large bags can also be utilized as a clean surface on which to lay material during sampling in order to prevent contamination through ground contact.
- Bags should be labelled at the time of sampling with as much relevant information as

Fig. 5.1. Pocket knife with locking blade and sharpener. (© Crown copyright. Forest Research.)

Fig. 5.2. Anvil secateurs. (© Crown copyright. Forest Research.)

possible but at a minimum with: the date of collection of the material, the species of tree from which the sample originated, the location of the tree (ideally using the appropriate national grid reference system, but otherwise stating the latitude and longitude of the location to at least three/four decimal places; three places will get you to within about 100 m and four to within about 10 m), a unique sample number (also to be recorded in the field notebook) and the name of the collector.

- Use of an indelible marker for labelling sample bags is recommended but in extremely wet conditions this method may fail and a label written in pencil on waterproof paper should be placed inside the bag instead.

- **A small pump-action spray bottle of disinfectant or packet of alcohol-impregnated wipes** should also be carried to allow disinfection of the tools used to collect samples after each use. This serves both to prevent cross-contamination between samples and potential cross-infection of trees because of the sampling process. Use of an alcohol-based disinfectant is less likely to damage tools and is useful for removing resin from cutting implements and hands when conifers are being sampled. Care should be taken when using these disinfectants and the procedures recommended by the manufacturer must be followed.

- As the size of trees to be sampled increases, **additional tools** will be required to remove larger diameter material, reach high branches, excise samples from extremely thick bark or remove soil to expose structural roots. Such tools include: anvil loppers (Fig. 5.4), pruning or bow saw (Fig. 5.5), high pruners (Fig. 5.6), pole-saw, hatchet, chisel (Fig. 5.7), trowel, rabbiting spade, mattock, etc.

5.3 Biosecurity

Unless preventive measures are taken, the process of sampling can result in the transfer of pathogens between trees or sites (further details in Chapter 12). Vehicles, tools and workwear should be scrupulously clean when arriving at a site and tools should be disinfected after each sample is taken, as noted above. When sampling is completed at a given site, plant debris and soil

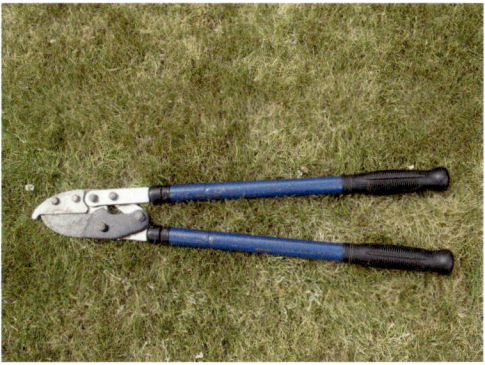

Fig. 5.4. Anvil loppers. (© Crown copyright. Forest Research.)

Fig. 5.3. Tree marking tape. (© Crown copyright. Forest Research.)

Fig. 5.5. Pruning or bow saw. (© Crown copyright. Forest Research.)

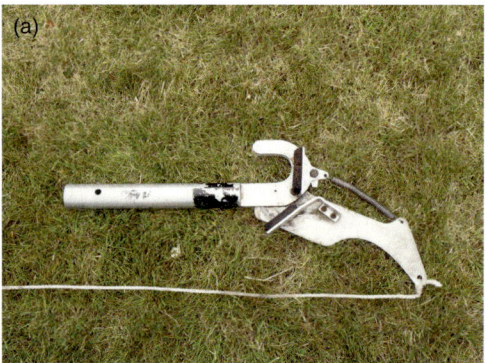

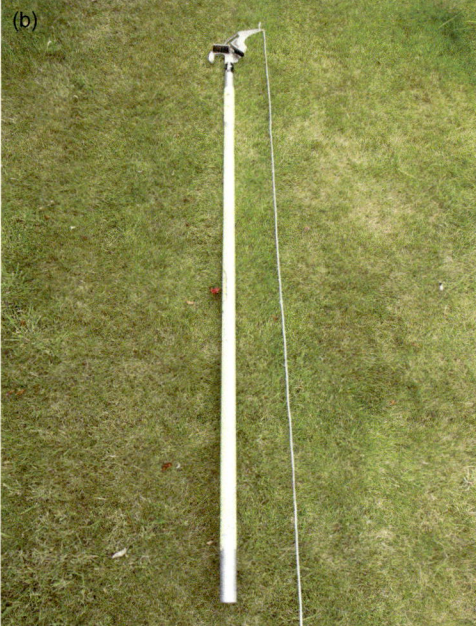

Fig. 5.6. High pruners on long pole. (© Crown copyright. Forest Research.)

should be removed from clothes and boots using a stiff brush (and clean water carried for the purpose, if necessary) and the soles of boots disinfected using a suitable alcohol or water-based product such as 70% IMS (industrial methylated spirit), 70% isopropyl alcohol or commercial products sanitizing spray (see full safety data sheet before using the product). Where this is not immediately possible, soiled clothing and boots should be placed in a large bag for later disinfection on an area of hardstanding where runoff to adjacent soil is precluded. For more information, please check: https://www.gov.uk/ guidance/prevent-the-introduction-and-spread-of-tree-pests-and-diseases (accessed 1 March 2023).

5.4   Soil

Soil should be collected from four evenly spaced points around the rooting zone of the tree(s) concerned. The cardinal points of the compass (north, east, south, west) can be used to define the four collection points. Soil should generally be removed close to the stem in younger trees but extend out to 50 cm and 150 cm from the stem base for older trees. Soil can also be taken from close to the stems of older trees but, if so, care should be taken to ensure that the main roots are not damaged. Checking the soil with a blunt-ended metal probe such as a peat probe (Fig. 5.8) will help to avoid excavating in areas where structural roots are close to the soil surface, saving time and effort.

Use of a nylon or composite plastic trowel to carry out any excavations can also minimize cutting injuries to the bark of any roots in the vicinity. Another option is to use a soil auger (Fig. 5.9), but care should be taken to avoid root damage with this method.

Any vegetation or organic layer should be removed prior to sampling. Depending on the type of soil, the appropriate tool should be selected (e.g. spade, trowel or mattock (Fig. 5.10)) to dig a hole 20–30 cm deep.

Using a trowel, soil from the hole should be placed into a plastic bag and any large stones removed, but fine roots or coarse woody debris do not need to be discarded. Soil from the four sampling points around the tree should total between 1 l and 3 l in volume and be combined in the same plastic bag. After enough soil has been collected, refill the holes and replace the displaced vegetation. Finally, any tools used should be wiped clean of soil and sprayed with a disinfectant.

5.5   Roots and Root Collar

On larger trees, the root collar and the upper surfaces of major surface lateral roots can often be sampled by removing a bark panel (see Section 5.8). Surface lateral roots decrease in diameter

Fig. 5.7. Chisel with guard and mallet. (© Crown copyright. Forest Research.)

Fig. 5.8. Peat probe. (© Crown copyright. Forest Research.)

Fig. 5.9 (a) Soil auger and (b) sampling. (© Crown copyright. Forest Research.)

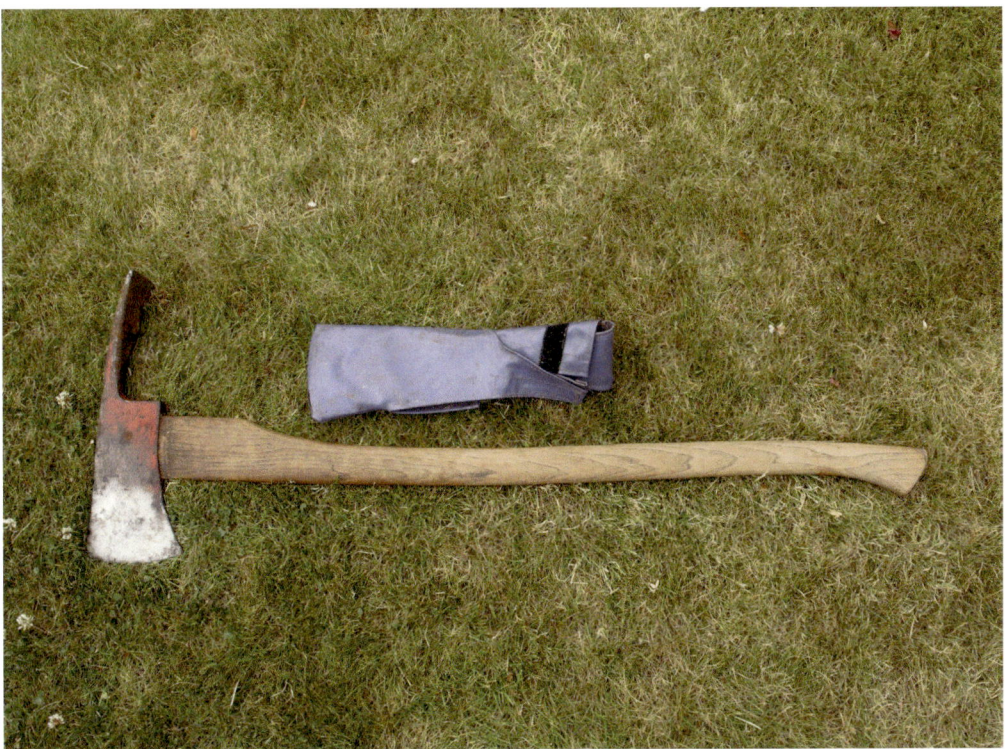

Fig. 5.10. Pulaski combined mattock and axe (with guard). (© Crown copyright. Forest Research.)

quite sharply with increasing distance from the stem, even in large trees. Exposing lateral roots to determine their condition and facilitate sampling is best achieved by starting at the root collar and working away from the stem using a trowel to loosen the soil over and around the roots (Fig. 5.11); a stiff nylon brush can then be used to sweep away the loosened soil from the root surface.

Using a nylon or composite plastic trowel can help to avoid cutting injury to the root bark during this operation. After exposure, excision of roots with a diameter of less than 7.5 cm can generally be achieved quite easily using loppers. If possible, avoid using a saw to cut roots as this often drags soil across the cut surfaces created during root detachment. As roots are rarely round in cross-section and frequently deeper than their width when viewed from above, this should be borne in mind when root excision is being considered. After removal, again use a stiff nylon brush to remove any remaining soil from the root surface and place in a plastic bag for transit.

Root systems of small trees (up to about 2 m in height) can be sampled in their entirety by digging around the margins of the root ball and gently lifting the soil. Pulling the base of the stem against resistance from roots in undisturbed soil should be avoided – if the root system cannot be extracted by gently lifting via the stem base then further loosening of the soil is required. Once extracted, adhering soil can be gently shaken off or brushed away but no force should be employed in this operation. If only the roots need to be retained, the tree stem should be severed a few centimetres above ground level (provided that no basal lesions extend beyond this point) and the root system along with any remaining soil placed in a sample bag.

Avoid handling the stem base of forest tree seedlings or young nursery stock as far as possible when lifting them for sampling as necrotic tissue at the ground-line is often extremely fragile and easily damaged. Very young seedlings should be handled by their foliage and lifted after loosening the soil around their roots. Better developed

Fig. 5.11. Exposing roots for sampling. (© Crown copyright. Forest Research.)

nursery stock can usually be lifted on a trowel blade and supported by holding the foliage during transfer to a sample bag.

5.6   Foliage, Shoots and Twigs

Shoots and twigs which display symptoms, or which bear symptomatic foliage, should be removed using secateurs, long-handled loppers or pole-mounted pruners, taking care to include several centimetres of healthy tissue proximal to any bark lesions if these are present (Fig. 5.12). Several shoots or twigs from the same tree can be placed in a single sample bag if required. Relatively long lengths of twigs and fine branches can be accommodated without damage in smaller sample bags if they are carefully coiled before inserting them, although this tends to be easier in the spring and summer when twigs are at their most flexible.

In the case of foliar diseases of broadleaved trees, entire shoots bearing both healthy and symptomatic foliage should be selected and bagged with the foliage still attached. Conifer needle diseases present different challenges because symptoms often develop on needles which have been retained for more than one year, so including information about the age of the affected foliage can be important. Ideally, a sample should not only include the section of branch bearing the symptomatic needles, but also all the internodes (internode – section of stem between two successive nodes) towards the distal end (tip) of the shoot and at least part of the internode proximal to (below) it.

If such a sample is too long to place in a bag, the internode bearing the affected foliage and half of the proximal and distal (if present) internodes should be taken; it is then essential to write on the sample bag the year in which the internode bearing the symptomatic needles was produced.

If cones and fruits are collected, they must come directly from the tree rather than from the ground to avoid (cross-)contamination.

For some conifers affected by needle cast diseases, the simple act of cutting a twig can cause all the symptomatic needles to fall. Therefore, make sure that you hold an open sample bag below the selected twig/branch before any cut is made to ensure that the needles of interest are caught and then place the shoot from which they fell into the same bag. Occasionally, sampling of individual infected needles, fascicles or short shoots from across the crown of a tree can be useful, e.g. if fungal fruit bodies are evident on scattered foliage. If so, place each sub-sample into a separate small bag of 5 × 7.5 cm and place them all inside the main sample bag for the tree under investigation.

Fig. 5.12. Samples of symptomatic branches need to include healthy tissue as this will allow finding of the leading edge of the lesion (a). The leading edge between diseased and healthy tissue is revealed when the outer bark is removed (b). (© Crown copyright. Forest Research.)

5.7   Branches

Whole branches which are accessible from ground level should be removed using appropriate hand tools (handsaw (Fig. 5.13), pole-saw, secateurs or loppers).

When using a handsaw, never attempt to remove a branch with a single cut. Particularly in the spring, this can result in tearing of the bark for a considerable distance below the branch insertion point causing damage to the tree. The correct sequence of pruning cuts during removal of a branch with a saw are:

- a shallow cut is made on the underside of the branch (about 25% of its diameter) to prevent subsequent bark tearing, followed by
- a topside cut just outside (i.e. distal to) the underside cut extending most of the way though the diameter of the branch and resulting in its removal, followed by
- a final pruning cut from the branch bark ridge to the branch collar to create a clean branch stub with optimal callusing potential.

Undercutting as described above cannot be carried with a pole-saw (Fig. 5.14). Instead, the sharp blade/sickle located just beneath the saw-blade of a properly specified pole-saw should be used to scribe an incision in the bark on the underside of the branch which serves the same purpose as the undercut described above. Do not use a pole-saw which is not equipped with a sickle.

If using a pole-saw, appropriate personal protective equipment should also be worn – a chainsaw helmet with visor is essential to avoid potential injury from a falling branch.

After removal, long branches may need to be cut into sections before bagging and the advice for dealing with larger whole trees given below should be followed.

5.8   Large Stems and Branches

Bark panels

Where samples from the main stem and structural branches of larger trees are required, the recommended method of sampling is the removal of bark panels from areas of interest such as bleeding lesions or cankers (Fig. 5.15).

When an area of bark for sampling has been identified, use a mallet and chisel (blade width 4 cm or 1.5 inches) or a hatchet (Fig. 5.16) to cut and remove a bark panel of approximately 6 × 4 cm down to the surface of the underlying wood.

Usually, this is most easily achieved by first cutting the outline of the panel through the outer bark down to the sapwood surface with the chisel, inserting the chisel or hatchet on one of the narrow edges of the inscribed area and then gently levering up the overlying bark (with the aid of the mallet if necessary) until the panel can be lifted away (Fig. 5.17). An ideal sample panel consists of approximately 70–80% lesion material and 20–30% healthy bark. Avoid removing the outer bark from the sample panel as far as possible because this will initiate oxidation of the phloem and potentially compromise subsequent efforts to isolate a pathogen or extract DNA from the material. If the outer bark does

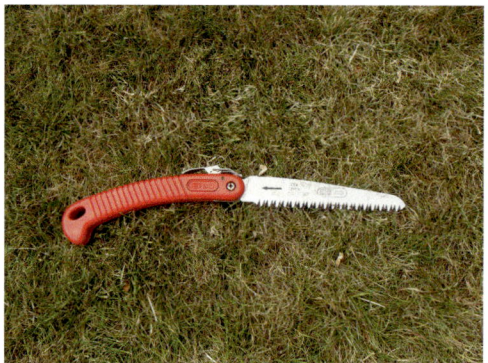

Fig. 5.13. Folding handsaw. (© Crown copyright. Forest Research.)

Fig. 5.14. Pole-saw head with sickle (and guard). (© Crown copyright. Forest Research.)

Fig. 5.15. Bark removed from stem lesion to show leading edge between diseased (left hand side) and healthy tissue (right hand side). (© Crown copyright. Forest Research.)

become detached from a panel, place it on top of the lesion and hold it in position with a rubber band or by wrapping in a paper towel. After removal, the bark panel should be sealed in a labelled, small, plastic bag.

Problems can occur when attempting to take bark panel samples during the dormant season as the cambium does not always separate from the underlying wood and attempts to remove bark alone will result in small, fragmented samples. In this case, a deeper cut which penetrates into the outer sapwood should be made when scribing the outline of the panel and the chisel or hatchet should be used to split the sapwood beneath the panel along its grain just below the bark in order to remove an intact area of bark backed by a few millimetres of sapwood. An advantage of this technique is that the inner surface of the sampled bark is not exposed to the air and oxidation of the phloem is therefore reduced. The technique can be employed to advantage when sampling from thin-barked tree species at any time of the year to keep the bark panel intact.

5.9 Sampling Whole Trees

Seedlings, transplants and small trees can be sampled in their entirety by digging them up and

Fig. 5.16. Hatchet and edge guard. (© Crown copyright. Forest Research.)

Fig. 5.17. Bark panel removed from stem lesion. (© Crown copyright. Forest Research.)

gently shaking loose soil from the roots. Soil can be collected into a separate bag if the tree is to be retained as a whole. If necessary, the stem can be cut into sections which will fit into the bag being used but it is preferable to carry bags of sufficient size to enclose trees of up to 1.5 m in height while leaving them intact.

Larger trees will usually need to be cut into sections using a hand saw or loppers in order to accommodate them in a large sample bag. While doing so, try to avoid making cuts through any bark lesions and leave them intact; ideally cuts should be made several centimetres above or below the extremities of any necrosis to minimize potential contamination of lesion tissue. A separate bag should be used as ground cover during cutting to prevent the cut ends of sections from contacting litter or soil before they are placed into the appropriate sample bag. Label each section (either directly or on a piece of tree marking tape tied securely to it) and include a diagram in your notebook to show the positions which each of the cut sections occupied on the tree.

5.10   Fungal Fruit Bodies

Small fungal fruit bodies are easily removed from the part of the tree on which they have been produced. In the case of larger fruit bodies, either all or part(s) of them can be collected, depending on their size and type (Fig. 5.18). The least degraded fruit bodies should be selected to maximize the chances of subsequent identification by isolation or examination of their morphology. Removal of fleshy fruit bodies can usually be achieved by hand, while a chisel or knife should be used to remove woody basidiocarps in whole or part.

Where decayed wood is associated with a fruit body, this should be removed (including the zone between decayed and sound wood, where possible) and placed in a separate sample bag. After removal, fruit bodies or parts removed from them should be wrapped in paper and placed in a sampling bag and transferred to the laboratory immediately. Otherwise, the sampled material should be placed in a labelled paper bag which will allow the specimen to dry slowly rather than rotting.

Fig. 5.18. Basidiocarps of *Armillaria mellea*. (© Crown copyright. Forest Research.)

5.11 Sample Processing

Humid chambers

Plastic food storage boxes, Petri dishes or plastic bags can all make suitable damp chambers, depending on the type of sample requiring incubation. Disinfect the containers before use, which can be done by washing them with water, drying and then spraying with 70% methylated spirit. Once dry, they can be used.

To provide humidity, dampened paper towelling/cotton muslin can be placed at the base of the chamber. Free water can also be used instead of dampened paper/cotton, in which case a raised stand made out of wire or plastic should be placed over the water and the samples placed on top of it to ensure they do not become waterlogged or submerged.

The quantity of plant material to be incubated in the damp chamber will depend on its size and consistency. Fragmented or multiple pieces of material should not be placed in a pile but laid out in a single layer. Although some overlapping is acceptable, try to avoid it.

Once the plant material is in place, place the lid on the damp chamber or Petri dish. If plastic bags are used, seal the plastic bag with adhesive tape or with a knot.

Label the damp chamber with the date, the sample number or reference number, the tree species and your name/initials.

Be careful when incubating plant material as it is frequently colonised by mites or other arthropods, which can quickly infest the incubator and be difficult and time consuming to eradicate.

See Protocol 3.1 for further details.

5.12 Isolation on Agar Media

Plant material should first be rinsed in water, then surface-sterilized using 70% methylated spirits or a 10% sodium hypochlorite solution (domestic bleach) and finally rinsed in sterile distilled water. This eliminates saprophytic organisms on the surface, which may outcompete or overgrow the target pathogen on an agar plate. When isolating from sample material, tools such as scalpels and forceps or tweezers should be sterilized to prevent any cross-contamination.

Saplings: Wash the lower stem and roots of sapling thoroughly with water to remove soil and debris. Carefully pare away the outer bark to expose the inner bark (phloem) beneath and look for the junction between uninfected tissue and discoloured infected tissue (lesion). Take small pieces of tissue (several cubic mm) from the edge of an exposed lesion and place 6–7 of these onto a plate of agar medium. If uncertain of the causal pathogen, use a range of agar media suitable for different types of agent.

Mature trees: isolate from cankers, lesions, bark panels. If the samples are fresh, they can be processed directly. If the samples are not fresh, first wash thoroughly with water and blot dry before processing. Remove thin sections of tissue (less than 1 mm thick) from the edges of any lesions found beneath the outer bark. Cut sections into approximately 5 mm × 5 mm pieces and place 6–10 of these pieces onto a plate of isolation agar media, ensuring they are evenly spaced (Fig. 5.19).

For each sample, prepare a duplicate of each plate and, if needed, use a range of agar media. As standard practice incubate plates from sample isolations (saplings or mature trees) at room temperature or at 20°C for 48–72 h before checking for any agents.

Fungal fruit bodies: Remove a section of tissue from the fruit body that is free of any vegetation or debris. If needed, excess moisture in the fungal fruit body can be reduced by pressing it between paper towels. Once enough moisture has been removed from the sample it can be cut

into small sections so that 6–7 pieces can be placed onto each plate of isolation agar. The plates should be incubated at 20°C for up to 72 h.

In all cases, label the plates on the underside with a permanent marker with the reference sample number and date of isolation and incubate the plates upside down.

See Protocol 3.2 for further details.

5.13   Soil Baiting

Soil baiting can be used to isolate soil-borne organisms and especially applies when *Phytophthora* species are of interest. Baits are selected based on susceptibility to the target organism. A range of baits can be used, such as seedlings or unripe fruit (Pérez-Sierra *et al.*, 2022).

See Protocol 5.1 for further details.

5.14   Preparation of Samples for DNA Extraction

Only a very small amount of plant material is needed for DNA extraction. Infected tissue is selected from the sample and cut using a sterile scalpel blade. Initially, cut pieces of approximately 0.5–1.5 cm² from the affected tissues and subsequently cut these into smaller pieces (<2–3 mm square). Carefully transfer these smaller pieces with sterile tweezers and place inside a sterile microcentrifuge tube. Label the tube with the sample reference number. Keep the sample tubes in the freezer until DNA can be extracted.

See Chapters 9 and 10 for further information.

5.15   Pathogenicity Test Method

Under-bark inoculation by wounding on seedlings/trees

Seedlings/trees should be potted in a compost mix in suitable containers which are kept in the inoculation area and watered on demand. Isolates of the pathogen (where possible include several isolates) to be inoculated are grown in pure culture on agar media for a length of time

Fig. 5.19. Agar plate showing the fungus has been consistently isolated. (© Crown copyright. Forest Research.)

that can vary from a few days to weeks or even months in order to ensure that actively growing isolates are used for inoculation.

To test the pathogenicity of isolates, first select the part of the plant to be inoculated (e.g. stem or branch) and disinfect the outer bark at the inoculation point with 70% methylated spirit. Generally, a wound is made as part of the inoculation process either by cutting a small flap in the bark with a sterile scalpel or by removing a disc of bark using a cork borer. When using a cork borer (approx. 5 mm diameter), remove the outer layer of bark and replace it with an agar disc (of the same diameter) taken from the pure culture. The agar disc is then placed mycelium-face down in the wound, which is covered with cotton wool moistened with sterile distilled water,

and the covering held in place with Parafilm® or cling film around the stem/branch. Finally, wrap the point of inoculation on the stem/branch with aluminium foil (Pérez-Sierra et al., 2022). Control trees are inoculated in the same way but using sterile agar discs.

Incubate the trees using the required conditions and monitor them closely until symptoms appear. It can be useful to add extra inoculated specimens into any trial, so these can be unwrapped to check the progress of any bark colonization without affecting the integrity of the entire trial. The time needed for symptoms to develop depends on the pathogen concerned and the season, but can vary from a couple of weeks to several months.

See Protocol 3.8 for further details.

Protocol 5.1

Soil baiting

Materials

- Pear or apple fruit
- Large knife
- Scalpel
- Cork borer
- Tweezers
- Industrial methylated spirits
- Sticky tape
- Marker pen

Baiting from soil (Pérez-Sierra *et al.*, 2022) using fruit to isolate *Phytopthora* species:

1. Select a visually healthy and undamaged semi-ripe pear (*Pyrus communis*) or green apple (Granny Smith or Golden delicious) (*Malus domestica*) fruit as the bait. Where possible obtain organic fruit (no pesticide application), wash it to remove any residues and dry carefully.

2. Initially sterilize the surface of the bait using 70% industrial methylated spirits.

3. Prepare the bait by cutting two to four holes, evenly spaced around the circumference, with a sterilized cork borer to an approximate depth of 2.5 cm (dependent on the size of the fruit being used).

4. Each hole is then stuffed with soil until completely full; small fragments of roots can be included, but any stones should be removed. Distilled water is pipetted into each hole, allowing the soil to soak.

5. The holes can then be sealed using a clear, sticky tape making sure all the holes are fully sealed to prevent any soil or water from falling out.

6. On the fruit, using a permanent marker write the sample reference number and date.

7. Soil baits should be kept within a container/dish to contain any leakages at room temperature (20°C) in daylight for up to 7 days. Usually symptoms will be observed after 4–5 days, depending on the *Phytophthora* species.

8. The baits should be checked for the formation of lesions (necrotic tissue) around the holes daily.

9. Once lesions are present, the baits can be processed. Using a sterile knife, split the bait into several sections.

10. Using a sterilized scalpel, remove small pieces from the lesion edge and plate onto the appropriate isolation media.

This method can be used on bark samples as well (Phillips and Burdekin, 1992; Pérez-Sierra *et al.*, 2022). The method is similar except, instead of using a cork borer to create holes, a flame-sterilized knife is used to cut a flap into the bait approximately 4 cm across and 1 cm thick, ensuring that one edge is still intact. Lifting the flap, thin sections of woody material can be placed underneath using sterilized tweezers. The flaps are then sealed in the same manner as above using tape. If required, a second flap can be made on the opposite side of the fruit. The bait should be labelled as appropriate and then stored in a container at room temperature (20°C) in daylight for up to 7 days but checked daily for the formation of lesions on or around the flaps. Once lesions are present, the baits can be processed as described above.

References

Gregory, S.C. and Redfern, D.B. (1998) *Diseases and Disorders of Forest Trees.* Forestry Commission Field Book 16. The Stationery Office, London.

Pérez-Sierra, A., Jung, M.H. and Jung, T. (2022) Survey and monitoring of *Phytophthora* species in natural ecosystems: methods for sampling, isolation, purification, storage, and pathogenicity tests. In: Luchi, N. (ed.) *Plant Pathology. Methods in Molecular Biology*, vol. 2536. Humana, New York, pp. 13–49.

Phillips, D.H. and Burdekin, D.A. (1992) *Diseases of Forest and Ornamental Trees.* Palgrave Macmillan, London.

Strouts, R.G. and Winter, T.G. (2000) *Diagnosis of Ill-health in Trees.* The Stationery Office, London.

6 Identification of Fungi Based on Morphological Characteristics

Aiga Ozolina[1]*, Michael Long[1] and Paul A. Beales[2]
[1]Fera Science Ltd, Sand Hutton, York, UK; [2]The Animal and Plant Health Agency, Sand Hutton, York, UK

6.1 Introduction

The morphology of a fungus refers to characteristics of its form, shape, and external and internal features that are visible either with the naked eye or microscopically and the relationships of their constituent parts, but not to its function (Bessey, 1964). Each characteristic described helps to build a picture, enabling identification of the fungus under examination.

Fungi may exhibit two recognizably different morphological stages, namely vegetative and reproductive (Fig. 6.1). The vegetative stage consists of filamentous, thread-like structures called hyphae (singular, hypha), with few distinguishing features, whereas the reproductive stage can be more variable and discernible. The mass of the vegetative structure is referred to as mycelium.

Fungi are exceptionally diverse in their morphological manifestations, from having relatively few structures such as *Rhizoctonia* spp. (a common disease of turfgrass – Fig. 6.2) that are typically only observed in vegetative hyphal forms, to the rust fungi (e.g. Fig. 6.3) that can produce up to five morphologically distinct fruiting bodies and spore types (Peterson, 1974; Farr

et al., 1989). Available morphological features for examination may be affected by environmental influences such as temperature, light period, humidity or even time of year, so consistency of isolation procedures and impacts of environmental variables should be minimized where practically possible (Cooke *et al.*, 2006; Madden *et al.*, 2007).

A brief literature review is usually a good starting point before carrying out a morphological examination of an infected plant or fungal culture. This helps to determine the range of known pathogens that infect the host and the typical morphological features (signs) that they can exhibit. It is important, however, to bear in mind that you may be observing one of the many undescribed fungi, or a new host–pathogen combination. Occasionally, for example when developing identification keys for fungi, it will also be necessary to carry out comparative morphological studies in which varying vegetative or reproductive structures from many fungi of a range of species are compared and grouped according to their similarities and differences.

This chapter will highlight fungal and fungus-like plant pathogens that are commonly encountered and identify important morphological

---

* E-mail: aiga.ozolina@fera.co.uk

© Crown copyright 2023. Reproduced with permission of the Controller of His Majesty's Stationery Office and Department for Environment, Food & Rural Affairs (Defra)
DOI: 10.1079/9781800620575.0006

Fig. 6.1. (a) Vegetative fungal hyphae grouped into a mycelium; (b) fruiting bodies of honey fungus pathogen *Armillaria mellea*. (From Shutterstock.)

features that are necessary for identification and diagnosing disease. Protocols are also included that cover processes for identification of fungi using different host tissue and diseases on specific hosts. It is not a comprehensive treatise of all groups of these pathogens, as these are described in more specific identification guides.

Fig. 6.2. *Rhizoctonia solani* causes diseases such as patch on turfgrass. (From Shutterstock.)

Fig. 6.3. Cedar apple rust caused by *Gymnosporangium juniperi-virginianae.* (From Shutterstock.)

6.2 Preparation of Plant Material for Microscopical Examination

Leaves, stems, flowers and fruits

In order to study the morphology of some fungi (principally obligate plant pathogens, i.e. require a host plant to complete their life cycle and do not grow in culture) (Fig. 6.4), structures need to be prepared or 'mounted' directly from the affected plant tissue for microscopical examination.

Adhesive tape method

This method captures external fungal structures and fixes them onto the surface of adhesive tape. It is generally used to observe the morphological features of hyphae, fruiting bodies and spores, and can be viewed with or without a stain. This approach is a rapid and temporary method and should be viewed immediately after preparation (Fig. 6.5). Protocol 6.1 describes this technique. This method will capture all material on the leaf surface, including numerous artefacts and contaminants (e.g. dust, dirt, leaf hairs). If fungal

Fig. 6.4. The obligate plant pathogen, oak powdery mildew (*Erysiphe alphitoides*). (From Shutterstock.)

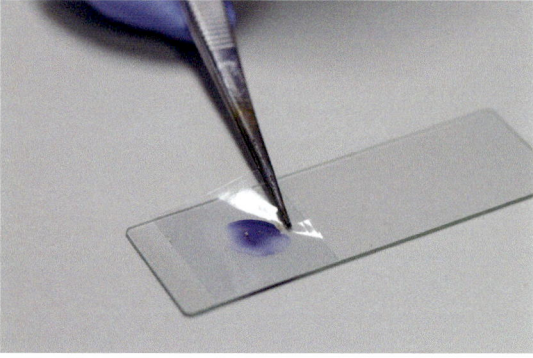

Fig. 6.5. Adhesive tape method ready for viewing under compound microscope. (Image courtesy Fera-Science Limited © Copyright Fera-Science Limited.)

structures such as fruiting bodies are clearly visible under a low-powered microscope, Protocol 6.2 indicates how to remove and view these.

Hand sectioning

Thin cross-sections of affected plant tissue are a useful and quick way to observe endophytic fungi, fruiting bodies and the association of internal and external fungal structures (Ruzin, 1999). With practice, it is possible to cut sections that are one cell thick. A microtome (consisting of a steel, glass or diamond cutting blade) can be used to cut ultrathin sections of consistent thickness and is especially useful when carrying out morphological examinations of plant pathogenic structures (Fig. 6.6). Protocol 6.3 describes how to produce thin sections using readily available laboratory equipment.

Roots

It can be difficult to prepare roots for microscopical examination owing to their dense and woody nature and pigmentation. As a rule, it is better to select finer roots (if present), as these are the easiest to prepare for light microscopy and often contain more useful morphological features than older and thicker roots (Fig. 6.7). Protocol 6.4 describes the microscopical examination of roots for pathogens.

6.3 Morphological Examination of Cultures

When describing a fungus, it is important that culture conditions are consistent (see Chapter 7) to ensure that morphological features described can be replicated (Fig. 6.8). Key factors include:

1. *Growth medium* – concentration, supplier of ingredients, age and the volume added to the Petri dish.
2. *Temperature* – this can be variable between fungi and a literature search may be required to

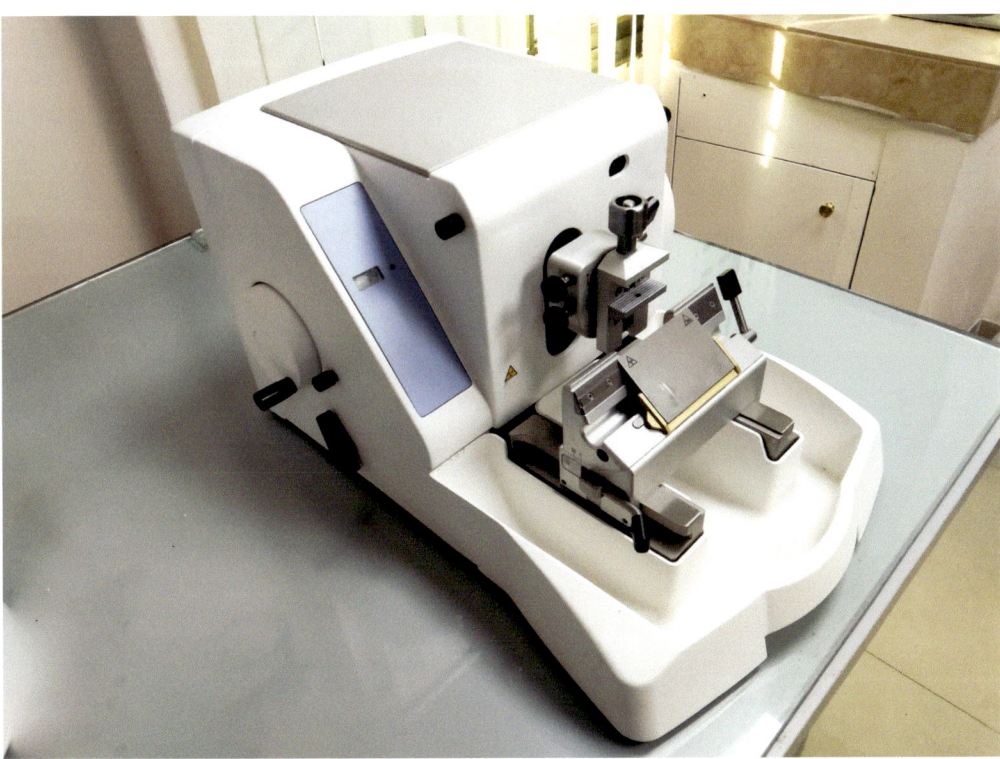

Fig. 6.6. Hand-operated rotary microtome for cutting thin microscopic sections of plant material. (From Shutterstock.)

Fig. 6.7. Finer roots are easier to prepare than older, woody roots. Fungal structures are frequently more prevalent in younger, finer roots. (From Shutterstock.)

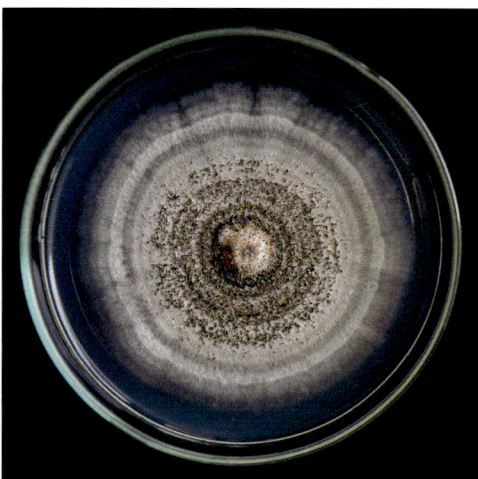

Fig. 6.8. Typical culture plate for isolating fungi. (From Shutterstock.)

determine the optimum value. Some fungi even have different optimal temperatures for growth and sporulation. If the fungus being cultivated is unknown, then it is advisable to try a range of temperatures to determine the optimum value.

3. *Humidity* – generally, most fungi grow above 85% relative humidity (RH) (Agrios, 2005). Within a sealed Petri dish, the RH can be considerably higher unless a breathable sealant tape is used.

4. *Light* – length of light/dark (day/night) regime should be specified and consistent. Many fungi have different growth patterns at varying light/dark durations, and these can be diagnostic. Also, the induction of some morphological features (e.g. resting spores) may only occur during the light or the dark. The type of light used should also be specified. Fluorescent lighting is commonly used in a controlled environment or incubator, as discussed in Chapter 7.

5. *Time* – cultures should be incubated under controlled conditions for a specified period before examination. A literature search may aid in determining the conditions for the isolate concerned; if not, further experimentation will be required to obtain the optimal time for morphological features of interest to develop. Typically, this may range from a few days to a few weeks, but it can be considerably longer.

The morphological examination of cultures is described in Protocol 6.5.

6.4 Identification of Some Common Groups of Fungal Plant Pathogens Using Morphological Features

Plasmodiophora brassicae (clubroot pathogen)

The clubroot pathogen is a protozoan pathogenic organism that causes disease of numerous genera of the Brassicaceae (mustard) family, including cabbage, cauliflower, broccoli, radish, oilseed rape and turnip (Snowdon, 1990b; Agrios, 2005; Averre, 2010) (Fig. 6.9). The disease has been recorded in Europe for many hundreds of years, but its effects can still be devastating, as it causes the galling or clubbing of roots, resulting in stunting and, in some cases, in wilting of the affected plants (Sherf and MacNab, 1986).

The organism produces dormant cysts that can remain viable in soil for many years (Walker, 1969). Identification in the field is usually rapid owing to the obvious appearance of clubbed roots. However, the detection of dormant cysts in the field is often necessary in deciding whether to plant a cruciferous crop. Protocol 6.6 gives a reliable method to determine presence or absence of this pathogen in soil.

Phytophthora species

Phytophthora species are not true fungi but belong to the kingdom Chromista (Brasier *et al.*, 2022). They attack a wide range of plants, including trees, herbaceous plants, and bulb and soft fruit plants, although grasses are believed not to be affected (Mehrotra and Aneja, 1990) (Fig. 6.10). The host range of a particular species may vary and can be wide, e.g. *Phytophthora cactorum* infects both woody and herbaceous hosts from over 150 genera, whereas *Phytophthora idaei* has a narrow host range and only infects raspberry plants (Erwin and Ribeiro, 1996; Brasier, 2009). Often referred to as water moulds, all phytophthoras have a motile zoospore stage that moves through water (Erwin, 1983). They are generally soil-borne organisms that infect the roots or crowns of susceptible host plants. However, some species can also infect aerial plant parts (leaves, green shoots, tree trunks) via deciduous (caducous) sporangia that can travel in the wind or wind-driven mists (e.g. *P. cactorum*) (Shaw, 1988). *Phytophthora* produces different sporing structures, all of which are important for morphological diagnosis. These include sporangia and associated sporangiophores, oogonia, antheridia, oospores and

Fig. 6.9. Clubroot infection of oilseed rape plants. (From Shutterstock.)

Fig. 6.10. Severe root and stem rot of cassava caused by *Phytophthora* sp. (From Shutterstock.)

chlamydospores. The identification of *Phytophthora* is described in Protocol 6.7, and the measurement of its growth temperatures in Protocol 6.8.

Downy mildews

Downy mildews are classified as water moulds and belong to the family Peronosporaceae. They are genetically closer to *Phytophthora* spp. than to true fungi (Agrios, 2005). Favouring damp and humid conditions, these organisms are significant biotrophic pathogens to a range of plants, including vegetables, ornamental plants, grapes and crucifers (Gäumann, 1923) (Fig. 6.11). Symptoms are typically observed on leaves, stems and, occasionally, on flowers. Leaf infection appears as chlorotic or necrotic spots on the upper surface and as a dense, felt-like, often greyish-coloured 'down' on the lower surface. As these are obligate pathogens, it is not possible to culture them on agar plates. Individual species frequently have a narrow host range, so knowledge of the host will aid species identification, which is discussed in Protocol 6.9. Some sporangia of downy mildews germinate directly, whereas others

Fig. 6.11. Cucumber downy mildew (*Pseudoperonospora cubensis*) infected leaf. (From Shutterstock.)

produce zoospores. To determine which mode of action occurs, a germination test can be carried out (see Protocol 6.10).

Powdery mildews

Powdery mildews are diseases caused by a diverse group of obligate pathogens in the order Erysiphales (Ing, 1990a, 1990b, 1990c, 1990d, 1991; Braun, 1995; Cook *et al.*, 1997). They can infect numerous plants (apart from gymnosperms) and all aerial parts, including leaves, stems, flowers and fruits (Hirata, 1969; Horst, 1983; Munro, 1986) (Fig. 6.12). The fungus typically appears as a white powdery deposit on the infected part of the plant; however, chlorosis, necrosis, dieback and even death of the host can result (Chase, 1987). Economically, if left untreated, powdery mildew can cause significant crop loss. For example, an infected vineyard can result in 100% of fruit unsuitable for winemaking, and up to 20% yield loss is common with a powdery mildew infection of wheat or barley (Pearson and Goheen, 1988; Chan *et al.*, 1990; Bowen *et al.*, 1991).

From the diagnostic viewpoint, powdery mildews are quite distinct from other plant pathogens owing to their characteristic white powdery appearance on infected plant tissue (Gorter, 1988). Symptoms are most commonly first seen on leaves on both upper and lower surface (caused by numerous asexual spores [conidia] borne on conidiophores), in contrast to a downy mildew, in which spores are produced on the lower leaf surface. Early leaf symptoms are commonly mild chlorosis at and around the point of infection. The host may respond to an attack, resulting in leaf flecking due to numerous small necrotic leaf spots or purplish rings (e.g. Rhododendron powdery mildew; Beales, 1997).

The asexual or anamorphic spore stage of powdery mildews is more noticeable to the casual observer than the sexual spore (ascospore) stage (Zaracovitis, 1965), which is formed in small (*c*.0.5–1.0 mm) fruiting bodies called asci that are housed in a protective structure known as a chasmothecia (Ialongo, 1992; Bélanger *et al.*, 2002). The sexual stage is often only evident during winter

Fig. 6.12. Powdery mildew infected roses – note white appearance on flowers and leaves due to numerous conidia of *Podosphaera pannosa*. (From Shutterstock.)

or the hot summer months, when it occurs in relatively small numbers compared with those of the asexual stage (Braun, 1987). Protocol 6.11 can be used with the anamorphic or teleomorphic fruiting stage.

Colletotrichum

The genus *Colletotrichum* consists of approximately 189 species organized into at least 11 species complexes (Baroncelli *et al.*, 2017), many of which are fungal pathogens. This group of fungal organisms has worldwide importance owing to its impact on a wide range of plants, which results in a direct reduction in crop quality and/or quantity (Sutton, 1980) (Fig. 6.13). Most frequently, the fungus is found in the asexual *Colletotrichum* stage, but also some species produce the teleomorphic (sexual) *Glomerella* stage (Sutton, 1992).

Virtually every crop grown throughout the world is susceptible to one or more species of *Colletotrichum* (Waller, 1992). Symptoms are typically postharvest rots, anthracnose spots and blights of aerial plant parts. Members of this genus can result in major economic losses, especially of fruits, vegetables and ornamentals

(Snowdon, 1990a). For example, the strawberry blackspot pathogen *Colletotrichum acutatum* species complex attacks strawberry fruits, resulting in blackened necrotic spots that spread over the fruits and causes them to rot (Simmonds, 1965; McGechan, 1977; Smith and Black, 1990); it is regarded as the second most important pathogen of strawberries (Cook, 1993).

The identification of *Colletotrichum* from morphological characteristics can be difficult, owing to the similarity of cultural features between species, many of which have only recently been described. Protocol 6.12 highlights some of the features that need to be recorded to identify the genus and separate some of the species or species complexes.

Monilinia

Commonly referred to as brown rot fungi, *Monilinia* spp. affect a wide variety of (mainly) fruit crops (particularly pome and stone fruits), causing blossom and twig blight and fruit rot (Pirone *et al.*, 1960; Snowdon, 1990a) (Fig. 6.14). Host symptoms include wilting of flowers and shoot death, along with browning and collapse of fruit, followed by copious amounts of spore production

Fig. 6.13. Anthracnose disease of chilli pods caused by *Colletotrichum capsici*. (From Shutterstock.)

Fig. 6.14. Rotten, mummified plums due to *Monilinia laxa* infection. (From Shutterstock.)

in damp conditions – often seen forming in concentric rings on fruit (Wormald, 1954). The protocol to help to identify some of the key species principally involves examination of macroscopic colony characteristics and introduces a polytomous key (Protocol 6.13).

Smut fungi

The terms 'smut' and 'bunt' fungi are used for obligate plant pathogens in the class Ustilaginomycetes (Bauer *et al.*, 1997). These produce serious diseases of a range of plants, including grasses and, until the 20th century, when stricter disease management controls were implemented, they caused significant losses to cereal crops (Wiese, 1987). These fungi inhabit stems, leaves and floral organs; in cereals, the kernel is replaced with a mass of black powdery spores; this reduces crop yield, but also taints the grain with an often foul-smelling, black residue (Martens *et al.*, 1984). Some smut fungi (*Tilletia* spp.) that infect cereals are known as bunts or covered smuts, and these destroy the contents of infected kernels, replacing them with spores of the fungus; they can also cause stunting of the host (Carris and Castlebury, 2006) (Fig. 6.15). Smut fungi often have a narrow host range, so knowledge of this will help with species identification,

which is described in Protocol 6.14. When a smut spore (teliospore) germinates, it produces sporidia or basidiospores, and various features of germinated teliospores can be used to aid identification of species (see Protocol 6.15).

Armillaria

Honey fungus (*Armillaria* spp.), which causes bootlace, shoestring or collar rot, is one of the most common pathogens affecting and killing a wide range of (principally) shrubby plants and trees (Pegler, 2000) (Fig. 6.16). It attacks the roots and spreads to the collar of plants, killing them in a matter of months or over a period of years (Pérez-Sierra, 2003). One of the key methods to diagnose honey fungus involves examining features on the plant; however, there are a number of saprophytic species, so accurate species identification is important. As these fungi produce a limited number of distinguishing morphological characteristics, molecular methods are most frequently used to separate this group of organisms, although mating compatibility tests and morphological characteristics can help to separate some of the pathogenic species from the saprobes. Species identification from cultural characteristics is described in Protocol 6.16.

Fig. 6.15. Bunted grains of wheat caused by *Tilletia tritici.* (From Shutterstock.)

Fig. 6.16. Clumps of honey fungus (*Armillaria mellea*) basidiocarps growing out from a decaying tree. (From Shutterstock.)

Sclerotia-forming fungi

Sclerotia-forming fungi produce at some stage in their life cycle a survival structure consisting of a compact mass of hardened fungal hyphae that contains food reserves. The majority of plant pathogenic forms are ascomycetes in the genera *Sclerotinia* and *Sclerotium* (e.g. *Sclerotinia sclerotiorum*, which has a wide host range and causes stem rots of brassicas, solanaceous plants and horticultural plants such as chrysanthemum (Fig. 6.17), and *Sclerotium cepivorum*, which causes root rots of onions and other alliums) (Schwartz and Mohan, 1996; Horst and Nelson, 1997). However, other plant pathogenic genera also produce sclerotia, including some species of *Botrytis*, *Ciborinia*, *Rhizoctonia*, *Colletotrichum* and *Verticillium* (Agrios, 2005). Sclerotia vary considerably in size, depending on the species, and size is a key morphological diagnostic characteristic, along with host plant and hyphal features (e.g. width). Isolating sclerotia from the surrounding substrate is therefore important in the diagnostic process. The identification of sclerotia-forming fungi is described in Protocol 6.17.

Rust fungi

Rust fungi are a diverse group of about 7000 biotrophic fungal pathogens classified in the division Basidiomycetes, order Pucciniales (Scott and Chakravorty, 1982). The name derives from the fact that many rusts produce reddish, yellow or brown 'rusty' coloured spores as part of their life cycle (Fig. 6.18), although symptoms can range from leaf chlorosis and distortion to galls (Peterson, 1974; Schumann and D'Arcy, 2010). These fungi affect most plants, including ornamentals, grasses and conifers, but the host range of each rust species is usually narrow (Agrios, 2005). Unusually within fungi, some rust species need two unrelated host plants to complete their life cycle; these are referred to as heteroecious rusts. Those that can complete their life cycle on a single host are referred to as autoecious rusts. Morphological identification is complicated, as some species can produce five different spore states, e.g. apple and pear rust (*Gymnosporangium* spp.), which produces spermatia, aeciospores, urediniospores, teliospores and basidiospores (Jones and Aldwinckle, 1990). Other species, such as Chrysanthemum white

Fig. 6.17. Cross-section of an oilseed rape stem infected with *Sclerotinia sclerotiorum*. Note presence of dark black sclerotia. (From Shutterstock.)

Fig. 6.18. Close-up of rust pustules showing spores emerging from leaf underside. (From Shutterstock.)

rust (*Puccinia horiana*) only produce two spore types (teliospores and basidiospores) (Horst and Nelson, 1997). It is also worth noting that many of the spore stages can occur at the same time on the host, and they may also develop sequentially in the same pustule. The identification of rust fungi is described in Protocol 6.18, and the use of keys to identify poplar rust in Protocol 6.19.

Fusarium

Fusarium spp. are an important group of plant pathogens which affect a wide range of hosts, from cereals to ornamental plants, and are commonly isolated from plant tissue (Wiese, 1987; Gleason *et al.*, 2009) (Fig. 6.19). They are generally opportunistic organisms, attacking stressed or damaged plant tissue. However, once established in the host, they can cause a range of symptoms from chlorosis to damping off or wilt, root decay, stem necrosis, cankers and death (Leslie and Summerell, 2006). The production of mycotoxins by some species can affect grain

Fig. 6.19. Fusarium head blight of wheat. (From Shutterstock.)

quality and pose a risk to human and animal health if the infected grains enter the food chain (Moss, 2008). Many species are saprophytic, so further pathogenicity testing may be required to identify pathogenic species. The identification of *Fusarium* from cultural characteristics is described in Protocol 6.20.

6.5 Diagnosis of Fungal Diseases from Some Common Host/Crop Species

Potato tubers

Potatoes (*Solanum tuberosum*) are affected by numerous pathogens, insects and physiological disorders. Many fungi and bacteria can cause soft or firm rots in potato tubers; symptoms are often present during harvest, but some of the problems will also spread through stored tubers (Fig. 6.20), and a few will only develop after prolonged storage (Brenchley and Wilcox, 1979). Significant problems often follow a wet growing season, particularly if the tubers are then lifted from wet soil. Some of the

pathogens causing tuber rots also produce symptoms on the aerial parts of the plant.

Common fungal causes of potato tuber rots are *Fusarium* spp. (dry rot) (Fig. 6.21 (a, b)) and *Boeremia* (*Phoma*) *foveata* (gangrene). Tubers with gangrene remain firm and dry, but cavities form within the tubers with a clear, dark margin between the healthy and diseased tissue (UNECE, 2014). Dry rot symptoms are usually observed following storage, with brown, slowly enlarging lesions as well as wrinkles (concentric rings) of the skin as the diseased tissues dry out (Wale *et al.*, 2008). White, blue and/or pink fungal growth may be seen on the affected areas and within cavities (Arora and Sagar, 2014). Gangrene and dry rot can sometimes be confused on symptoms alone (Brenchley and Wilcox, 1979), therefore laboratory confirmation of the cause is important. Other potato tuber rots caused by fungal and fungus-like pathogens include rubbery rot (*Geotrichum candidum*), pink rot (*Phytophthora erythroseptica*), watery wound rot (*Pythium* sp.) and the less common late blight (*P. infestans*) and early blight (*Alternaria solani*).

Currently, there are two fungi which cause rot-like symptoms in potato tubers which are

Fig. 6.20. Potatoes rotting in storage. (From Shutterstock.)

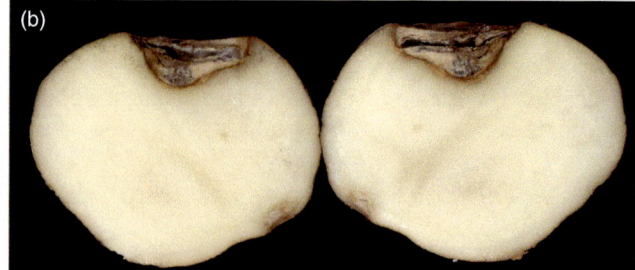

Fig. 6.21. (a) Fusarium dry rot of potato; (b) cross-section of tuber showing dry rot lesion caused by *Fusarium* sp. (From Shutterstock.)

listed as quarantine organisms in many countries. *Synchytrium endobioticum* (potato wart disease) and *Thecaphora solani* (potato smut) both cause galls to form on the surface, leading to the whole tuber becoming heavily distorted and unmarketable (OEPP/EPPO, 2017; EFSA Panel on Plant Health, 2018a, 2018b).

Potatoes can also be infected with a variety of pathogens that change the appearance of the surface of tubers. These diseases generally only affect the skin, and do not cause internal rotting. Two blemish diseases that are frequently encountered on potato are powdery scab (caused by *Spongospora subterranea*) and common scab (caused by *Streptomyces scabiei*) (Fig. 6.22). These differ by powdery scab having subglobose sandy to green spore balls within scab lesions, while common scab does not (Wale *et al.*, 2008). *Spongospora subterranea* vectors potato mop-top virus, which causes the economically damaging disease spraing, meaning that there is a lower tolerance for powdery scab in seed potatoes (Carnegie *et al.*, 2007).

Other common blemish diseases are silver scurf (*Helminthosporium solani*) and black dot (*Colletotrichum coccoides*). Both pathogens cause dotting on the tuber surface, but the dots of *C. coccoides* are larger and are produced within lesions which are less well defined and more limited in size than those of *H. solani* (UNECE, 2014). Furthermore, potatoes can get the tuber blemish diseases black scurf (*Rhizoctonia solani*), skin spot (*Polyscytalum pustulans*) and violet root rot (*Helicobasidium purpureum*). Visual aids for symptom recognition (for example, AHDB Potatoes, 2018) can help decide if the problem is likely to be fungal.

Fig. 6.22. Potato tuber affected with blemish common scab disease. (From Shutterstock.)

The testing of potato tubers for both rot and blemish diseases is covered in Protocol 6.21.

Strawberries

Strawberries (*Fragaria* spp.) are host to a range of fungal and fungus-like pathogens. These include the oomycetes *Phytophthora fragariae*, *P. cactorum* and *Pythium* spp., which all infect via the roots and form oospores within the root tissue. *P. fragariae* and *P. cactorum* cause root rot, which can lead to wilting and stunting of both young and mature leaves and stems, however *P. fragariae* causes rat-tailing (unbranching) of roots, and a characteristic reddening of the root stele, which can extend into the crown (giving the disease its common name red core/red stele) (Fig. 6.23). Contrastingly, infection by *P. cactorum* characteristically gives strawberry crowns an intense brown staining and causes disintegration

Fig. 6.23. Wilting strawberry plants due to a *Phytophthora* sp. root infection. (From Shutterstock.)

of vascular tissue (Maas, 1998; HDC, 2004). *Pythium* spp. infection is known to lead to severe stunting of above ground parts, yield loss and destruction of the roots (Maas, 1998).

Furthermore, the soil-borne fungi *Verticillium dahliae* and *V. albo-atrum sensu lato* cause wilting of outer leaves, significant yield loss and eventual plant death (MAFF, 1970; Maas, 1998). While no crown or root rotting occurs (HDC, 2004), *V. dahliae* and *V. albo-atrum* can be distinguished by the former producing chlamydospores and dark microsclerotia within roots, whereas *V. albo-atrum* produces dark, swollen resting mycelium (Maas, 1998).

Other common fungal diseases of strawberries include root rotting by *Rhizoctonia* spp., leaf spot by *Mycophaerella fragariae*, grey mould of fruit and leaves from *Botrytis cinerea* infection, and strawberry blackspot disease affecting leaves, petioles and fruit, caused by *Colletrotrichum acutatum* species complex (HDC, 2004) (Fig. 6.24).

As a range of strawberry plant parts can be infected by fungal pathogens, there are a broad range of procedures that can be utilized for examination and testing, such as incubation chambers, isolations onto selective media, root preparations and root floats, as outlined in Protocol 6.22.

Turfgrass

The health of turf can be affected by a range of abiotic and biotic factors. Disease plays a major role in determining the success or failure of a turfgrass stand, and fungi are recognized as the most important cause of turfgrass diseases (Vargas, 1994). There are numerous fungal and oomycete turfgrass pathogens which can cause root rots, leaf blights, leaf spots, smuts, rusts, powdery and downy mildews, as well as fairy rings (Fig. 6.25). Disease complexes are important in turfgrasses and especially likely when plant roots are damaged before periods of environmental adversity, when roots as well as foliage become more susceptible to colonization by facultatively parasitic and saprotrophic fungi (Smiley, 1983).

Guidance on how to assess and test turf samples for the presence of a range of fungal plant pathogens is given in Protocol 6.23.

Fig. 6.24. Strawberry blackspot disease affecting strawberry pseudocarp. (From Shutterstock.)

Fig. 6.25. Darkened ring of turf grass caused by fairy ring fungus. Basidiocarps (mushrooms) of the fungus often appear at the edge of the ring. (From Shutterstock.)

Fruit crops

Fungal diseases can have a significant economic impact on various fruit crops. For example, *Botrytis cinerea*, which affects a wide range of hosts, is one of the most important pre- and post-harvest pathogens in fresh fruits and vegetables (Hua *et al.*, 2018) (Fig. 6.26).

Fruit can be infected by fungal pathogens during the growing season or post-harvest. Weather conditions play a significant role in facilitating fungal infections during the growing season, for example, wet cloudy weather around the apple blossoming period encourages outbreaks of apple scab caused by *Venturia inaequalis* (Fig. 6.27) and grey mould in strawberries takes a heavy toll if there is wet weather during fruit ripening (Wormald, 1955). Fruit deterioration can occur post-harvest and can be caused by physiological changes, physical damage, chemical injury or pathological decay (Snowdon, 1990a).

Some fungal plant pathogens affect a narrow range of hosts, for example *Phyllosticta citricarpa* (the cause of citrus black spot), which affects *Citrus*, *Poncirus* and *Fortunella* and their hybrids (EPPO, 2020). Others, such as *Alternaria alternata* and *Colletotrichum gloeosporioides sensu lato*, can cause disease in a broad range of fruit and vegetable hosts. It is worth noting that some of the fungi which can cause fruit rots in a wide range of hosts, like *Penicillium* and *Mucor* species (Snowdon, 1990a), are often encountered as saprophytes in other types of plant samples but should not be overlooked when identifying the cause of fruit decay.

To find out the cause of deterioration in fruit, a good place to start is finding out if a pathogen is involved. A procedure for examination and testing of fruit samples for the presence of a range of fungal plant pathogens is described in Protocol 6.24.

6.6   Use of Identification Keys

Keys to identify fungi generally fall into two categories: diagnostic and synoptic. The former are designed to make identification convenient and

Fig. 6.26. *Botrytis cinerea* grey mould affecting mature blackberries. (From Shutterstock.)

Fig. 6.27. Apple fruit affected by apple scab disease (*Venturia inaequalis*). (From Shutterstock.)

reliable, whereas the latter aim to reflect the scientific classification of fungi (Winston, 1999). For example, a diagnostic key may use host and symptom type to identify a possible pathogen, while a synoptic key would take features of the fungus affecting the same plant to identify its taxonomic rank.

Dichotomous (also known as sequential or pathway) keys are designed in such a way as to give two choices at each branching point, whereas a polytomous key will give several options after each statement. There are advantages and disadvantages to both types. A dichotomous key can give a final answer quickly but can cause problems if one of the early questions cannot be answered, e.g. owing to lack of a feature, or is answered incorrectly. A polytomous key can help to allow identification even if some features are lacking but can be time consuming to complete or provide inconclusive results.

The protocols that follow give examples of both dichotomous and polytomous keys to enable the separation of indigenous poplar rusts in the UK from a non-indigenous species (Protocol 6.19). A second polytomous key can be found in Protocol 6.13 to enable the separation of *Monilinia* spp.

Preparation from plant material for microscopical examination
Leaves, stems, flowers and fruits
Protocol 6.1

Adhesive tape method

Materials

- Clear adhesive tape
- Microscope slides
- Coverslips
- Mounting solution (lactoglycerol or water ± stain)
- Heat source
- Dissecting and compound microscope

Method

1. Carefully cut a 2–3 cm square piece of clear adhesive tape and hold with forceps to prevent fingerprints on the sticky surface.

2. Press on to plant surface with enough pressure to adhere to the tape and remove the structure of interest, but not too hard to remove plant material, and peel off carefully.

3. Add 1–2 drops of mounting solution with stain (e.g. cotton blue in lactoglycerol), clear lactoglycerol or water to a microscope slide.

4. Place the tape sticky side down on to the stain (note no coverslip is required) (Fig. 6.28).

5. Press down gently to remove air bubbles, and then warm very carefully to avoid boiling which may melt the adhesive tape. Allow slide to cool.

6. View directly under a compound microscope.

Fig. 6.28. Adhesive tape method to remove fungal structure direct for viewing under microscope. (Image Courtesy Fera-Science Limited © Copyright Fera-Science Limited.)

Protocol 6.2

Direct removal of fungal fruiting bodies from plant surfaces

Materials
- Microscope slides
- Coverslips
- Mounting solution (lactoglycerol or water ± stain)
- Mounted needle
- Heat source
- Dissecting and compound microscope

Method

1. Place a small drop of lactoglycerol (with or without stain) onto the surface of a clean glass microscope slide.

2. Keeping the sample material under the dissecting microscope, carefully pick off a specimen of several suspect fungal fruiting bodies with the tip of a fine alcohol-flamed mounted needle or new sterile hypodermic needle (Fig. 6.29). It may be necessary to cut a thin section of a fruiting body if details of spore attachment are not clear. This can be done with either a sharp single-edged razor blade or a scalpel.

3. Examine the slide under dissecting microscope to confirm that structures were successfully removed and are correctly orientated for examination (excessive amounts of host material can also be teased away at this stage and discarded).

4. Gently lower a suitably sized glass coverslip onto the slide and then press down lightly to expel any air bubbles. If necessary, holding the slide by one end, gently warm the slide over the flame of a spirit burner until any air pockets start to expand, remove from the flame and place on the bench top to cool. This removes any further pockets of air and assists in clearing of tissues by the mountant. It can also cause the spores to expand. This is helpful when viewing rust spores.

5. Transfer the slide to a compound light microscope for viewing.

Fig. 6.29. Removing fungal structures from plant surface. (Image Courtesy Fera-Science Limited © Copyright Fera-Science Limited.)

Protocol 6.3

Hand sectioning

Materials
- Dissecting and compound microscope
- Razor blade/scalpel
- Microscope slides
- Coverslips
- One carrot
- Cutting board
- Fine hairbrush
- Mounting solution (lactoglycerol or water) with or without stain (e.g. cotton blue)
- Clear nail varnish
- Tissue paper

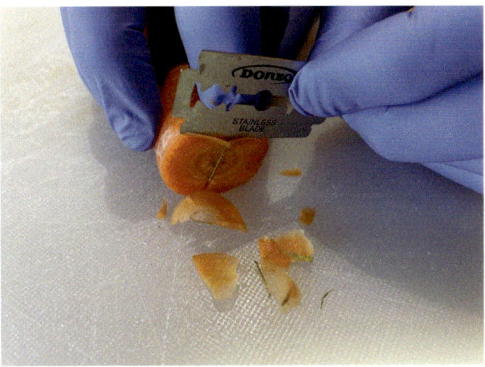

Fig. 6.30. Hand sectioning using carrot to support plant tissue of interest. (© UK Crown copyright.)

Method

1. Using a low-powered dissecting microscope, select an area of interest (e.g. fungal fruiting body).

2. Where possible, cut the plant tissue with a new scalpel/razor blade to an area two to three times greater than the specific area you wish to examine microscopically and insert into a slot cut into one end of the carrot.

3. Hold the tissue section of interest firmly within the carrot. Cut through the tissue with the blade in one clean movement (Fig. 6.30). Repeat this process until numerous thin sections are prepared. These can be floated in water in a Petri dish to aid selection (stage 4).

4. Still viewing through a dissecting microscope, use a fine brush to separate the carrot tissue and transfer gently thin sections of plant tissue containing the area of interest onto a glass slide. Wipe any excess liquid with tissue paper.

5. Add 1–2 drops of a mounting solution to the slide, cover with a coverslip and view under a compound microscope.

6. If a more permanent preparation is required, the coverslip can be sealed with three coats of clear nail varnish (or equivalent propriety sealant) painted around the edges (allow each coat to dry fully between each application). Note: Semi-permanent slide mounts are most effective when lactoglycerol or a stain is used (water tends to evaporate more quickly). Ensure that no bubbles are present in the mounting solution (gentle warming will reduce any bubbles prior to sealing) and that the solution extends to the edge of the coverslip but not beyond (use a small piece of paper tissue to remove excess from the outside the perimeter of the coverslip before application of sealant).

Roots
Protocol 6.4

Microscopical examination of roots for pathogens

Materials

- Dissecting and compound microscope
- Safety glasses
- Forceps
- Microscope slides
- Coverslips (large)
- Potassium hydroxide crystals/pellets
- Boiling tube + heat source, e.g. gas burner
- Mounting solution (stain/lactoglycerol or water)

Method

1. Carefully wash roots under flowing cold water (refer to Chapter 12 if handling quarantine pathogens). If the soil is particularly heavy or dry and adhering to the roots, soak in warm water for 24 h to loosen soil particles. If the soil is sandy, ensure that all sand particles are removed from the roots to be examined as a single grain can cause difficulty in mounting on the microscopic slide.

2. If fine, non-woody roots are present, it may be possible to examine these directly without softening under the microscope.

    2.1. Carefully remove roots using forceps under a dissecting microscope, and place on a microscope slide. Crush by placing a second microscope slide on top of the sample and apply gentle pressure.

    2.2. Remove the second slide, add mountant solution (stain/lactoglycerol or water) and place coverslip over the sample. View using a compound microscope.

3. If fine roots are not present, or they are too tough to crush directly, the roots will need to be softened.

    3.1. Select roots from different parts of the root ball and add to a boiling tube with approximately 2.5 cm of water and add two potassium hydroxide pellets (note: use gloves/forceps when adding potassium hydroxide crystals).

    3.2. Wearing safety glasses, warm boiling tube gently at a 45° angle with the open end away from your face, gently agitating continually and remove from heat source before it boils. Repeat this step for approximately 2 min depending on the root density.

    3.3. Pour the roots and liquid out of the tube into an empty Petri dish, rinsing out the tube with water as necessary. If required, the roots can be left to soak up to 2 h to further soften.

    3.4. Place a few of the softened roots onto a microscope slide and place a second slide on top. Apply a small amount of pressure to crush the roots and remove secondary slide. Add appropriate mountant solution and carefully place coverslip over sample. View with a compound microscope.

Note

Some fungal structures form less frequently than others and therefore more than one root preparation may be required. For example, when examining strawberry roots for *Phytophthora* spp. oospores, up to seven slides often need to be examined.

Morphological examination of cultures Protocol 6.5

Cultural characteristics

Following incubation under controlled conditions (Protocol 3.1), various cultural characteristics can be helpful in identifying fungi.

Materials
- Compound and dissecting microscopes
- Petri dish

Method

1. Examine culture under low-power magnification using a dissecting microscope (up to ×50 magnification).

An initial examination which reduces delicate structures disturbance (e.g. spore chains) enables the selection of a suitable area for more detailed study and can be done repeatedly to identify changes, study growth characteristics and calculate culture growth rate. This can be carried out with the lid of the Petri dish on or off.

Record in Table 6.1:
- Rate of culture growth, presence/absence of pigmentation.
- Hyphae type, e.g. sparse, compact, aerial, fluffy, pigmentation. Note whether this changes over time.
- Presence/absence of fruiting bodies (note: it may be necessary to scrape away some of the hyphae to observe any obscured fruiting bodies (e.g. pycnidia); it is also advisable to examine the culture plate from the underside to look for fruiting bodies immersed within the agar). It may be necessary to cut a thin section of a fruiting body if details of spore attachment are not clear. This can be done with either a sharp single-edged razor blade or a scalpel.
- Presence/absence of conidiophores. Identify whether these are short or long.

- Spores: catenulate (in chains), length (short or long) or single.

2. Examine culture under compound microscope.

Make a slide mount of culture and add an appropriate stain (note this can be repeated at different times to identify morphological changes).

Examine initially under low magnification (×100) and increase as required until the morphological feature can be observed and measured. Record:

2.1. Hyphae
- General appearance (e.g. width, coralloid, thick walled).
- Degree and angle of branching (e.g. slight, extensive).
- Presence or absence of septa and clamp connections.
- Pigmentation.
- Ornamentation.

2.2. Fruiting bodies
- Type (e.g. pycnidium, perithecium, apothecium, acervulus, sporangium, oogonium).
- Size (measure breadth and length of any features, such as perithecial beak).
- Other features (e.g. papillation, antheridia, appendages, ornamentation).

2.3. Exposed spore-bearing structures (e.g. conidiophore/sporangiophore)
- Length.
- Length of foot cell.
- Spore-forming type (e.g. blastic/phialidic).
- Any additional features (e.g. swollen or twisted base).

2.4. Spores
- Length, breadth and length-to-breadth ratio.
- Ornamentation (side wall/end wall if visible).
- Type (sexual, asexual, zoospore, chlamydospore, arthrospore, other).

Table 6.1. General culture identification. Fill in as fully as possible.

General details	
Date of culture	
Reference number	
Rate of growth (mm day^{-1})	
Cultural characteristics	
Hyphae (note changes over time)	General appearance
	Colour:
	Ornamentation:
	Aerial: Present ❑ Absent ❑
	Width (µm):
	Branching pattern:
	Septa: Present ❑ Absent ❑
	Clamp connections: Present ❑ Absent ❑
Fruiting bodies	Present ❑ Absent ❑
	Type:
Exposed spore-bearing structures (e.g. conidiophores/ sporangiophores)	Present ❑ Absent ❑
	Length (µm):
	Length of foot cell (µm):
	Spore-forming type:
	Additional features:
Spores	Present ❑ Absent ❑
	Type:
	Length (l) (µm):
	Breadth (b) (µm):
	l:b ratio:
	Ornamentation:

Identification of some common groups of fungal plant pathogens using morphological features
Protocol 6.6

Testing soil for the clubroot pathogen (*Plasmodiophora brassicae*)

Materials

- Oilseed rape seeds (*Brassica napus* var. 'Broadleaf Essex', or other highly susceptible variety)
- Clubroot-infested soil for positive control, which can be produced by finely chopping infected plant roots into potting compost. Note: Set up positive control soil using the method described below, before setting up test soil
- Plant pots and saucers
- Sterile compost (John Innes No. 2)
- Seedling compost

Method

1. Plant 250 susceptible oilseed rape seeds into seedling compost and incubate under conditions suitable for germination. Seedlings are suitable for use as bait plants 10 days post emergence.

2. Collect 1 kg of test soil made up from 100 collection points over *c.* 1 ha from the location of interest (Melville and Hawken, 1967).

3. If the test soil is heavy clay, mix thoroughly with 1 kg sterile compost (John Innes No. 2).

4. Plant seedlings in pots as follows:

4.1. Five seedlings in one pot containing sterile potting compost (John Innes No. 2). Label as negative control.

4.2. Divide test soil sample evenly between five pots and plant five seedlings in each pot. Label as test soil.

4.3. Plant five seedlings in a pot containing positive control material. Label as positive control.

5. Place a saucer under each of the pots and water thoroughly and carefully using distilled water or tap water with a pH value lower than or near neutral (pH 7.0). Note – ensure water does not splash between pots when watering.

6. Incubate in a glasshouse at 21 °C for 6 weeks. Note: Ensure that compost remains well watered throughout the test period.

7. After 6 weeks, remove plants in the following order (negative control, positive control and test plants), carefully wash roots to remove soil, and examine and record signs of swellings or clubbing on roots of plants (see Fig. 6.31). If symptoms are observed in the negative control, or there are no symptoms in the positive control, the test must be repeated.

Severe

 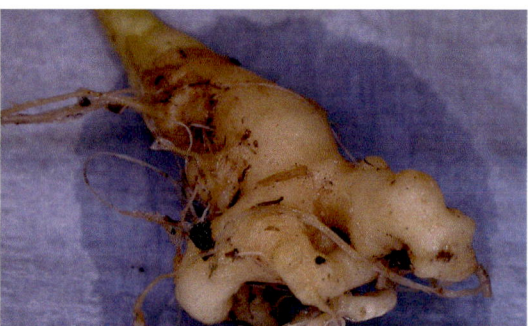

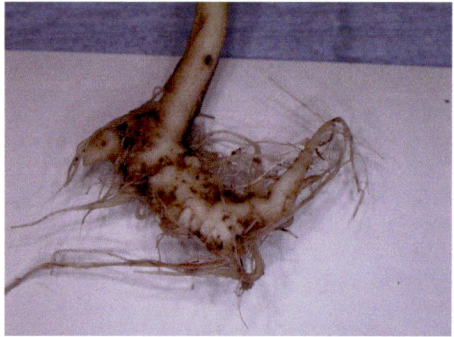

Moderate

Slight

Fig. 6.31. Various degrees of infection by the clubroot pathogen (*Plasmodiophora brassicae*) on test roots of *Brassica napus* var. 'Broadleaf Essex'. (© Crown copyright – courtesy of Fera.)

Protocol 6.7

Identification of *Phytophthora*

> **Materials**
>
> - Dissecting and compound microscope
> - Scalpel/sharp scissors
> - Microscope slides and coverslips
> - Cotton blue (0.1%) in lactoglycerol

Method

1. Following isolation from plant material, water or soil onto culture medium (Protocols 3.2, 3.3(a) and 3.3(b)), cut a small piece of agar (1 × 1 cm) from the culture plate (locate using a dissecting microscope) or select a suitable piece of tissue/agar from the float, stain with cotton blue in lactoglycerol and mount for viewing.

2. Examine under a compound microscope and record the following features in Table 6.2:

2.1. Hyphae
- Young hyphae should be aseptate (coenocytic, i.e. without cross-walls).
- Width.
- Note general appearance (e.g. coralloid, swollen, nodular, tuberculate, coiled).
- Presence/absence of hyphal swellings (globose swellings on hyphae near a septum).

2.2. Sporangiophores
- Branching pattern (unbranched, irregularly branched or sympodial; sympodial sporangiophores emerge from the base of the previous sporangium and can be simple, compound, umbel or proliferate through the base of the empty sporangium).
- Note presence/absence of swellings just below the sporangia.

2.3. Chlamydospores
- Record presence or absence, and whether they are numerous or sparse.

- Note whether chlamydospores are terminal or within (intercalary) hyphae.
- Diameter.
- Presence or absence of pigmentation.
- Wall thickness (thin/thick).

2.4. Hyphal swellings
- These are often confused with chlamydospores and are distinguished by the lack of delimitation by septa from the hyphae; note presence or absence and diameter.

2.5. Sporangia
- Sporangial caducity: note whether the sporangia are shed from the sporangiophore at maturity (caducous) or remain attached (non-caducous); species that are truly caducous possess a plug of wall material at the point of detachment.
- If caducous, the length of the sporangial pedicel can be a useful morphological feature to record (short <5 μm, medium 6–15 μm, long >16 μm).
- Shape can be variable or distorted (this is sometimes due to contaminated cultures, environmental factors or if the primary isolation is from a plant where pesticide has been applied previously). Sporangial shape is normally relatively consistent and is a useful morphological feature. The most common sporangial shapes are ovoid, limoniform, obturbinate, ellipsoid, pyriform and globose.
- Record length (l), breadth (b), l:b ratio.
- Papillation – sporangia are classified into three groups according to the degree of thickening at the apical end, namely papillate (conspicuous thickening), semi-papillate and non-papillate (inconspicuous thickening) (Waterhouse,

1963). Staining sporangia with 0.1% cotton blue in lactoglycerol (Waterhouse, 1963) can help to distinguish the level of papillation; measure exit pore size – these are generally narrow (c.5.7 µm) for papillate and semi-papillate sporangia and wider (c.12 µm) for non-papillate species (Erwin and Ribeiro, 1996).

2.6. Zoospores

Zoospores of *Phytophthora* spp. mature within sporangia. They are biflagellate with a longer smooth anterior whiplash flagellum and a shorter ornamented (hairy) shorter posterior flagellum. A deep longitudinal groove is also present on one side of the zoospore. Zoospores are not generally used to distinguish species, as morphological features are difficult to observe with compound microscopes.

Sexual spore features

The sexual spore form of *Phytophthora* is the oospore that is formed by fusion of the female oogonium and the male antheridium. *Phytophthora* spp. can be divided into two groups depending on whether distinct fertile mating types are formed. These are referred to as:

- Heterothallic species, in which separate mating types (designated A1 and A2) need to pair to produce the sexual oospore both *in planta* and *in vitro*.
- Homothallic species, which are self-fertile so do not require the pairing of separate mating types to initiate oospore formation. Oospores occur in a single culture or *in planta*.

2.7. Antheridia
- Measure size and record shape (round or square).
- Size (short <10 µm) or elongated (>20 µm) (Erwin and Ribeiro, 1996).
- Amphigynous or paragynous – these refer to whether the maturing oogonium develops through the antheridium (amphigynous) or appress laterally (paragynous). This is a major morphological diagnostic feature.

2.8. Oogonia
- Record shape (round ± tapered base).
- Numerous or sparse in culture.
- Smooth or ornamented wall.
- Size.

2.9. Oospores
- Note whether the oospore fills (plerotic), almost fills (partially plerotic) or does not fill (aplerotic) the oogonial space.
- Wall thickness (distance from the outer to innermost wall).
- Diameter.

Table 6.2. Morphological characteristics of *Phytophthora*. Fill in as fully as possible.

General details	
Location	
Date of collection	
Reference number	
Host species	
Cultural characteristics	
Hyphae	Appearance:
	Septate ❑ Aseptate ❑
	Width (µm):
	Hyphal swellings: Present ❑ Absent ❑
Sporangiophores	Branching:

Table 6.2. Continued.

	Unbranched ❑ Irregularly branched ❑
	Sympodial ❑ Compound ❑ Umbel ❑ Proliferating ❑
Chlamydospores	Present ❑ Absent ❑
	Numerous ❑ Sparse ❑
	Terminal ❑ Intercalary ❑
	Diameter (μm):
	Pigmentation: Present ❑ Absent ❑
	Wall: Thick ❑ Thin ❑
Hyphal swellings	Present ❑ Absent ❑
	Size (μm):
Sporangia	Caducous ❑ Non-caducous ❑
	Caducous: Pedicel length: Short ❑ Medium ❑ Long ❑
	Shape:
	Length (l) (μm):
	Breadth (b) (μm):
	l:b ratio
	Papillate ❑ Semi-papillate ❑ Non-papillate ❑
Sexual spore features	Heterothallic ❑ Homothallic ❑
Antheridia	Size (μm):
	Shape: Round ❑ Square ❑
	Amphigynous ❑ Paragynous ❑
Oogonia	Shape:
	Numerous ❑ Sparse ❑
Oospores	Plerotic ❑ Partially plerotic ❑ Aplerotic ❑
	Wall thickness (μm):
	Diameter: (μm):
Growth temperatures	Minimum (°C):
	Optimum (°C):
	Maximum (°C):

Protocol 6.8

Measuring *Phytophthora* growth temperatures

Growth temperature maximums, minimums and optimal for *Phytophthora* can be used in addition to the morphological features to aid identification. One of the most useful temperature measurements is growth maximum, as this can help to separate some morphologically similar *Phytophthora* spp. (e.g. *Phytophthora drechsleri* grows well at 35°C, whereas *P. cryptogea* does not (Ho, 1992)).

Materials

- Scalpel
- Agar culture plates
- Petri dish
- Incubator

Methods

1. Using a flame-sterilized scalpel, cut small (0.5 cm square) pieces of agar from the edge of an actively growing culture at room temperature of the *Phytophthora* to test.

2. Place each plug in the centre of a Petri dish containing suitable medium for growth (e.g. V8 juice, carrot agar).

3. Incubate at a range of temperatures and record minimum, optimal and maximum growth for test *Phytophthora*.

Suggested incubation temperatures are (after Erwin and Ribeiro, 1996):

To determine minimum: <5°C and 5–10°C.

To determine optimum: 10–25°C, 25–30°C and >30°C.

To determine maximum: 25–27°C, 28–35°C and >35°C.

4. Identification of species. The following sources will help in species identification: Waterhouse (1963), Ho (1981, 1992), Stamps *et al.* (1990), Erwin and Ribeiro (1996), Bush *et al.* (2006) and Gallegly and Hong (2008).

Protocol 6.9

Identification of downy mildews

Materials

- Dissecting and compound microscopes
- Microscopic slides and coverslips
- Cotton blue in lactoglycerol

Method

1. Record host and symptoms (range, type of tissue affected, any chlorotic or necrotic spots, extent of spotting).
2. Using a dissecting microscope, examine affected plants for visible signs of the pathogen (e.g. dense or sparse 'downy' appearance on the underside of leaves) and record observations.
3. Prepare a microscope slide using both the adhesive tape method (Protocol 6.1) and by sectioning the tissue (Protocol 6.3). Cotton blue in lactoglycerol is a good stain to help to observe key morphological features of downy mildews.
4. Observe under a compound microscope ($\times100$–$\times1000$ magnification) and record the following characteristics in Table 6.3:
   - **4.1.** Sporangiophores
     - Arising singly or in multiples through stomata.
     - Colour.
     - Overall shape (club shaped, straight or curved).
     - Shape of distal, basal end (bulbous, swollen, straight).
     - Callose plug present/absent (internal structure similar to a septum).
     - Branching pattern (+ number of branches) of lower and ultimate branchlets (single, pairs, amount branching, shape (straight, curved, sigmoid)), tip shape (e.g. truncate).
     - Length before first branch, total length, length of final branchlet.
   - **4.2.** Apical vesicles
     - Present/absent.
     - Size.
   - **4.3.** Sporangia
     - Colour.
     - Shape.
     - Size, length (l), breadth (b), l:b ratio.
     - Pore diameter.
     - Papilla thickness.
     - Pedicel present/absent, size and shape, pedicel scar present/absent.
   - **4.4.** Zoospores
     - Present/absent.
   - **4.5.** Oogonia
     - Present/absent.
     - Shape.
     - Colour.
     - Diameter.
     - Wall features (e.g. smooth/wrinkled), thickness.
   - **4.6.** Oospores
     - Plerotic/aplerotic.
     - Colour.
     - Size.
     - Wall thickness.
   - **4.7.** Haustoria
     These are observed as various shaped bulbous extensions into host cells and can be seen on cross-sectional microscopical preparations.
     - Shape (lobed, globose).
     - Number per host cell.
     - Size (also note whether they fill or partially fill the host cell).

Table 6.3. Downy mildew characteristics. Fill in as fully as possible.

General details	
Location	
Date of collection	
Reference number	
Host species	
Host symptoms	
Morphological characteristics	
Sporangiophores	Number arising through stomata: 1 ❑ >1❑ Undetermined ❑
	Colour:
	Overall shape: Club ❑ Straight ❑ Curved ❑
	Distal end: Bulbous ❑ Swollen ❑ Straight ❑
	Basal end: Bulbous ❑ Swollen ❑ Straight ❑
	Branching pattern (lower and ultimate branches): Single ❑ Pairs ❑ Shape: Straight ❑ Curved ❑ Sigmoid Tip shape:
	Number of branches:
	Branch length/s (µm): Length before first branch (µm): Final branch length (µm): Total length (all branches) (µm):
Apical vesicles	Present ❑ Absent ❑
	Size (µm):
Sporangia	Colour:
	Shape: Size: Length (l) (µm): Breadth (b) (µm): l:b ratio:
	Pore diameter (µm):
	Papilla thickness (µm):

Continued

Table 6.3. Continued.

	Pedicel: Present ❏ Absent ❏ Shape: Size (μm): Scar: Present ❏ Absent ❏
Zoospores	Present ❏ Absent ❏
Oogonia	Present ❏ Absent ❏
	Shape:
	Colour:
	Diameter (μm):
	Wall features: Wall thickness (μm):
Oospores	Plerotic ❏ Aplerotic ❏
	Colour:
	Size (μm):
	Wall thickness (μm):
Haustoria	Shape:
	Number per cell:
	Fills host cell ❏ Partially fills host cell ❏
	Size (μm):
Sporangial germination	Present ❏ Absent ❏

Protocol 6.10

Germination of downy mildew spores

Material
• Distilled water
• Camel hair brush
• Microscope slides and coverslips
• Compound microscope
• Incubator

Method

1. Carefully dust fresh downy mildew sporangia (e.g. from young lesions on plant, or recently formed spores) onto a drop of distilled water added to microscope slides.

2. Incubate slides in two humid chambers, one at 15°C and the other at 20°C.

3. Examine slides under a compound microscope every 2 h and record evidence of sporangial germination.

4. The following sources will help to identify downy mildew species: Wilson (1914), Gäumann (1923), Yerkes and Shaw (1959), Constanantinescu (1979), Spencer (1981), Waterhouse and Brothers (1981), Skidmore and Ingram (1985), Barreto and Dick (1991), Choi *et al.* (2005) and Shin and Choi (2006). The CMI Descriptions of Pathogenic Fungi and Bacteria (from the Commonwealth Mycological Institute, Kew, UK) and CABI Crop Protection Compendium online databases are also useful as general sources for pathogen identification.

Protocol 6.11

Identification of powdery mildews

Materials

- Clear adhesive tape, e.g. Crystal Clear
- Dissecting needle
- 10 ml pipette with bulb
- Compound microscope with up to ×100 objective lens and eyepiece graticule
- Dissecting microscope up to ×50 magnification
- Distilled water
- Potassium sulphate KH_2SO_4
- Microscope slides and coverslips
- Parafilm or equivalent sealing tape
- Petri dishes
- Small weigh boats (3 cm square)
- Cotton blue in lactoglycerol (0.1%)
- Lactic acid

Method

- Plant material should be fresh (preferably less than 48 h old), unless herbarium specimens are being examined.
- Record host species in Table 6.5.
  Examine upper and lower surfaces with the dissecting microscope for signs of powdery mildew. These can include hyphae, conidiophores and spores (conidia). Select a suitable area in between the centre and the edge of an infection – at the centre of infection the sporulation may be too dense and at the very edge spores may not yet have formed).
- Record in Table 6.5 whether the mycelium is over the surface of the leaf (exogenous; in most powdery mildew genera) or is mainly underneath the leaf surface (endogenous; in *Phyllactinia* and *Leveillula* species only). Note whether the conidiophores are emerging from the mycelium on the leaf surface or through stomata; conidiophores emerging through stomata indicate endogenous mycelium. If exogenous, note whether it is only on the upper (epiphyllous), lower (hypophyllous) or on both leaf surfaces (amphigenous). Note whether spores are forming singly (non-catenate) or in chains (catenate).

- Remove surface structures using the adhesive tape method (see Protocol 6.1). Place on microscope slide sticky side down, over a drop of cotton blue in lactoglycerol. To examine the mother cell/base of the foot cell, place the adhesive tape sticky side up on the slide, but this is more difficult to manipulate.
- View structures using a compound microscope: first, locate the structures at ×100 magnification, then increase the magnification to ×400 and measure and fill in as much detail as possible in Tables 6.4 and 6.5. Use images below for guidance on powdery mildew features.

Conidia

- Identify whether spores mature singly (non-catenate) or in chains (catenate). (*Neoerysiphe*, *Sawadaea*, *Podosphaera* and *Golovinomyces* species all form catenate conidia; *Phyllactinia*, *Leveillula* and *Erysiphe* species are non-catenate). Be aware of pseudo-chains; they may form in non-catenate genera if undisturbed (for illustration, see Fig. 6.32).
- Observe the shape of the conidia and any obvious ornamentation (e.g. you may see striations on wrinkled conidia). (See Fig. 6.33.)
- Measure the length and width of the smallest and largest conidium in at least five traverses of the specimen.
- Record the ranges and calculate the mean values.
- Measure length and width of five conidia to determine the spore size range.
- Calculate the average length to width ratio using the five measurements from step above.
- Check for the presence or absence of fibrosin bodies. These appear as small refractive structures, very irregular in shape, resembling glass shards within the conidia. Granular inclusions may look like fibrosin bodies, but granular inclusions are somewhat

Anatomy of non-catenescent conidiophore
(Conidiophore produces a single **mature**, non-catenate, conidium)

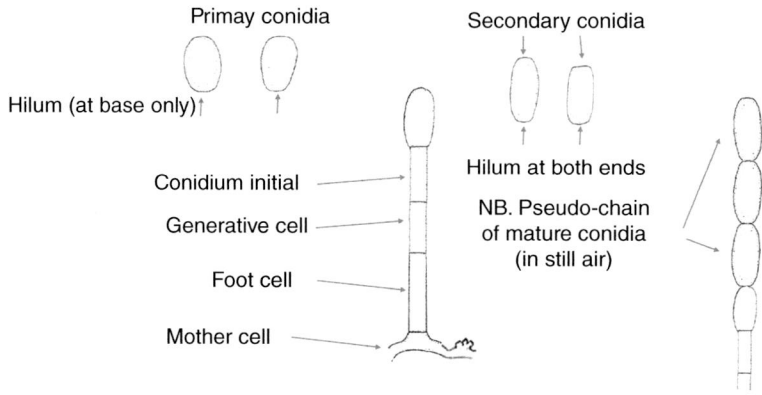

Anatomy of catenescent conidiophore

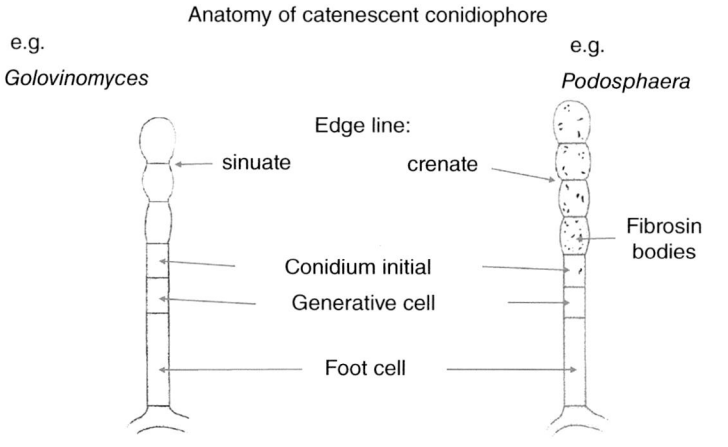

(*Neoerysiphe* similar but has lobed hyphal appressoria)

Fig. 6.32. Anatomical representation of non-catenescent and catenescent powdery mildew conidiophores. (Drawing courtesy of Roger Cook.)

nebulous and not so irregular in shape. Slide preparation using water can make it easier to detect the presence of fibrosin bodies. Fibrosin bodies are present in the catenescent conidia of *Podosphaera* and *Sawadaea* species and absent in *Blumeria*, *Neoerysiphe* and *Golovinomyces* species.

Conidiophores

- Measure the length and width of the shortest and longest conidiophore foot cells; the length of the foot cell is measured from the base of the conidiophore to the first septum and the width across the middle of the foot cell.
- Measure the length and width of at least five, preferably more, shortest and longest conidiophore foot cells to determine the size range.
- Measure the length of at least five conidiophores (up to and including the first expanding or immature conidium).
- Record the various shapes of the conidiophore base (foot cell) (as many as possible).

Conidial shapes

(as illustrated in Taxonomic Manual by Braun and Cook, 2012)

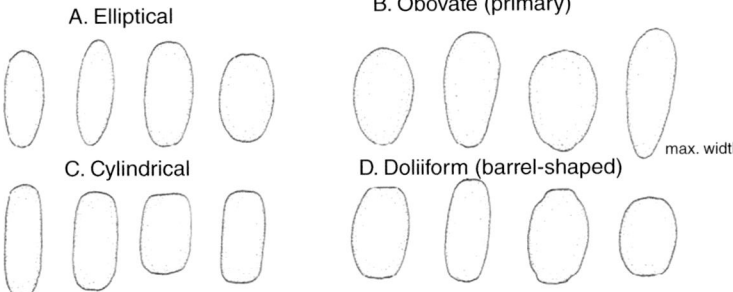

A. Elliptical

B. Obovate (primary)

C. Cylindrical

D. Doliiform (barrel-shaped)

max. width

Fig. 6.33. Powdery mildew conidial shapes. (Adapted by Roger Cook from Braun and Cook, 2012. *Taxonomic Manual of the Eysiphales (Powdery Mildews)*.)

These can be straight and cylindrical or expanding rapidly in width, curved, flexuous or with a marked twist or kink.

- Be aware of hyperparasites such as *Ampelomyces* which can be found forming fruiting bodies within swellings of powdery mildew structures such as conidiophores and conidia, especially those of *Podosphaera* species.

Hyphal appressoria

- These may be unlobed (inconspicuous or nipple-shaped, e.g. *Podosphaera, Golovinomyces*, or lobed (*Erysiphe, Neoerysiphe, Sawadaea* spp.). Appressoria are usually visible using the adhesive tape method.
- Record shape as inconspicuous, nipple-shaped or lobed, and whether appressoria are formed singly or in pairs.

Conidial germination

Assessing the morphological features of germinated conidia may be useful to help distinguish between anamorph genera.

Check under stereoscopic microscope for recent infections bearing turgid-looking conidiophores and conidia. (Do not use if the structures are collapsed/collapsing or heavily parasitized). Tap the leaf over a plastic Petri dish lid to release conidia which immediately stick to the plastic surface.

Reverse the lid over its base containing paper towel dampened with saturated solution of KH_2SO_4.

View under stereoscopic microscope to check that a good deposit of conidia has been obtained.

Incubate in the dark at $+/-20°C$.

If possible, view conidia under the compound microscope after 5 h to check germination has started and then after 48 h when full development should have been completed to observe whether 50% germination has occurred. Conidia are classified as germinated when the length of the germ tube is as long as the width of the spore. Observe the following features using a compound microscope and record in Table 6.5.

- Position of germ tube (observe up to 100 germinated conidia or as many as possible); record the number of germ tubes emerging from an end (hilar) or from the side (lateral). Note that hilar germ tubes actually emerge from the 'shoulder', rather than from the centre of a hilum even though they may appear to do so optically depending on the orientation of the conidium.
- Length of germ tubes.
- Shape of appressoria on the tips of germ tubes.

Sexual morphs (chasmothecia)

Under the dissecting microscope locate a chasmothecium (see Fig. 6.34; note it may be necessary

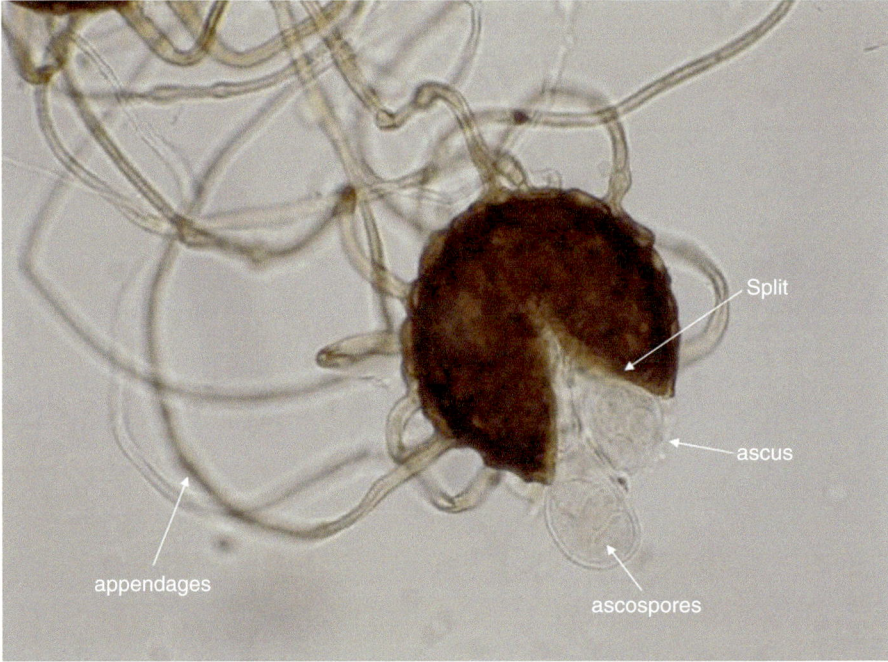

Fig. 6.34. Powdery mildew chasmothecium. (Image courtesy of Roger Cook.)

to move some of the hyphae aside to reveal embedded fruiting bodies) and, using a dissecting needle, carefully remove it, place it on a slide and add a drop of water and a glass coverslip. The adhesive tape method may also be suitable. Examine under a compound microscope and record the following features in Table 6.6.

- Chasmothecium diameter.
- Chasmothecium wall: differentiated/simple, presence/absence of layers (if present, note whether layers are easily separated or not), are the peridial cells large (*Podosphaera* sect. *Sphaerotheca* subsect. *Magnicellulata*, e.g. *P. xanthii* in the *Taxonomic Manual* (Braun and Cook, 2012).
- Asci.
  In at least five chasmothecia, record the number of asci per chasmothecium (crush chasmothecia gently to reveal asci). *Podosphaera* species have only one ascus per chasmothecium. To examine mature asci, select the

most mature chasmothecia: the largest, darkest chasmothecia, with fully developed appendages (appendages fully develop before the ascospores are mature).

Measure the length and width of the asci in at least five chasmothecia, and obtain range and mean values, and record in Table 6.4.

Record the shape of the asci: elliptical, ovate, obovate; are the asci stalked or sessile?

Appendages: Present/absent

Type: See Fig. 6.35 mycelioid (e.g. *Erysiphe* sect. *Erysiphe*), branched, flexuous, stiff (*Erysiphe* sect. *Microsphaera*), hooked, flexuous, stiff with or without bristles (*Erysiphe* sect. *Uncinula*), hooked and branched (*Sawadaea*), needle shaped (*Phyllactinia*).

Colour: light/dark in part or throughout.

Branching: unbranched or branched (*Erysiphe* sect. *Microsphaera*).

Distribution: equatorial, all over, basal.

(a)

Chasmothecial appendages

| Mycelioid | Branched dichotomously | Hooked | Hooked and branched |

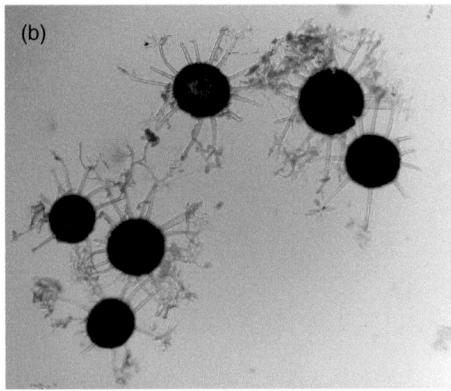

Fig. 6.35. (a) Chasmothecial appendages. (Figure used with permission from Roger Cook from *Taxonomic Manual of the Eysiphales (Powdery Mildews)* (Braun and Cook, 2012)). (b) Oak powdery mildew chasmothecia. (© UK Crown copyright.)

Powdery mildew identification tables (for an expected known powdery mildew)

Table 6.4. Measurements.

	Conidia						Foot cells				Sh[a]	Conidiophore length[b]	Ascus				Ascospores			
	Length		Width		L/W		Length		Width				Length		Width		Length		Width	
	min	max	min	max	min	max	min	max	min	max			min	max	min	max	min	max	min	max
1																				
2																				
3																				
4																				
5																				
…																				
Mean											NA									
Range	…… – ……		…… –		…… –		…… –		…… –		–	… – …	…… – ……		…… –		…… – ……		…… –	

[a]Shape of footcell, e.g. straight (str), curved (cur), cylindrical (cyl), expanding upwards (exp), flexuous (fl), kinked (k), etc.
[b]Length from base up to (including) the first expanding conidium.

Table 6.5. Characteristics of the asexual morph for the identification of an unknown powdery mildew. Fill in as fully as possible.

Conidial germination	
Time for 50% of conidia to have germinated	5 h ❑ 48 h ❑
Position of germ tube (%):	Hilar
	Lateral
Length of germ tubes (mainly):	Very short ❑ Fairly short ❑ Variable ❑ Long ❑ Longitubus (very long, undifferentiated) ❑
Shape of appressoria on germ tubes:	Inconspicuous ❑ Nipple-shaped ❑ Lobed ❑
Other notable features:	
General details	
Sample reference number/Host	
Powdery mildew mycelium	
Exogenous ❑ Endogenous ❑ Epiphyllous ❑ Hypophyllous ❑ Amphigenous ❑	
Conidia	
Spore formation	Singly ❑ Chains ❑
Conidia shape	
Conidia length, smallest spore (mm)	
Conidia length, largest spore (mm)	
Conidia width, smallest spore (mm)	
Conidia width, largest spore (mm)	
Mean conidial length (mm) from Table 6.4	
Mean conidial width (mm) from Table 6.4	
Conidia length:width ratio	
Fibrosin bodies present?	Yes ❑ No ❑
Conidiophores	
Conidiophore foot cells size range	Smallest (mm):
Conidiophore foot cells size range	Smallest (mm):
	Largest (mm):

Continued

Table 6.5. Continued.

Mean foot cell length (mm)	
Mean foot cell width (mm)	
Mean length of conidiophore (mm)	
Shape of conidiophore foot-cell	
Hyperparasites observed?	Yes ❑ No ❑
Hyphal appressoria	
Appressoria type	Inconspicuous ❑ Nipple-shaped ❑ Lobed ❑

Table 6.6. Characteristics of the sexual morph (if present) for the identification of a powdery mildew. Fill in as fully as possible.

Sexual characteristics	
Chasmothecium	
Diameter (μm)	
Peridial wall	Differentiated ❑ Simple ❑
	Layers: Present ❑ Absent ❑
	Peridial cells large? Yes ❑ No ❑
Asci – number per chasmothecium:	
Asci – average length (μm), from Table 6.4:	
Asci – average width (μm), from Table 6.4:	
Asci – shape:	Elliptical ❑ Ovate ❑ Obovate ❑ Stalked ❑ Sessile ❑
Appendages	Present ❑ Absent ❑
Type:	
Colour:	
Branching pattern:	
Distribution:	
Other notable features:	

Protocol 6.12

Identification of *Colletotrichum* spp.

> **Materials**
>
> - 4% Potato dextrose agar (PDA)
> - Microscope slide and coverslip
> - Damp chamber
> - Dissecting and compound microscopes
> - Scalpel
> - Forceps

Method

1. Carry out primary isolation from host tissue onto PDA as described in Protocol 6.2.

2. Aseptically transfer a small piece of culture from the growing margin of the colony onto PDA and incubate for 10 days at 20°C with a 12 h light/dark cycle.

3. To induce appressoria production, cut a 1 × 1 cm square of agar from just behind the edge of the growing colony (three-quarters of the agar square should contain hyphae) and place on a sterile microscope slide. Lay a sterile coverslip directly onto the agar square and place the whole slide in a damp chamber. Incubate for up to 1 week at 20°C, examining every day for the presence of appressoria.

4. For the appressoria, record in Table 6.7:
  4.1. Colour.
  4.2. Size.
  4.3. Shape (common shapes are lobed, crenate, clavate, circular).

4.4. Margin shape (e.g. smooth, irregular).

5. Examine culture plates and record in Table 6.7:
  5.1. Colony colour.
  5.2. Presence/absence of sclerotia (and shape if present).
  5.3. Presence/absence of setae.
  5.4. Presence/absence of chlamydospores.
  5.5. Conidia:
   - Shape (e.g. straight, falcate, fusiform, cylindrical, curved).
   - Size: length (l) and breadth (b), l:b ratio.

6. The following sources can be used to help to identify species: Sutton (1980), Gunnell and Douglas Gubler (1992) and Guerber and Correll (2001). The CMI Descriptions of Pathogenic Fungi and Bacteria (from the Commonwealth Mycological Institute, Kew, UK) can also be used as a general source; in this case, those of specific interest are for *Colletotrichum musae* (No. 222), *Colletotrichum gloeosporioides* species complex (No. 315), *Colletotrichum lindemuthianum* (No. 316), *Colletotrichum capsici* (No. 317), *Colletotrichum coccodes* (No. 131), *Colletotrichum graminicola* (No. 132), *Colletotrichum falcatum* (No. 133) and *Colletotrichum acutatum* species complex (No. 630). Caution with morphological species identification is needed due to the recent changes to genus *Colletotrichum* and similarities between species within species complexes. Morphological features should be used in conjunction with molecular methods for species identification. Helpful sources on some species complexes include Damm *et al.* (2012) and Weir *et al.* (2012).

Table 6.7. *Colletotrichum* characteristics. Fill in as fully as possible.

General details	
Location	
Date of collection	
Reference number	
Host species	

Continued

Table 6.7. Continued.

Cultural characteristics	
Colony	Colour:
Sclerotia	Present ❑ Absent ❑
	Shape:
Setae	Present ❑ Absent ❑
Chlamydospores	Present ❑ Absent ❑
Conidia	Shape: Straight ❑ Falcate ❑ Fusiform ❑ Cylindrical ❑ Curved ❑
	Length (l) (μm):
	Breadth (b) (μm):
	l:b ratio:
Appressoria	Colour:
	Size:
	Shape:
	Margin shape:

Protocol 6.13

A polytomous key to identify some *Monilinia* spp. from stone or pome fruits

Materials

- Scalpel or dissecting needle
- Potato dextrose agar (PDA) plates (4%, 12.5 cm³ per 9 cm Petri dish)
- Dissecting microscope

Method

1. Carry out primary isolation from infected fruit or plant as described in Protocol 6.2.

2. Aseptically select a small piece of actively growing fungus using a dissecting needle or scalpel from the edge of a colony and subculture onto PDA.

3. Incubate at 22°C under a 12 h light/dark cycle for 10 days.

4. Record the following cultural features in Table 6.8:

    4.1. Colony colour: upper surface of plate grey (A), yellow (B) or cream/white (C).

    4.2. Growth rate: mean colony diameter >80 mm – fast (D), 70–80 mm – medium (E) or <70 mm – slow (F).

    4.3. Sporulation: upper surface of colony, viewed with a dissecting microscope, showing abundant (G) or sparse (H) sporulation.

    4.4. Concentric rings of sporulation: upper surface of colony, viewed with a dissecting microscope, showing concentric rings present (I) or absent (J).

    4.5. Colony margin: colony, when examined from the underside of the plate, showing margin lobed (K) or non-lobed (L).

    4.6. Rosetting: upper surface of colony 'rosetted', i.e. showing mycelium in distinct layers (petals) on top of each other, with the appearance of an open rose flower (M) or not (N).

    4.7. Black arcs: lower surface of colony showing black arcs or rings associated with the 'petals' of a rosetted isolate (O), black dotted areas or brown arcs or rings (P) or no black arcs or rings absent (Q).

5. Based on the morphological features A–Q identified in step 4, use the following synoptic key to identify species:

Monilinia fructicola	A, D, (E), G, I, (J), L, (M), N, (P), Q
Monilinia laxa	A, (C), (E), F, H, J, K, M, (N), O
Monilinia fructigena	B, (C), (D), E, (F), (G), H, (I), J, L, N, Q

Note: Bracketed characters indicate that the feature is sometimes observed.

Table 6.8. Key to differentiate key *Monilinia* spp. Fill in as fully as possible.

General details	
Location	
Date of collection	
Ref. no.	
Cultural characteristics	
Colony colour – upper surface	Plate: Grey (A) ❑ Yellow (B) ❑ Cream/white (C) ❑

Continued

Table 6.8. Continued.

Growth rate: mean colony diameter	>80 mm – fast (D) ❏ 70–80 mm – medium (E) ❏ <70 mm – slow (F) ❏
Sporulation: upper surface of colony, viewed with a dissecting microscope, showing abundant or sparse sporulation	Abundant (G) ❏ Sparse (H) ❏
Concentric rings of sporulation: upper surface of colony, viewed with a dissecting microscope, showing concentric rings present or absent	Rings: Present (I) ❏ Absent (J) ❏
Colony margin: colony, when examined from the underside of the plate, showing margin lobed or non-lobed	Lobed (K) ❏ Non-lobed (L) ❏
Rosetting: upper surface of colony 'rosetted', i.e. showing mycelium in distinct layers (petals) on top of each other, with the appearance of an open rose flower or not	Yes (M) ❏ No (N) ❏
Black arcs	Lower surface of colony showing black arcs or rings associated with the 'petals' of a rosetted isolate (O) ❏ Black dotted areas or brown arcs or rings (P) ❏ No black arcs or rings absent (Q)❏

Transpose selected key code letter below:																
A	B	C	D	E	F	G	H	I	J	K	L	M	N	O	P	Q

Protocol 6.14

Identification of smut fungi

Materials

- Microscope slides and coverslips
- Sterile distilled water
- Compound microscope

Method

1. Record host details.

2. Note distribution of sori (fungal fruiting bodies), e.g. are they restricted to ovaries or are they also associated with leaves/stems. Note whether all host ovaries are infected or just some.

3. Record shape of sori and whether they are enclosed within a fungal membrane or are loose.

4. Produce a microscope slide for the examination of spores. Smut spores can usually just be dusted onto a slide (e.g. by making a small hole in an intact sorus) and mounted in sterile distilled water.

5. Record the following smut spore characteristics in Table 6.9:

5.1. Size.

5.2. Colour: pale yellow, orange, reddish brown, chestnut brown or black; spores opaque, semi-opaque or translucent.

5.3. Surface ornamentation, e.g. reticulate (net-like), echinulate (warty), verrucose (spiny), cerebriform (brain-like) and whether conspicuous or not.

5.4. Spines:

- Sharply pointed, conical or curved.
- Size range.
- Presence/absence of sheath (hyaline or coloured).
- Arrangement (dense or coarse).
- Presence/absence of sterile cells between spores.

Table 6.9. Characteristics of smut species. Fill in as fully as possible.

General details	
Location	
Date of collection	
Reference number	
Host species	
Cultural characteristics	
Sori	Distribution: Restricted to ovaries ❑ Leaves ❑ Stems ❑
	Enclosed in membrane? Yes ❑ No ❑
	Shape:
Spores	Size (µm):
	Colour: Pale yellow ❑ Orange ❑ Reddish brown ❑ Chestnut brown ❑ Black ❑ Other:

Continued

Table 6.9. Continued.

	Opaque ❏ Translucent ❏
	Surface ornamentation: Conspicuous ❏ Non-conspicuous ❏ Reticulate ❏ Echinulate ❏ Verrucose ❏ Cerebriform ❏ Other:
	Spines: Sharply pointed ❏ Conical ❏ Curved ❏ Arrangement: Dense ❏ Sparse ❏ Size range: Sheath: Present ❏ Absent ❏
Sterile cells	Present ❏ Absent ❏
Germinated spores	Present ❏ Absent ❏
Promycelium	Number per spore:
	Size (µm):
	Septate ❏ Non-septate ❏
	Number of septa:
	Branching pattern:
Basidiospores (primary sporidia)	Present ❏ Absent ❏
	Size (µm): Shape:
Secondary sporidia	Size (µm): Shape:
Sporidial fusion	Present ❏ Absent ❏

Protocol 6.15

Germination of smut teliospores

Materials

- Centrifuge tube
- Distilled water
- Centrifuge
- Pipette
- 10% Bleach (0.3–0.5% active NaOCl)
- Sterile spreader
- Agar plates with antibiotics (AWA)*
- Electrical tape or polyethylene bags

---

*AWA prepared using Agar Technical No. 3 (2%) from Oxoid; antibiotics penicillin and strepto-mycin (60 mg penicillin G (Na salt) and 200 mg streptomycin sulphate per litre of agar). After autoclaving, add the antibiotic suspension to the cooled agar before pouring plates.

Method

1. Dust spores into a centrifuge tube and add 5 ml distilled water.

2. Incubate the teliospore suspension overnight at 21°C to hydrate the teliospores and make fungal and bacterial contaminants more susceptible to subsequent surface sterilization.

3. Pellet the teliospores by centrifuging at $1200 \times g$ for 3 min.

4. Remove the supernatant with a pipettor with a plugged tip, or a disposable Pasteur pipette, taking care not to disturb the pellet. Pipette the supernatant into a suitable disposable waste bottle for subsequent quarantine disposal if appropriate.

5. Resuspend the pellet in 5 ml of 10% domestic bleach (replace the centrifuge tube cap and immediately invert the tube three times to ensure that the bleach contacts all inner surfaces).

6. Immediately centrifuge at $1200 \times g$ for 1 min, then quickly and aseptically remove the supernatant and resuspend the pellet in 1 ml sterile distilled water (SDW) to wash the pellet.

NB: Some teliospores can be killed if the total time in the bleach exceeds 2 min.

7. Centrifuge at $1200 \times g$ for 5 min, aseptically remove the supernatant and then add another 1 ml SDW to wash the pellet again.

8. Centrifuge at $1200 \times g$ for 5 min, aseptically remove the supernatant and resuspend the final pellet in 1 ml SDW.

9. Transfer 200 µl of the teliospore suspension aseptically onto individual 2% AWA plates and spread with a sterile spreader. (NB: Three-day-old plates are recommended as these quickly absorb the suspension; excessive surface water can inhibit teliospore germination. Alternatively, prepare the agar plates on the day of use, but pour the liquid agar when cool and do not replace the lids fully until the agar has set.)

10. Incubate the AWA plates at 21°C with a 12 h light cycle (e.g. TLD 18W/83 Philips white light tubes). Leave for about 5 days before sealing plates with electrical tape or placing the plates inside clear polyethylene bags.

11. After 7 days, examine the plates and record in Table 6.10:

- Presence or absence of germinated teliospores.
- Promycelium number per spore, size, septation (non-septate/septate and number of septa), branching pattern.
- Presence of a tuft of filiform basidiospores (primary sporidia), size, shape.
- Presence/absence of secondary sporidia (these generally derive from small colonies forming around germinated teliospores).
- Size and shape of all observed morphological features.
- Any evidence of sporidial fusion.

12. Identification of species. Using the recorded features, the following sources can be used for species identification: Ainsworth and Sampson (1950), Mathre (1996), Piepenbring *et al.* (1998) and OEPP/EPPO (2004). The CMI Descriptions of Pathogenic Fungi and Bacteria (from the Commonwealth Mycological Institute, Kew, UK) are also useful as a general source.

Table 6.10. Morphological characteristics of smut spores.

General details	
Location	
Date of collection	
Reference number	
Host species	
Cultural characteristics	
Germinated teliospores	Present ❏ Absent ❏
Promycelium	Number per spore: Size: Septation: Non-septate ❏ Septate ❏ Number of septa: Branching pattern:
Presence of a tuft of filiform basidiospores (primary sporidia)	Absent ❏ Present ❏ Size: Shape:
Presence/absence of secondary sporidia (these generally derive from small colonies forming around germinated teliospores)	Absent ❏ Present ❏
Size and shape	
Any evidence of sporidial fusion	Absent ❏ Present ❏

Protocol 6.16

Identification of *Armillaria* spp.

Materials

- 3% Malt extract agar (MEA)
- 1.5% MEA amended with tannic acid (0.5% w/v) (TAA)
- Hazelwood malt agar (HMA) slopes (after Rishbeth, 1986)
- Dissecting and compound microscopes

Method

1. Examine plant material for diagnostic features of *Armillaria* infection. These include a white mycelial sheet that has a strong mushroom-like odour under the surface of the bark near the soil line of shrubby plants. Examine surrounding area of affected plant *in situ* for large macro (mushroom) basidiocarps (these are not always present). A further confirmatory test in the field is to examine the roots, soil and stem base for thick black mycelial cords (rhizomorphs).

2. To confirm presence of *Armillaria* or identify species from morphological features, carry out primary isolation from leading edge of white mycelial sheets (include host woody material) (as described in Protocol 6.2) on to MEA. Incubate plates for 25°C for up to 2 weeks.

3. Aseptically transplant onto TAA and MEA plates and HMA slopes from the growing margin of the colony.

4. Incubate TAA plates and HMA slopes at 25°C and MEA plates at 30°C for 14 days.

5. Measure and record growth rate on MEA plates as mm week^{-1}. Refer to Table 6.11 to help identify species using difference of growth rate.

6. Examine TAA plates and record the following features in Table 6.12:

    6.1. Colony form (e.g. woolly, flat).

    6.2. Presence/absence radial growth within colonies.

    6.3. Colony colour.

    6.4. Presence/absence rhizomorphs:
- Abundant/infrequent.
- Branching frequency.
- Length.

    6.5. Presence/absence of brown water-soluble pigment in medium.

7. Examine HMA slopes and record the following in Table 6.12:

    7.1. Colony appearance and colour (pay attention to margin of colony).

    7.2. Amount of growth on woody host surface.

    7.3. Features of rhizomorphs (if present):
- Length: long (i.e. reach bottom of tube) or short.
- Shape.
- Increase in width along length.
- Colour.
- Shape of branches.
- ± Aerial features (hairy, flattened, notched).
- Wrinkled or not.

8. Identification. The following are useful sources to aid species identification: Gibson (1961), Philips (1981), Rishbeth (1986), Bérubé and Dessureault (1989), Pérez-Sierra (2003) and Mwangi *et al.* (2007).

Table 6.11. Identification of *Armillaria* spp. based on growth rate on malt extract agar (MEA) (after Rishbeth, 1986).

Species	Growth rate on 3% MEA (mm week^{-1})
Armillaria mellea	0.9 ± 0.2
Armillaria ostoyae	2.4 ± 0.2
Armillaria bulbosa	3.3 ± 0.3
Armillaria tabescens	5.3 ± 0.5

Table 6.12. Characteristics of *Armillaria* spp. Fill in as fully as possible.

General details	
Location	
Date of collection	
Reference number	
Host species	
Mycelial sheet under bark	Present ❑ Absent ❑
Basidiocarps (mushrooms)	Present ❑ Absent ❑
	Conform to *Armillaria* spp.: Yes ❑ No ❑
Rhizomorphs	Present ❑ Absent ❑
Cultural characteristics	
On TAA[a] plate	
Growth rate (mm week^{-1})	
Colony	Form:
	Colour:
	Radial growth within colony: Present ❑ Absent ❑
	Brown pigment in medium: Present ❑ Absent ❑
Rhizomorphs (in culture)	Present ❑ Absent ❑
	Abundant ❑ Infrequent ❑
	Branching frequency:
	Length:
On HMA[b] slope	
Colony	Form:
	Colour:
	Radial growth within colony: Present ❑ Absent ❑
	Brown pigment in medium: Present ❑ Absent ❑
Rhizomorphs (in culture)	Present ❑ Absent ❑
	Length: Long ❑ Short ❑
	Width along length (µm):
	Colour:
	Branch shape:
	Wrinkled: Yes ❑ No ❑
	Aerial: Yes ❑ No ❑
Growth on woody host surface	Nil ❑ Sparse ❑ Abundant ❑

[a]TAA, malt extract agar (1.5%) amended with tannic acid (1.5% w/v).
[b]HMA, hazelwood malt agar slopes.

Protocol 6.17

Identification of sclerotia-forming fungi from plant tissue and soil

---

Materials

- Dissecting needle/scalpel
- Incubation box (damp chamber)
- Forceps
- Small (50 ml) and large (2000–3000 ml) beaker
- Plastic tray
- Various sized sieves (2–0.180 mm)
- Dissecting and compound microscopes
- Bleach (NaOCl) (concentrated domestic bleach)
- Sterile distilled water
- Potato dextrose agar (PDA) plates amended with penicillin and streptomycin (60 mg penicillin G (Na salt) and 200 mg streptomycin sulphate per litre of agar)

Method

1. Sclerotia from plant tissue:

    1.1. Sclerotia are commonly easily visible on plant tissue and are often associated with obvious fungal hyphae. These can be on stems, the underside of flowers, leaves or roots. If no sclerotia are visible, incubation in a damp chamber (see Protocol 3.2) for up to 2 weeks will usually trigger their production.

    1.2. Using a dissecting needle (or forceps), carefully remove individual sclerotia and place in a small beaker. Proceed to step 3.

2. Sclerotia from soil: due to the small size of many sclerotia and their similarity in colour to some soil particles, a sieving process can be used. A quantitative assessment of the sclerotia present can be made if the initial weight of soil is known (1 kg is the usual upper limit per test).

    2.1. Air-dry soil on a plastic tray at room temperature so that it is suitable for sieving (note that heavy clay soils will become too hard if left to completely dry, so it is important to monitor drying of soil until it will pass easily through the sieve).

    2.2. Sieve through a 2 mm clean sieve into a beaker containing 1000 ml tap water. Wash sieve with flowing tap water and examine the wash under a dissecting microscope for large black sclerotia.

    2.3. Stir sieve wash for 30 s and allow sediment to settle.

    2.4. Filter through suitable sieve to capture sclerotia of interest (see Table 6.13). Note sclerotia usually float away from the soil/sludge so only pour the water layer through the sieve. For a general check of all sclerotia, filter through a graduated series of filters beginning with a 1 mm sieve, followed by 0.5 mm and 0.180 mm sieves.

    2.5. Add a further 1000 ml tap of tap water to the filtrate and repeat steps 2.3–2.4.

3. Examine sieve under a dissecting microscope and remove sclerotia using forceps. If a quantitative assessment is required, count the number of individual sclerotia found and calculate number per gram of soil.

4. Select viable sclerotia (this can often be done by gently squeezing a sclerotium between the thumb and forefinger). If the sclerotia are spongy and do not crumble or collapse, they are more suitable for germination. Surface sterilize in 10% sodium hypochlorite solution (bleach) for 1 min followed by washing in sterile distilled water.

5. Using a surface-sterilized scalpel, cut each sclerotium in half and place onto PDA plates. Incubate at 18–20°C for up to 2 weeks, observing every few days for hyphal growth. (Note that some sclerotia may require up to a month at 4°C to break dormancy and for germination to be triggered.)

    5.1. Record the following morphological features:
    - Primary hyphal width.
    - Length of tip cell.
    - Presence/absence of clamp connections on primary hyphae.
    - Features of secondary cells.

    5.2. If no sclerotia have formed, leave plates for a further 2 weeks then record colony colour and sclerotia formation (size and distribution on plates).

6. Compare with the morphological features in Table 6.14:

Table 6.13. Size range of sclerotia for some common sclerotia-forming plant pathogenic fungi (adapted from Clarkson and Whipps, 2002).

Pathogen	Size of sclerotia (mm)
Sclerotinia sclerotiorum	1–10
Sclerotium cepivorum	0.2–0.5
Sclerotium rolfsii	0.5–2
Rhizoctonia solani	*c*.1

Note: Sclerotia of *Botrytis* spp. are extremely variable in size, ranging from 10 μm to 18 mm. Sclerotia of *Verticillium dahliae* are in the region of 50 μm in size.

Table 6.14. Morphological features of some common sclerotia-forming fungi.

Species	Colony colour	Sclerotia (on plate)	Primary hyphal width	Tip cell	Secondary cell features	Clamp connections
Sclerotinia sclerotiorum	White to faintly grey	Form at edge up to 1 mm	9–18 μm	>300 μm One or more branches before 1st septum	Narrower than primary hyphae	Absent
Sclerotium cepivorum	White to brownish grey	Distributed over plate, occasionally in diurnal rings, 0.2–0.5 mm	9–18 μm	300–400 μm One or more branches before 1st septum	Narrower than primary hyphae	Absent
Sclerotium rolfsii	White	Distributed over plate, near spherical, 1–2 mm	4.5–9 μm	<350 μm	Adpressed to primary hyphae	Present (primary hyphae)
Rhizoctonia spp.	Colourless to brown	In crust or distributed over plate	5–12 μm	<250 μm	Branches at distal end constricted at point of origin and septate shortly above. Branching *c*.90°	Absent

Protocol 6.18

Identification of rust fungi

> **Materials**
>
> - Dissecting and compound microscopes
> - Microscope slides and coverslips
> - Mounting agent (lactoglycerol ± 0.1% cotton blue or water)
> - Dissecting needle
> - Scalpel
> - Forceps

Method

1. Record host details.

2. Examine infected tissue under a dissecting microscope and identify affected area. In general, rusts in their spermatial state may be difficult to detect. However, they often form in groups on the upper leaf surface among discoloured spots or hypertrophied tissue (swollen host cells). Rusts in their aecial and urediniospore stages and, in many cases, in the teliospore stage, sporulate on the lower leaf surface (or around the stem). These stages are usually visible with the naked eye, with spores (aeciospores urediniospores or teliospores) emerging from either rusty- or black-coloured pustules. The teliospore stage of certain rusts (e.g. *Cronartium ribicola* – white pine blister rust) is observed within galls on the host plant. The basidiospore stage is not generally used for morphological identification, owing to the lack of distinguishing morphological characteristics between species.

3. Using methods described in Protocols 6.1, 6.2 and 6.3, make a slide preparation from the affected region. Rust spores can be mounted in water or lactoglycerol.

4. Observe under a compound microscope. If spores appear collapsed, slight warming of slide from the underside will help them swell. This is often an important step when viewing teliospores.

- Spermogonium: prepare a cross-section of the pustule and observe the shape and structure of the spermogonium and its position within the host tissue.

- Aecium: this is a cup-shaped structure with a well-developed peridium (the wall of the aecium). Record surface ornamentation of aeciospores and any distinguishable surface patterns. This is typically verrucose (warty) for most species, although Sato and Sato (1982) describe a further seven distinct types, namely: aciculate, nailhead, echinulate, coronate, tabulate, annulate and reticulate. Record presence or absence of spores forming in chains. Aeciospores often form earlier in the year than urediniospores.

- Uredinium: record structure of peridium if present (e.g. well developed, obscure, ± paraphyses), surface ornamentation of urediniospores (typically echinulate – spiny), and any distinguishable surface patterns, spore shape (ellipsoid, globoid, obovoid, helmet-shaped), size, presence or absence of germ pores and their distribution (scattered over spore, equatorial, unizonate or bizonate – distributed over one or two distinct areas – or basal). Urediniospores are usually unicellular and form singly on pedicels. Record the shape and wall thickness of any paraphyses.

- Telium: genera are often distinguished based on this state. Record position of telia within host tissue (e.g. throughout mesophyll, in epidermal cells, sub-epidermal, erumpent cushions, hair-like extensions). Teliospores are generally very different in appearance from all other rust spore stages. Record colour, number of cells (unicellular, two celled or multicellular) and arrangement (if there is more than one cell), number of germ pores per cell, length and breadth and presence/absence of chains.

- Basidium: these are produced from teliospores. The morphological features of neither basidia nor basidiospores are generally used in classification of rust fungi.

5. Record as many features as visible from the following structures in Table 6.15:
6. Identification of species. Using the above morphological data, a number of good reference sources are available to help identify species. A selection of these are: Grove (1913), Arthur *et al.* (1929), Arthur (1934), Wilson and Henderson (1966), Laundon (1967a, 1967b), Peterson (1974), Henderson and Bennell (1979, 1980), Henderson (2000, 2004), Cummins and Hiratsuka (2003), Hernandez *et al.* (2011) and Termorshuizen and Swert (2011).

Table 6.15. Characteristics of rust fungi. Fill in as fully as possible.

General details	
Location	
Date of collection	
Reference number	
Host species	
Symptoms	
Cultural characteristics	
Spermogonium	Shape:
	Position within host tissue:
Aecium	
Surface ornamentation:	Verrucose ❑ Aciculate ❑ Nailhead ❑ Echinulate ❑ Coronate ❑ Tabulate ❑ Annulate ❑ Reticulate ❑
Spores forming in chains	Yes ❑ No ❑
Uredinium	
Peridium	Present ❑ Absent ❑
Surface ornamentation	
Spore size (µm)	Length ❑ Breadth ❑
Spore shape	Ellipsoid ❑ Globoid ❑ Obovoid ❑ Helmet-shaped ❑ Other:
Germ pores	Present ❑ Absent ❑ Scattered ❑ Equatorial ❑ Unizonate ❑ Bizonate ❑ Basal ❑ Other:
Telium	
Position within host tissue	
Spore colour	
Number of cells in spore, arrangement	
Size (µm)	Length ❑ Breadth ❑
Chains	Present ❑ Absent ❑
Germ pores	Present ❑ Absent ❑ Number per cell: Arrangement: Scattered ❑ Equatorial ❑ Unizonate ❑ Bizonate ❑ Basal ❑ Other:

Protocol 6.19

Use of keys to identify poplar rust (*Melampsora* spp.)

The following protocol is a demonstration of two keys (dichotomous and polytomous) and how they can be used to identify rust species. This example will enable separation of four rust species found on *Populus* spp. trees, namely *Melamspora populnea*, *M. alli-populina*, *M. larici-populina* and *M. medusae*.

Materials

- Dissecting microscope
- Compound microscope
- Glass slides and coverslips
- Scalpel
- Lactoglycerol mountant
- Spirit burner

Method

1. Using a single-edged razor blade, cut a series of fine transverse sections through the pustule and leaf. If this is performed under the dissecting microscope then very fine sections can be cut.

2. Place one small drop of lactoglycerol onto a slide.

3. Carefully transfer the sections onto this slide.

4. Gently lower a coverslip onto the specimen and press down carefully to expel any trapped air.

5. Gently warm the slide over a flame to expel any further trapped air.

6. Examine the sections under the compound microscope.

7. Record in Table 6.16 the presence or absence of the following selected characters and assign the correct letter:

   7.1. Shape of spores – ellipsoid/clavate (A), ovoid/obovoid (B), globose (C), pyriform (D).

   7.2. Distribution of spines – all over surface (E), apical smooth patch (F), equatorial smooth patch (G).

   7.3. Shape of uredoparaphyses – capitate (H) or clavate (I).

   7.4. Uredoparaphyses wall thickness – even (J) or uneven (K).

Table 6.16. Poplar rust characteristics. Fill in as fully as possible.

General details	
Location	
Date of collection	
Ref. no.	
Cultural characteristics	
Spore shape	Ellipsoid/clavate (A) ❑ Ovoid/obovoid (B) ❑ Globose (C) ❑ Pyriform (D) ❑
Distribution of spines	All over surface (E) ❑ Apical smooth patch (F) ❑ Equatorial smooth patch (G) ❑
Shape of uredoparaphyses	Capitate (H) ❑ or Clavate (I) ❑
Uredoparaphyses wall thickness	Even (J) ❑ or Uneven (K) ❑

Transpose selected key code letter below:

A	B	C	D	E	F	G	H	I	J	K

Key 1. Polytomous key (based on characters A–K defined in Table 6.16 above)

Species	Characters
M. medusae	C, D, G, H, J
M. populnea	B, (C), E, I, (J), K
M. alli-populina	A, F, H, J
M. larici-populina	A, F, J, (K), I

Note: Bracketed characters indicate that the feature is sometimes observed.

8. Using either of the simple keys below (dichotomous or synoptic), identify the fungus to the correct species.

Key 2. Dichotomous key

Morphological features	*Melampsora* rust species
Spines distributed all over spore (no smooth/bald patches)	*M. populnea*
Spines not distributed all over spore (smooth/bald patch(es) present) and/or	*M. medusae*
Equatorial smooth/bald patch	
Apical smooth/bald patch and	*M. alli-populina*
Uredoparaphyses capitate with even wall thickness	
Apical smooth/bald patch and	*M. larici-populina*
Uredoparaphyses clavate with uneven wall thickness	

Protocol 6.20

Identification of *Fusarium*

Fusarium species cannot reliably be identified directly from plant material. It is therefore necessary to isolate onto appropriate media. Further pathogenicity testing (see Protocol 3.7) may be required to eliminate saprophytic species of *Fusarium* spp. or to identify special forms.

> **Materials**
>
> - Potato dextrose agar (PDA), Spezieller Nährstoffarmer Agar (SNA) and tap water agar (TWA) plates
> - Dissecting and compound microscopes
> - Scalpel
> - Microscope slides and coverslips
> - Cotton blue in lactoglycerol

Method

1. Using Protocol 6.2, carry out primary isolation onto PDA.

2. Incubate for 4 days at 18–20°C with 12–16 h light.

3. Using aseptic techniques, subculture from the edge of a single actively growing colony on to PDA, SNA and TWA plates.

4. Incubate for 7–10 days under the same conditions as step 2.

5. Record any pigmentation observed on reverse of PDA plate. Note mycelium colour and record as white, grey, pink, red or violet.

6. Remove the lid of the SNA plate and examine under low power (dissecting microscope). Record presence or absence of conidia in chains, dry heads or slimy droplets, and whether they are growing on long or short conidiophores.

7. Make a slide mount to include the sporulating area on SNA or TWA plates (cut an agar block approximately 1 cm square and mount in water).

8. Record the following features in Table 6.17:

    8.1. Colony characteristics: appearance, pigmentation (e.g. brown, vinaceous, violet, pink), fast or slow growing.

    8.2. Spore type:

- Macroconidia: these are large, usually banana-shaped spores of *Fusarium* spp. Record presence/absence, shape (pay particular attention to the foot cell, apical cell and degree of swelling of the central region of the spore), formation in sporodochia (cushion-like masses of spores enclosed in a mass of slime) or dry heads (individual macrocondia clearly visible).
- Microconidia: smaller spores of *Fusarium*. Record presence/absence, shape, formation in chains, dry heads or slimy droplets.
- Presence/absence of chlamydospores: these are thick-walled resting spores.

    8.3. Conidiogenous cells (cells from which conidia form):

        8.1. Size.

        8.2. Shape.

        8.3. Type of cell:

- Mono-phialides: single conidiogenous locus on the cell.
- Poly-phialides: more than one conidiogenous locus per cell.
- Polyblastic cells: several conidiogenous loci per cell (these resemble poly-phialides, but unlike poly-phialides, only one spore is formed from each locus, which is plugged after spore release. The conidiogenous cell then produces a further locus adjacent to or near the previous one before producing a further spore).

9. Identification of species. The following sources will be useful in determining species: Gordon (1952), Toussoun and Nelson (1968), Booth (1971, 1977), Gerlach and Nirenberg (1982), Burgess and Liddell (1983), Nelson *et al.* (1983), Summerell and Nelson (2003) and Leslie and Summerell (2006).

Table 6.17. Characteristics of *Fusarium*.

General details	
Location	
Date of collection	
Reference number	
Host species	
Pathogenicity test	Not done ❑ Positive ❑ Negative ❑
Cultural characteristics	
Pigmentation	Present ❑ Absent ❑ Colour:
Growth rate	Fast ❑ Slow ❑
Mycelium colour	White ❑ Grey ❑ Pink ❑ Red ❑ Violet ❑
Conidia (under dissecting microscope)	Forming singly ❑ Forming in chains ❑
	Dry heads ❑ Slimy droplets ❑
Conidiophores (under dissecting microscope)	Short ❑ Long ❑
Macroconidia	Present ❑ Absent ❑
	Shape:
	Sporodochia: ❑ Dry heads ❑
Microconidia	Present ❑ Absent ❑
	Shape:
	Chains: Present ❑ Absent ❑
	Dry heads ❑ Slimy droplets ❑
Chlamydospores	Present ❑ Absent ❑
Conidiogenous cells	Size (µm):
	Shape
	Mono-phialide ❑ Poly-phialide ❑ Polyblastic ❑

Diagnosis of fungal diseases from some common host/crop species
Mycological examination and testing
Protocol 6.21

Mycological examination and testing of potato tubers

Materials
• Paper towel
• Trays
• Sharp knife
• Dissecting and compound microscopes
• Clear-sided plastic boxes
• Extra absorbent paper towel
• Clear adhesive tape
• Cotton/trypan blue (0.1%) in lactoglycerol
• Microscope slides and coverslips
• Hypodermic needles/mounted needle
• Potato dextrose agar (PDA)
• Razor blades/scalpel
• 10% bleach solution
• Forceps
• Plastic bags
• Petri dishes
• Petri's mineral solution

Method

1. If potato tubers have any mud or dirt on them, wash a representative sample under a tap and lie them out on paper towel on a tray to dry.

2. If assessing for rots, cut tubers open with a sterile knife and record:

- if it is a dry or soft rot;
- how deeply it extends into the tuber;
- the colour of rot margins;
- if there are any cavities;
- if there are numerous brown specks within the tuber tissue, which may indicate *Thecaphora solani* infection; and
- any distinct smell (e.g. sweet/vinegary/fishy).

3. Under the dissecting microscope, examine each tuber for fungal structures, recording their:

- type (mycelium, conidiophores, sporodochium, pycnidium, acervulus, etc.);
- position;
- occurrence;
- relationship to the diseased areas; and
- size, measuring at least five spores/spore-producing structures.

Compare morphology and initial tentative diagnosis to reference material and literature.

4. Examine for surface lesions or blemishes such as scabbing or dotting on the skin, and for any signs of potato wart disease, *Synchytrium endobioticum* (warty, cauliflower-like protuberances in the eye tissue; in severe cases the whole tuber may be replaced by the warty proliferation). Record:

- Colour of blemishes/lesions.
- Margin shape (if regular or irregular).
- Occurrence.

5. Assess for dark *Rhizoctonia* mycelium, branched at right angles, particularly around the eyes.

6. Black dot/silver scurf (*Colletotrichum coccodes/Helminthosporium solani*)

    6.1. If tubers have black dots on their surfaces, incubate 2–3 tubers in a plastic box on damp paper towel (see Protocol 3.1).

    6.2. After 3 days, make tape preps and examine for any conidiophores branched at irregular angles, typical of *Helminthosporium* (silver scurf), under a compound microscope (see Protocol 6.2). *Helminthosporium* conidiophores and spores should also be visible under a dissecting microscope.

7. Common/powdery scab (*Streptomyces scabei/Spongospora subterranea*)

    7.1. If tubers have scabs on the surface, examine each tuber carefully under the dissecting microscope for clumps of sandy subglobose spore balls within the scabs.

    7.2. Dislodge any mud in or over scabs with a sterile needle.

    7.3. For any suspect spore balls, take them off the tuber with a needle and place them onto a slide with lactoglycerol. Examine under a compound microscope and compare morphology to reference material and Fig. 6.36 (a, b).

8. Skin spot (*Polyscytalum pustulans*)

   8.1. If tubers have dark purple to black pimples on the surface, place 2–3 tubers in a damp chamber.

   8.2. Incubate at approx. 6°C for 10 days.

9. If fungal structures cannot be identified confidently, take a small amount of inoculum using a needle (leaving at least a little bit of fungal material in case of unsuccessful culture) and place it in the centre of a PDA agar plate.

10. If no fungal structures were observed, set up isolations and incubations.

   10.1. Cut out pieces of tissue from the internal leading edge of infection with a sterile razor blade/scalpel.

   10.2. Soak four pieces of plant material for 2 min in 10% bleach solution, and place onto a PDA plate with sterile forceps. Incubate for up to 7 days (see Protocol 3.1).

   10.3. If there are no severe soft rots, place 2–3 tubers in a plastic box on a piece of damp paper towel and incubate at 18–25°C for at least 7–10 days (see Protocol 3.1).

11. *Phytophthora/Pythium*

   11.1. If internal rot is wet and creamy (such as in Fig. 6.37), cut out pieces from the internal leading edges of infection.

   11.2. Place them in a clean empty Petri dish. Pour in some Petri's mineral solution to allow the pieces of tissue to float. Check daily for signs of *Phytophthora* or *Pythium*, e.g. sporangia.

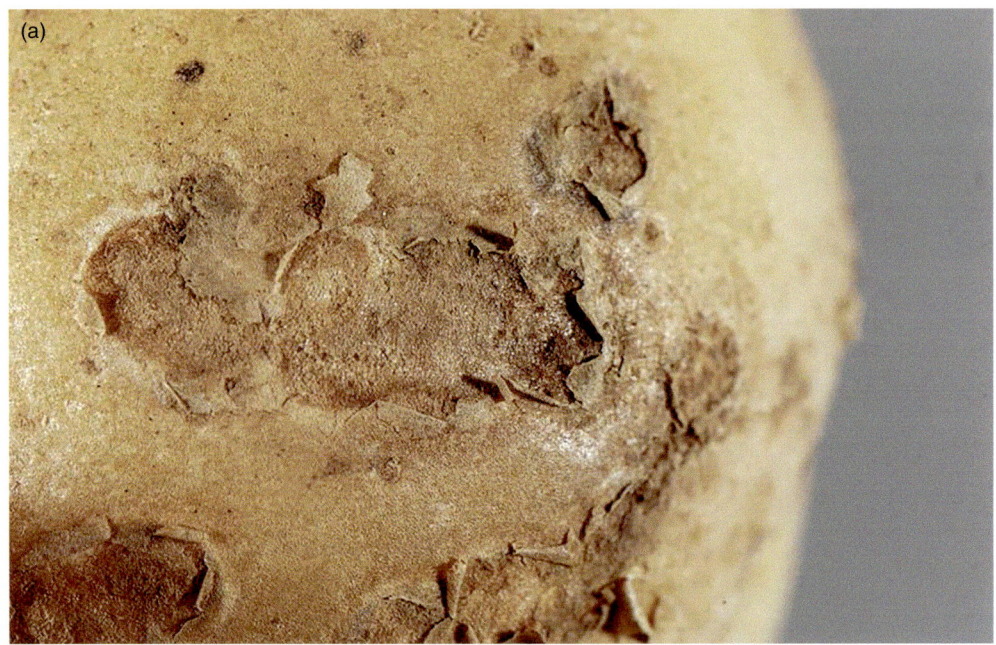

Fig. 6.36. (a) *Spongospora subterranea* (powdery scab) on tuber surface; (b) spore balls at ×400 magnification. (Image Courtesy Fera-Science Limited © Copyright Fera-Science Limited.)

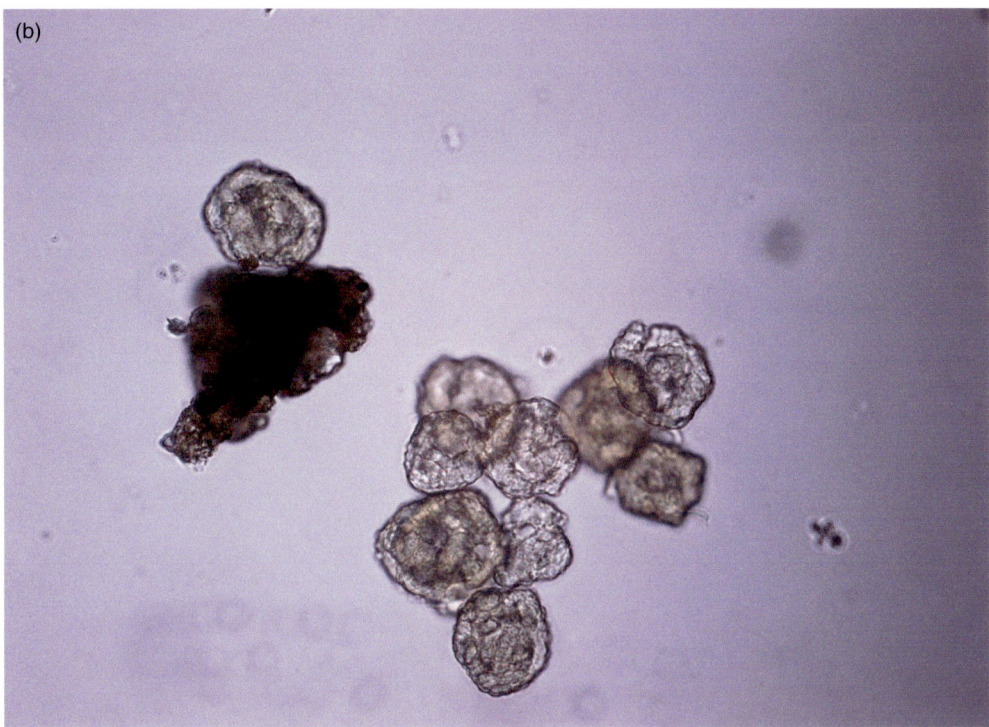

Fig. 6.36. Continued.

Fig. 6.37. Soft and creamy internal potato tuber rotting from *Pythium* infection (watery wound rot). (Image Courtesy Fera-Science Limited © Copyright Fera-Science Limited.)

Record findings in Table 6.18:

Table 6.18. Potato diseases checklist. Fill in as fully as possible.

General details	
Location	
Date of collection	
Reference number	
Rot/blemish characteristics	
Rots	Present ❑ Absent ❑ Dry ❑ Soft ❑ Wet and creamy ❑ Depth into tuber (cm): Colour of rot: Colour of rot margins: Cavities: Present ❑ Absent ❑ Internal brown specks: Present ❑ Absent ❑ Smell:
Fungal structures	Present ❑ Absent ❑ Type (mycelium, conidiophores, sporodochium, pycnidium, acervulus, etc.): Shape: Position on/in the tuber: Occurrence: Relationship to the diseased areas:
	Spore type:
	Spore sizes (µm): | | | | |
	Spore type:
	Spore sizes (µm): | | | | |
Surface blemishes	Absent ❑ Surface lesion(s) ❑ Warty protuberances ❑ Black pimples ❑ Other: Colour: Margin: Regular ❑ Irregular ❑ Coalescing ❑ Occurrence:
Surface black dotting	Present ❑ Absent ❑ Branched conidiophores: Present ❑ Absent ❑
Surface scabbing	Present ❑ Absent ❑ Spore balls: Present ❑ Absent ❑
Rhizoctonia mycelium	Present ❑ Absent ❑

Protocol 6.22

Strawberry sample mycological examination and testing procedures

Materials

- Single-edged razor blade/scalpel
- Scissors
- KOH (potassium hydroxide) pellets
- Boiling tube
- Flame heating device (e.g. gas burner or spirit burner)
- Petri dishes
- Forceps
- Cotton/trypan blue (0.1%) in lactoglycerol
- Microscope slides and coverslips
- Dissecting and compound microscopes
- Petri's mineral solution
- Plastic bags
- Sterile distilled water
- 10% bleach solution
- Hypodermic needles/mounted needle
- Clear-sided plastic boxes
- Extra absorbent paper towel

Method

1. If strawberry roots have any soil on them, wash off soil under a tap.
2. Examine the sample material and record extent and position of damage outlined below:
    2.1. Tissues appear dried and dead.
    2.2. Discoloration of foliage; chlorosis/necrosis/coloured deposits/curling (and whether young or mature leaves are affected).
    2.3. Spots or holes on leaves; symmetrical, irregularly shaped, association with chlorotic or discoloured haloes.
    2.4. Malformation or stunting of leaves or shoots.
    2.5. Wilting.
    2.6. Stem browning or canker formation.
    2.7. Stunted/contorted/blackened/unbranched (rat-tailed)/rotted root system.
    2.8. Firmness of fruits.
    2.9. Sticky exudates/gummosis.
    2.10. Lesions, vascular staining (such as reddening), rotting of the crown.
3. **Root preparation assessment**
    3.1. Using a razor blade/scalpel working away from the body and fingers, slice or scrape roots lengthwise and look for any red steles or rotten areas. Start at the root tips and work back towards the crown.
    3.2. Use flame-sterilized scissors to cut out symptomatic areas of the roots.
    3.3. Place symptomatic roots in a boiling tube and cover with approx. 2.5 cm of water.
    3.4. If roots are not soft or decaying, boil in KOH.

    - Add two pellets of KOH using forceps to a boiling tube. Check for any cracks in the tube.
    - Wearing safety glasses, boil for approx. 2 min using a spirit/gas burner by holding the boiling tube at an angle pointing away from your body. Gently agitate the tube to prevent large air bubbles forming. Remove it from the heat if it looks like it might boil over.
    - Pour roots in KOH into a Petri dish, rinsing out once into the Petri dish with tap water.
    - Pick out symptomatic roots with forceps and place on a slide in a single layer in straight rows for ease of examining.
    - Stain with a drop of cotton/trypan blue in lactoglycerol, place a long coverslip on top and gently tap to squash the roots flat. Warm with a burner to get rid of bubbles (see Fig. 6.38).

    3.5. Examine seven slides of roots under a compound microscope. For *Phytophthora* oospores (Fig. 6.39), record the length and width of five oogonia and five oospores, and calculate mean. See features of common strawberry pathogens below:

Organism	Characteristics
Phytophthora fragariae (Red core)	Oospores typically form around the stele of infected roots (EPPO, 2022)
	Lens-shaped or irregular/funnel-shaped oogonia 30–80 μm with a single oospore 28–37 μm in diameter
Phytophthora cactorum (Crown rot)	Smaller oospores, typically 20–26 μm, form throughout root cortex
	Globose oogonia and oospores (Waterhouse and Waterston, 1966)
Pythium	Oospores smaller than 20 μm diameter
	Oogonium wall often barely visible
Olpidium	Hexagonal resting sporangia
Verticillium dahliae	Microsclerotia, composed of a compact mass of thick-walled, black/brown cells and chlamydospores
Verticillium albo-atrum	Swollen, dark resting mycelium that lack lateral septa
Rhizoctonia	Hyphae have right-angled branching and are constricted at branch points
	Hyphae 3.6–3.7 μm thick

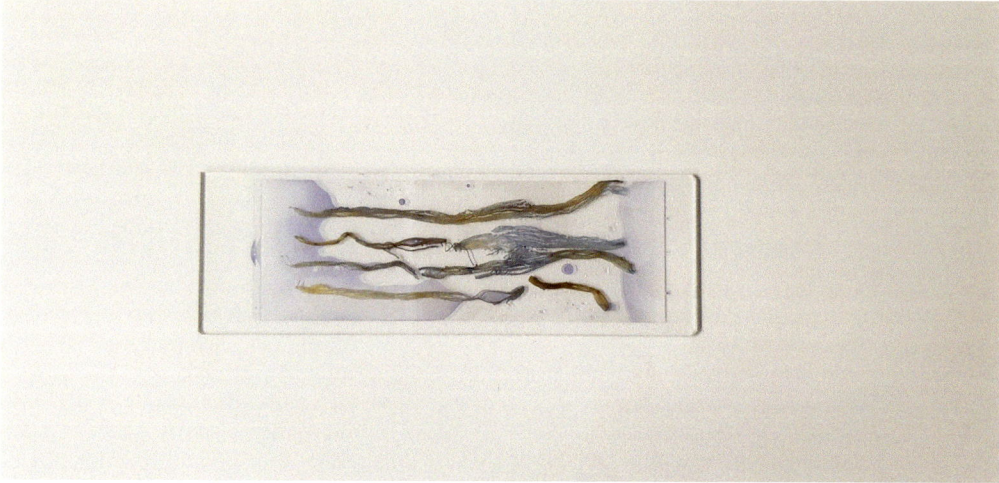

Fig. 6.38. Microscope slide with strawberry roots stained with trypan blue. (Image Courtesy Fera-Science Limited © Copyright Fera-Science Limited.)

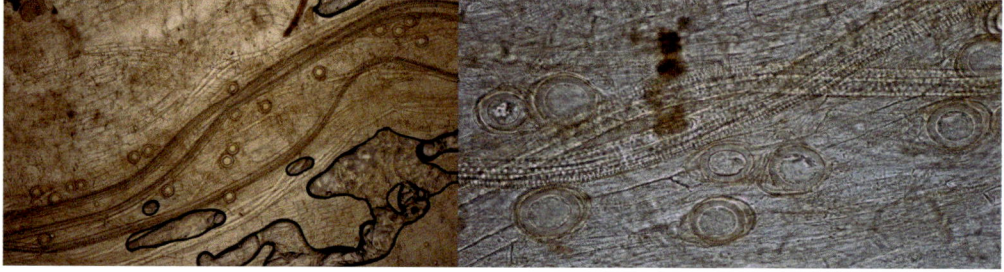

Fig. 6.39. *Phytophthora fragariae* oospores in diseased roots (×200 and ×400). (Image Courtesy Fera-Science Limited © Copyright Fera-Science Limited.)

4. Root floats

4.1. With sterilized scissors, cut off several pieces of symptomatic roots (4–10 if possible) and place in an empty Petri dish.

4.2. Add enough Petri's mineral solution to cover the roots and replace the Petri dish lid. Incubate at 21°C for up to 7 days.

4.3. Check roots daily if sporangia of *Phytophthora* or *Pythium* are attached, using the compound microscope at low magnification. If no sporangia are present after 2 days, gently pour off and replace the Petri's mineral solution.

5. Isolations onto agar

If looking for *Phytophthora* or *Pythium* spp., do not use bleach, but surface decontaminate by rinsing in a small plastic bag with three changes of sterile distilled water.

5.1. Leaves, stems, crowns:
- Cut out symptomatic tissue from the leading edge using sterilized scissors/scalpels/razor blades.
- Surface decontaminate by placing in 10% bleach in a Petri dish for 2 min.

5.2. Stems:
- Cut down stem pieces lengthwise to reveal vascular tissue using scissors/scalpels/razor blades.
- Place four small pieces in 10% bleach in a Petri dish for 2 min.

5.3. Crowns:
- Using secateurs, cut the crown open to reveal the core and vascular tissue.
- With a scalpel/razor blade, cut out four pieces of crown tissue at the leading edge.

- Place in 10% bleach for up to 2 min and rinse with three changes of sterile water.

5.4. Roots:
- Rinse off enough remaining soil to be able to select a few symptomatic roots, tear them away and then cut them to 2–3 cm lengths using scalpels/scissors/razor blades.
- Rinse root pieces thoroughly with sterile water.

5.5. Fungal structures:
- Using a sterile needle, pick off fruiting bodies/mycelium/spores/sclerotia, etc.
- Immerse structures such as sclerotia in 10% bleach for 3–5 min.

6. Use flame-sterilized forceps to place or immerse the selected plant/fungus material into the agar, placing typically four pieces per plate. Use PDA for most fungal pathogens, but if specifically testing for *Phytophthora* spp., use PARP-H and for *Pythium* spp. use PARP.

7. Incubate for 5–7 days as per Protocol 3.1.

8. Damp chambers

8.1. Set up plastic boxes of damp paper towel as described in Protocol 3.1.

8.2. Take symptomatic plant material, e.g.
- necrosis on leaves;
- reddening/darkening of crowns;
- vascular staining/rotting/dieback of stems.

8.3. Incubate for 10 days as in Protocol 3.1.

Record findings in Table 6.19.

Table 6.19. Mycological examination of strawberry. Fill in as fully as possible.

General details	
Location	
Date of collection	
Reference number	
Host *Fragaria* species	
Visual symptoms	
Leaves	Chlorosis ❏ Necrosis ❏ Coloured deposits ❏ Curling ❏ Malformation ❏ Stunting ❏ Wilting ❏ No symptoms ❏ Leaves affected: Young ❏ Mature ❏

Continued

Table 6.19. Continued.

	Spots ❑/Holes ❑: Symmetrical ❑ Irregular ❑
Stems	Browning ❑ Cankers ❑ Sticky exudates ❑ Malformation ❑ Stunting ❑ Wilting ❑ No symptoms ❑
Roots	Stunting ❑ Contorting ❑ Blackening ❑ Rat-tailing (unbranched) ❑ No symptoms ❑
Fruit	Soft ❑ Firm ❑
Crown	Lesions ❑ Rotting ❑ No symptoms ❑ Vascular staining: Reddening ❑ Browning ❑
Microscopic characteristics	
Oospores	Present ❑ Absent ❑ Location: Around stele ❑ Throughout cortex ❑
	Height (µm): Width (µm):
	Mean (height) (µm): Mean (width) (µm): <20 µm ❑ 20–26 µm ❑ >26 µm ❑
Oogonia	Present ❑ Absent ❑ Shape: Lens-shaped/irregular ❑ Globose ❑ Barely visible ❑
	Height (µm): Width (µm):
	Mean (height) (µm): Mean (width) (µm): <30 µm ❑ 30–80 µm ❑
Sporangia	Absent ❑ Hexagonal ❑ Other:
Microsclerotia	Present ❑ Absent ❑
Dark resting mycelium	Present ❑ Absent ❑
Chlamydospores	Present ❑ Absent ❑
Hyphae	Dematiaceous ❑ Hyaline ❑ Absent ❑ Branched ❑ Not branched ❑ Thickness (µm): Constricted at branch points ❑ Not constricted ❑
Fungal/fungus-like structures present on any of 7 slides	Present ❑ Absent ❑

Protocol 6.23

Mycological examination and testing of turf samples

There are numerous fungal and fungus-like pathogens which can cause disease in turfgrass plants. A good sample of turf for mycological examination and testing is a core from the affected area, approximately 10 cm deep × 10 cm in diameter, and includes the leading edge between symptomatic and healthy turf.

Materials

- Dissecting and compound microscopes
- Forceps
- Microscope slides and coverslips
- Clear adhesive tape
- Mounted or disposable sterile needles
- Mounting solution (lactoglycerol with or without stain, or water)
- Cotton or trypan blue stain
- Gas or spirit burner
- Scalpel or single-edged razor blades
- 1 ml Pipette
- Potato dextrose agar (PDA) plates with or without penicillin and streptomycin (P&S)
- Clear-sided boxes
- Absorbent paper towel
- Petri's mineral solution
- Petri dishes
- 70% IMS or equivalent
- Sterile distilled water

Method

1. Examine the sample and note the symptoms, e.g. rotting roots, leaf spots.

2. To check the overall health of the turf sample:

   2.1. Cut the turf sample in half vertically (include any leading edge) and check for black layer – it can be seen either as a continuous layer or black streaks. It smells of hydrogen sulfide and is often the result of waterlogging and asphyxiation, causing the turf to turn yellow and thin out. Black layer is often associated with too much thatch.

   2.2. Check the thatch level – this is a hard layer under the grass and above soft soil. Thatch that is less than 2 cm deep suggests good drainage, but a thatch layer deeper than 2 cm indicates poor drainage.

   2.3. Carry out a drainage test: using a pipette, add 1 ml of cotton or trypan blue in water to the surface of the soil/thatch. If drainage is good, the coloured solution should be visible 9 cm down the soil within a few minutes.

3. Carefully separate several whole plants from the sample (symptomatic plants as well as some healthy plants for comparison), wash the soil off, then examine them and set up further tests as detailed below.

4. Roots and stem base:

   4.1. Check if the roots look healthy or if they show signs of disease (e.g. rotting, blackening).

   4.2. During examination under a dissecting microscope (up to ×50 magnification), check the roots and stem bases of the plants for the presence of dark runner hyphae of *Rhizoctonia*, *Gaeumannomyces* (take-all) or *Magnaporthiopsis* mycelium. All three of these genera include species of grass root pathogens.

   4.3. If suspected *Rhizoctonia*, *Gaeumannomyces* or *Magnaporthiopsis* mycelium is observed, isolate the fungus in culture for accurate diagnosis: cut four pieces of roots with visible fungal mycelium and plate out onto PDA + P&S following surface disinfection in 10% bleach solution for 2 min.

   4.4. Alternatively, pick off a small section of the mycelium with a sterile needle and place it in the centre of the PDA + P&S plate. Repeat for 2–3 plates (as these are isolations of fungi directly from roots, there is a possibility that saprotrophic fungi may overgrow the plates).

   4.5. If there is not a lot of mycelium on the roots, float some of the roots in Petri's mineral solution for a few days – this will allow the fungus to produce more hyphae which then can be picked off for plating out.

   4.6. Check the isolation plates every few days and identify the pathogen by comparing the colony morphology to descriptions in literature.

4.7. Select several symptomatic roots, prepare and microscopically examine them as described in Protocol 6.4.

4.8. During microscopic examination, look out for *Pythium* oospores and *Rhizoctonia* or take-all mycelium.

4.9. Other organisms that may be present in grass roots are *Olpidium* spp. (look out for characteristic angular oospores) and *Polymyxa graminis* (look out for resting spores resembling a bunch of grapes within root cells); these organisms may infect roots without causing discernible damage, however, when environmental conditions are unfavourable for plant growth, they may exacerbate the problem.

5. Several fungi, including *Fusarium*, *Colletotrichum*, *Drechslera/Helminthosporium* species, can cause root and stem base rots as well as foliar blights. If roots and/or stem bases are rotting but no structures of fungal pathogens are observed during examination under a dissecting microscope:

5.1. Cut out four sections from the leading edges using a clean razor blade or scalpel.

5.2. Surface-disinfect in 10% bleach solution for 2 min.

5.3. Plate the plant material onto a PDA plate, incubate and assess every few days.

5.4. Also set up an incubation chamber as described in Protocol 3.1.

6. Leaves and stems. There are numerous fungal or fungus-like plant pathogens that can cause leaf and stem diseases in turfgrasses, including rust fungi, powdery mildew (*Blumeria graminis*), downy mildew (*Sclerophthora macrospora*), smuts (mostly species of *Urocystis*, *Ustilago* and *Entyloma*), *Fusarium* spp., *Ascochyta* spp., *Septoria* spp. and many others.

6.1. Examine the leaves and stems as described below, recording the symptoms and signs of fungi observed.

6.2. Examine healthy tissue under a dissecting microscope so that the appearance of non-infected tissue can be recognized.

6.3. Examine symptomatic material for any fungal structures.

6.4. If fungal structures are present, record their type (mycelium, conidiophores, sporodochium, pycnidium, acervulus, apothecium, perithecium, etc.), position, occurrence and relationship to the diseased areas. Decide whether the fungal structures are consistently associated with the symptoms.

6.5. Examine the fungal structures using a compound microscope, measure and record the morphological features of spores (including measurements of at least five spores) and spore-producing structures.

6.6. Compare morphology and initial tentative diagnosis to reference literature.

6.7. If a fungal specimen cannot be identified with confidence, proceed to culture the suspect fungus to make the identification easier. Take a small amount of inoculum using a sterile needle (leaving at least a little bit of the original fungal material on the sample in case the first culture is not successful) and place it in the centre of a PDA or PDA + P&S plate. Note that several fungal or fungus-like pathogens encountered on turfgrasses cannot be cultured, e.g. powdery mildew, downy mildew, rusts and smut fungi.

7. If no fungal structures were observed during the initial microscopic visual examination or no fungal structures present were identified as a primary pathogen:

7.1. Set up isolations from leaves or stems showing a leading edge of infection as described in Protocol 3.2.

7.2. Set up an incubation chamber using symptomatic leaf and/or stem tissue as described in Protocol 3.1.

7.3. Assess isolation plates and damp chamber every few days for signs of fungi and compare morphology and initial tentative diagnosis of sample to reference information.

Record findings in Table 6.20.

Table 6.20. Turf sample assessment checklist. Fill in as fully as possible.

General details	
Location	
Date of collection	
Reference number	
Host species	
Symptoms	Root rot ❑ Blackened roots ❑ Stem base necrosis ❑ Leaf spots ❑ Leaf yellowing ❑
Black layer	Present ❑ Absent ❑
Thatch level	Less than 2 cm ❑ More than 2 cm ❑
Drainage test	Good drainage ❑ Poor drainage ❑
Fungal structures	Present ❑ Absent ❑ Location on host: Roots ❑ Stem base ❑ Leaves ❑ Position of structure on host (superficial, erumpent, partially erumpent, raised, sunken): Relationship to the diseased areas:
	Type of fungal structures (mycelium, conidiophores, sporodochium, pycnidium, acervulus, etc.): Shape: Pigmentation:
	Spore type (conidia, ascospores, etc.): Spore colour: hyaline ❑ other ❑ Spore shape:
	Spore sizes (length by width, μm):

Protocol 6.24

Mycological examination and testing of fruit samples

Materials

- Dissecting and compound microscopes
- Forceps
- Microscope slides
- Coverslips
- Mounted or disposable needles
- Mounting solution (lactoglycerol with or without stain, or water)
- Gas or spirit burner
- Scalpel or single-edged razor blade
- Potato dextrose agar (PDA) plates
- Clear-sided boxes
- Absorbent paper towel
- Petri's mineral solution
- Petri dishes
- 70% IMS or equivalent

Method

1. Examine the sample material with naked eye and record in Table 6.21 the extent and position of disease symptoms. You may come across anthracnose lesions, blights, blotches, rots, moulds or scabs.

2. Place under dissecting microscope and select lowest magnification (×10).

- Examine healthy tissue under the dissecting microscope so that the appearance of non-infected tissue can be recognized.
- Then examine diseased material for any fungal structures, recording their position, occurrence, shape and relationship to the diseased areas. Decide whether any fungal structures are consistently associated with the symptoms.

3. Having detected fungal structures, increase magnification (up to about ×50) to determine their morphology (mycelium, conidiophores, sporodochia, pycnidia, acervuli, perithecia, etc.), pigmentation and position of structure on host (superficial, erumpent, partially erumpent, raised, sunken, etc.).

3.1. Fruiting bodies:

- Place a small drop of lactoglycerol (with or without stain) onto a clean glass microscope slide. Keeping the sample material under the dissecting microscope, carefully pick off a specimen of several suspect fungal fruiting bodies with the tip of a fine needle.
- Examine the slide under the dissecting microscope to confirm that structures were successfully removed and are correctly orientated for examination (excessive amounts of host material can also be teased away at this stage and discarded).
- Gently lower a suitably sized glass coverslip onto the slide and then press down lightly to expel any air bubbles. If necessary, holding the slide by one end, gently warm the slide over the flame of a spirit burner until any air pockets start to expand, remove from flame and place on the bench top to cool. Transfer the slide to a compound light microscope for examination.

3.2. Mycelium, conidiophores with or without spores, or unknown structures:

- Cut off a small piece of clear adhesive tape (up to 30 mm long) and place adhesive side down onto sample material, and then gently press the tape down. Specimen should have adhered to tape.
- Lift off tape and then place the desired portion adhesive side down onto a small drop of lactoglycerol on a glass microscope slide. Press down gently to remove air bubbles, and then warm very carefully to avoid boiling which may dissolve adhesive. Allow slide to cool and transfer it to the compound microscope for examination.

4. Examine the mounted material with a compound microscope:

- Select a low power objective (e.g. ×10) and locate the specimen.

Table 6.21. Fruit sample assessment checklist. Fill in as fully as possible.

General details	
Location	
Date of collection	
Reference number	
Host species	
Symptoms	Anthracnose ❏ Blight ❏ Blotches ❏ Rot ❏ Mould ❏ Scab ❏ Any notable odour:
Fungal structures	Present ❏ Absent ❏ Position of structure on host (superficial, erumpent, partially erumpent, raised, sunken): Relationship to the diseased areas:
	Type of fungal structures (mycelium, conidiophores, sporodochium, pycnidium, acervulus, etc.): Shape: Pigmentation:
	Spore type (conidia, ascospores, etc.): Spore colour: hyaline ❏ other ❏ Spore shape:
	Spore sizes (length by width, μm):

- When the fungal structures are located, record their morphology and dimensions. If a fruiting body is examined, attempt to expel its contents by gently pressing down on the coverslip. This can be done with extreme care on the microscope stage while observing under ×10 objective or after removing the slide from the microscope, placing it on the bench and tapping it very gently with a pencil.
- Re-examine under ×40 objective and describe morphology of spores and spore-producing structures.

5. If possible, at this stage identify the fungus/fungi to genus level.

6. Check literature for fungal diseases described for the host. If the fungus observed is consistently associated with the symptoms, and if the symptoms are typical, final diagnosis can be made at this stage. In addition to common fungal plant pathogens (e.g. *Colletotrichum, Fusarium, Botryosphaeria, Phomopsis* species and many others), there also fungi which are considered saprotrophic in other types of samples that can cause post-harvest diseases in fruit (and vegetables), for example, *Rhizopus, Penicillium, Aspergillus* and *Cladosporium* species.

7. If a fungal specimen cannot be identified with confidence, then proceed to culture the suspect fungus to make the identification easier. Take a small amount of inoculum using a flamed mounted needle or a disposable needle and place it in the centre of a PDA plate, leaving at least a little bit of the original fungal material on the sample in case the first culture is not successful.

8. If no fungal structures were observed during the initial microscopic visual examination or any fungal structures present were not identified as a primary pathogen:

- Set up isolations: cut out pieces of tissue with a leading edge of infection, surface-disinfect the selected material pieces for approximately 2 min in 10% bleach solution and plate onto a PDA plate.

- If the sample is not too deteriorated or already overgrown with fungi, set up an incubation box – a dry chamber for juicy fruit and/or soft, wet rots; a damp chamber if the sample is relatively dry. Expose any leading edges by cutting the fruit open through the lesion or symptomatic area and place the sample in the incubation box.

- If a *Phytophthora* or *Pythium* infection is suspected, set up a float in Petri's mineral solution – select small pieces of the leading-edge tissue (internal, where possible), place the selected material in a clean, empty Petri dish and add some Petri's mineral solution to cover the plant material pieces and allow them to float in the solution. Check the floated pieces of plant material daily for signs of *Phytophthora* or *Pythium*.

9. Assessment of incubation and isolation plates.

- Check the incubation box and isolation plates every few days for the presence of fungal structures. Examine, record and identify any fungal growth, referring to literature as required.

- Decide if any fungus observed is likely be the primary cause of symptoms or not, based on consistency of findings and evidence in literature.

- If no fungal growth is observed after 10 days, it is most likely that no fungal plant pathogens are present in the sample.

References

Agrios, G.N. (2005) *Plant Pathology*, 5th edn. Academic Press, San Diego, California.

AHDB Potatoes (2018) Diseases and defects of potatoes. Poster. AHDB Sutton Bridge Crop Storage Research, Lincolnshire, UK.

Ainsworth, G.C. and Sampson, K. (1950) *The British Smut Fungi (Ustilaginales).* Commonwealth Agricultural Bureaux, Farnham Royal, UK.

Arora, R.K. and Sagar, V. (2014) Tuber borne diseases of potato. In: Dinesh Singh, P. Chowdappa and Pratibha Sharma (eds) *Diseases of Vegetable Crops: Diagnosis and Management.* Indian Phytopathological Society, New Delhi, pp. 1–57.

Arthur, J.C. (1934) *Manual of the Rusts in the United States and Canada*. Purdue Research Foundation, Lafayette, Indiana.

Arthur, J.C., Kern, F.D., Orton, C.R., Fromme, F.D., Jackson, H.S. *et al.* (1929) *The Plant Rusts (Uredinales).* Wiley and Sons, New York.

Averre, C.W. (2010) *Club-Root of Cabbage and Related Crops*. Vegetable Disease Information Note 17 (VDIN-0017), College of Agriculture and Life Sciences, Plant Pathology Extension, North Carolina State University, Raleigh, North Carolina.

Baroncelli, R., Talhinhas, P., Pensec, F., Sukno, S.A., Le Floch, G. and Thon, M.R. (2017) The *Colletotrichum acutatum* species complex as a model system to study evolution and host specialization in plant pathogens. *Frontiers in Microbiology* 8:2001.

Barreto, R.W. and Dick, M.W. (1991) Monograph of *Basidiophora* (oomycetes) with the description of a new species. *Botanical Journal of the Linnean Society* 107, 313–332.

Bauer, R., Oberwinkler, F. and Vanky, K. (1997) Ultrastructural markers and systematics in smut fungi and allied taxa. *Canadian Journal of Botany* 75, 1273–1314.

Beales, P.A. (1997) The epidemiology and biology of rhododendron powdery mildew. PhD thesis, University of Hertfordshire, Hatfield, UK.

Bélanger, R.R., Bushnell, W.R., Dik, A.J. and Carver, T.L.W. (2002) *The Powdery Mildews: A Comprehensive Treatise.* APS Press, St Paul, Minnesota.

Bérubé, J.A. and Dessureault, M. (1989) Morphological studies of the *Armillaria mellea* complex: two new species, *A. gemina* and *A. calvescens. Mycologia* 81, 216–225.

Bessey, E.A. (1964) *Morphology and Taxonomy of Fungi*. Hafner Press, New York.

Booth, C. (1971) *The Genus Fusarium*. Commonwealth Agricultural Bureaux, Farnham Royal, UK.

Booth, C. (1977) *A Laboratory Guide to the Identification of the Major Species of Fusarium*. Commonwealth Mycological Institute, Kew, UK.

Bowen, K.L., Everts, K.L. and Leath, S. (1991) Reduction in yield of winter wheat in North Carolina due to powdery mildew and leaf rust. *Phytopathology* 81, 503–511.

Brasier, C.M. (2009) *Phytophthora* biodiversity: how many *Phytophthora* species are there? In: Goheen, E.M. and Frankel, S.J. (eds) *Phytophthoras in Forests and Natural Ecosystems*. General Technical Report PSW-GTR-221. US Department of Agriculture Forest Service, Albany, California, pp. 101–115.

Brasier, C., Scanu, B., Cooke, D. and Jung, T. (2022) *Phytophthora*: an ancient, historic, biologically and structurally cohesive and evolutionarily successful generic concept in need of preservation. *IMA Fungus* 13, 12.

Braun, U. (1987) A monograph of the *Erysiphales* (powdery mildews). *Beiheifte zur Nova Hedwigia* 89, 1–700.

Braun, U. (1995) *The Powdery Mildews (Erysiphales) of Europe*. Gustav Fisher Verlag, Jena, Germany.

Braun, U. and Cook, R.T.A. (2012) *Taxonomic Manual of the Erysiphales (Powdery Mildews)*. Centraalbureau voor Schimmelcultures, Utrecht, The Netherlands.

Brenchley, G.H. and Wilcox, H.J. (1979) *Potato Diseases.* HMSO, London.

Burgess, L.W. and Liddell, C.M. (1983) *Laboratory Manual for Fusarium Research.* Fusarium Research Laboratory, Department of Plant Pathology and Agricultural Entomology, University of Sydney, New South Wales, Australia.

Bush, E.A., Stromberg, E.L., Hong, C., Richardson, P.A. and Kong, P. (2006) *Illustration of Key Morphological Characteristics of Phytophthora Species Identified in Virginia Nursery Irrigation Water*. Plant Health Progress, Plant Management Network, St Paul, Minnesota. Available at: http://www.plantmanagementnetwork.org/pub/php/diagnosticguide/2006/va/ (accessed 2 September 2011).

Carnegie, S., Saddler, G., Davey, T. and Browning, I. (2007) *The importance of potato mop top virus (PMTV) in Scottish seed potatoes*. Project report. British Potato Council, Spalding, UK.

Carris, L.M. and Castlebury, L.A. (2006) Nonsystemic bunt fungi – *Tilletia indica* and *T. horrida*: a review of history, systematics, and biology. *Annual Review of Phytopathology* 44, 113–133.

Chan, K.C., Boyd, W.J.R. and Khan, T.N. (1990) Distribution, severity and economic importance of powdery mildew of barley in Western Australia. *Australian Journal of Experimental Agriculture* 30, 379–385.

Chase, A.R. (1987) *Compendium of Ornamental Foliage Plant Diseases.* APS Press, St Paul, Minnesota.

Choi, Y.J., Hong, S.B. and Shin, H.D. (2005) A re-consideration of *Pseudoperonspora cubensis* and *P. humuli* based on molecular and morphological data. *Mycological Research* 109, 841–848.

Clarkson, J.P. and Whipps, J.M. (2002) Control of sclerotial pathogens in horticulture. *Pesticide Outlook* June, 97–101.

CMI Descriptions of Pathogenic Fungi and Bacteria (various) Commonwealth Mycological Institute, Kew, UK. Various dates and publication numbers.

Constanantinescu, O. (1979) Revision of *Bremiella* (Peronosporales). *Transactions of the British Mycological Society* 72, 510–515.

Cook, R.T.A. (1993) Strawberry black spot caused by *Colletotrichum acutatum*. In: Ebbels, D. (ed.) *Plant Health and the European Single Market*. BCPC Monograph No. 54. British Crop Protection Council (BCPC), Farnham, UK, pp. 301–304.

Cook, R.T.A., Inman, A.J. and Billings, C. (1997) Identification and classification of powdery mildew anamorphs using light and scanning electron microscopy and host range data. *Mycological Research* 101, 975–1002.

Cooke, B.M., Jones, D. and Kaye, G.B. (2006) *The Epidemiology of Plant Diseases*, 2nd edn. Springer, Dordrecht, The Netherlands.

Cummins, G.B. and Hiratsuka, Y. (2003) *Illustrated Genera of Rust Fungi*, 3rd edn. APS Press, St Paul, Minnesota.

Damm, U., Cannon, P., Woudenberg, J.H.C. and Crous, P.W. (2012) The *Colletotrichum acutatum* species complex. *Studies in Mycology* 73, 37–113.

EFSA Panel on Plant Health (2018a) Pest categorisation of *Thecaphora solani*. *EFSA Journal* 16, 5445.

EFSA Panel on Plant Health (2018b) Scientific opinion on the pest categorisation of *Synchytrium endobioticum*. *EFSA Journal* 16, 5352.

EPPO (2020) PM 7/017 (3) *Phyllosticta citricarpa* (formerly *Guignardia citricarpa*). *OEPP/EPPO Bulletin* 50, 440–461.

EPPO (2022) *Phytophthora fragariae.* EPPO datasheets on pests recommended for regulation. Available at: https://gd.eppo.int/taxon/PHYTFR/datasheet (accessed 8 February 2023).

Erwin, D.C. (1983) *Phytophthora: Its Biology, Taxonomy, Ecology, and Pathology.* APS Press, St Paul, Minnesota.

Erwin, D.C. and Ribeiro, O.K. (1996) *Phytophthora Diseases Worldwide.* APS Press, St Paul, Minnesota.

Farr, D.F., Bills, G.F., Chamuris, G.P. and Rossman, A.Y. (1989) *Fungi on Plants and Plant Products in the United States.* APS Press, St Paul, Minnesota.

Gallegly, M.E. and Hong, C. (2008) *Phytophthora: Identifying Species by Morphology and DNA Fingerprints.* APS Press, St Paul, Minnesota.

Gäumann, E. (1923) Beiträge zu einer Monographie der Gattung *Peronospora* Corda. *Beiträge zur Kryptogamenflora der Schweiz* 5, 1–360.

Gerlach, W. and Nirenberg, H. (1982) *The Genus Fusarium – A Pictorial Atlas.* Mitteilungen aus der Biologischen Bundesanstalt für Land- und Forstwirtschaft, Berlin-Dahlem No. 209, Berlin.

Gibson, I.A.S. (1961) A note on variation between isolates of *Armillaria mellea* (Vahl ex Fr.) Kummer. *Transactions of the British Mycological Society* 44, 123–128.

Gleason, M.L., Daughtrey, M.L., Chase, A.R., Moorman, G.W. and Mueller, D.S. (2009) *Diseases of Herbaceous Perennials.* APS Press, St Paul, Minnesota.

Gordon, W.L. (1952) The occurrence of *Fusarium* species in Canada II. Prevalence and taxonomy of *Fusarium* species in cereal seed. *Canadian Journal of Botany* 30, 209–251.

Gorter, G.J.M.A. (1988) Identification of South African Erysiphaceae with a key to the species. *Phytophylactica* 20, 113–119.

Grove, W.B. (1913) *The British Rust Fungi (Uredinales).* Cambridge University Press, Cambridge, UK.

Guerber, J.C. and Correll, J.C. (2001) Characterization of *Glomerella acutata*, the teleomorph of *Colletotrichum acutatum. Mycologia* 93, 216–229.

Gunnell, P.S. and Douglas Gubler, W. (1992) Taxonomy and morphology of *Colletotrichum* species pathogenic to strawberry. *Mycologia* 84, 157–165.

HDC (2004) *Crop Walkers Guide: Strawberry.* Horticultural Development Council, East Malling, UK.

Henderson, D.M. (2000) *A Checklist of the Rust Fungi of the British Isles.* British Mycological Society, Cambridge, UK.

Henderson, D.M. (2004) *The Rust Fungi of the British Isles: A Guide to Identification by their Host Plants.* British Mycological Society, Cambridge, UK.

Henderson, D.M. and Bennell, A.P. (1979) British rust fungi: additions and corrections. *Notes from the Royal Botanic Garden Edinburgh* 37, 475–501.

Henderson, D.M. and Bennell, A.P. (1980) Supplement to British rust fungi: additions and corrections. *Notes from the Royal Botanic Garden, Edinburgh* 38, 184.

Hernandez, J.R., Cline, E., Palm, M.E., Farr, D.F. and McCray, E.B. (2011) *Rust Fungi on Fabaceae (legumes).* Systematic Mycology and Microbiology Laboratory, Agricultural Research Service, US Department of Agriculture, Beltsville, Maryland. Available at: http://nt.ars-grin.gov/taxadescriptions/keys/LegumeRustsIndex.cfm (accessed 2 September 2011).

Hirata, K. (1969) Notes on host range and geographic distribution of the powdery mildew fungi II. *Transactions of the Mycological Society of Japan* 10, 47–72.

Ho, H.H. (1981) Synoptic keys to the species of *Phytophthora. Mycologia* 73, 705–714.

Ho, H.H. (1992) Keys to the species of *Phytophthora* in Taiwan. *Plant Pathology Bulletin (Taiwan)* 1, 104–109.

Horst, R.K. (1983) *Compendium of Rose Diseases.* APS Press, St Paul, Minnesota.

Horst, R.K. and Nelson, P.E. (1997) *Compendium of Chrysanthemum Diseases.* APS Press, St Paul, Minnesota.

Hua, L., Yong, C., Zhanquan, Z., Boqiang, L., Guozhengh, Q. and Shiping, T. (2018) Pathogenic mechanisms and control strategies of *Botrytis cinerea* causing post-harvest decay in fruits and vegetables. *Food Quality and Safety* 2, 111–119.

Ialongo, M.T. (1992) Taxonomic study of some species of the genus *Erysiphe. Mycotaxon* 44, 251–256.

Ing, B. (1990a) An introduction to British powdery mildews – 1. *The Mycologist* 4, 46–48.

Ing, B. (1990b) An introduction to British powdery mildews – 2. *The Mycologist* 4, 88–90.

Ing, B. (1990c) An introduction to British powdery mildews – 3. *The Mycologist* 4, 125–128.

Ing, B. (1990d) An introduction to British powdery mildews – 4. *The Mycologist* 4, 172–177.

Ing, B. (1991) An introduction to British powdery mildews – 5. *The Mycologist* 5, 24–27.

Jones, A.L. and Aldwinckle, H.S. (1990) *Compendium of Apple and Pear Diseases.* APS Press, St Paul, Minnesota.

Laundon, G.F. (1967a) Terminology in the rust fungi. *Transactions of the British Mycological Society* 50, 189–194.

Laundon, G.F. (1967b) The taxonomy of the imperfect rusts. *Transactions of the British Mycological Society* 50, 349–353.

Leslie, J.F. and Summerell, B.A. (2006) *The Fusarium Laboratory Manual*. Blackwell Publishing, Ames, Iowa.

Maas, J.L. (1998) *Compendium of Strawberry Diseases*, 2nd edn. American Phytopathological Society, St Paul, Minnesota.

Madden, L., Hughes, G. and van den Bosch, F. (2007) *The Study of Plant Disease Epidemics*. APS Press, St Paul, Minnesota.

MAFF (1970) *Strawberries*. Ministry of Agriculture, Fisheries and Food: Bulletin 95. HMSO, London.

Martens, J.W., Seaman, W.L. and Atkinson, T.G. (1984) *Diseases of Field Crops in Canada*. Canadian Phytopathological Society, Harrow, Ontario, Canada.

Mathre, D.E. (1996) Dwarf bunt: politics, identification and biology. *Annual Review of Phytopathology* 34, 67–85.

McGechan, J.K. (1977) Black spot of strawberry. *Agricultural Gazette of New South Wales* 88, 26–27.

Mehrotra, R.S. and Aneja, K.R. (1990) *An Introduction to Mycology*. New Age International, New Delhi, India.

Melville, S.C. and Hawken, R.H. (1967) Soil testing for club root in Devon and Cornwall. *Plant Pathology* 16, 145–147.

Moss, M.O. (2008) Fungi, quality and safety issues in fresh fruits and vegetables. *Journal of Applied Microbiology* 104, 1239–1243.

Munro, J.M. (1986) Infection studies with powdery mildew from *Rhododendron* and *Erysiphe cruciferarum*. *Transactions of the British Mycological Society* 86, 686–687.

Mwangi, L.M., Lin, D. and Hubbes, M. (2007) Identification of Kenyan *Armillaria* isolates by cultural morphology, intersterility tests and analysis of isozyme profiles. *European Journal of Forest Pathology* 19, 399–406.

Nelson, P.E., Toussoun, T.A. and Marasas, W.F.O (1983) *Fusarium Species. An Illustrated Manual for Identification*. Pennsylvania State University Press, University Park, Pennsylvania.

OEPP/EPPO (2004) PM 7/29 *Tilletia indica*. *Bulletin OEPP/EPPO Bulletin* 34, 219–228.

OEPP/EPPO (2017) PM 7/28 (2) *Synchytrium endobioticum*. *Bulletin OEPP/EPPO Bulletin* 47, 420–440.

Pearson, R.C. and Goheen, A. (1988) *Compendium of Grape Diseases*. APS Press, St Paul, Minnesota.

Pegler, D.N. (2000) Taxonomy, nomenclature and description of *Armillaria*. In: Fox, R.T.V. (ed.) *Armillaria Root Rot: Biology and Control of Honey Fungus*. Intercept, Andover, UK, pp. 81–93.

Pérez-Sierra, A. (2003) Systematics, diagnostics and epidemiology of the fungal genus *Armillaria*. PhD thesis, University of London, London.

Peterson, R.H. (1974) The rust fungus life cycle. *The Botanical Review* 40, 453–513.

Philips, R. (1981) *Mushrooms and Other Fungi of Great Britain and Europe*. Pan Books, London.

Piepenbring, M., Bauer, R. and Oberwinkler, F. (1998) Teliospores of smut fungi: teliospore connections, appendages, and germ pores studied by electron microscopy; phylogenetic discussion of characteristics of teliospores. *Protoplasma* 204, 202–218.

Pirone, P.P., Dodge, B.O. and Rickett, H.W. (1960) *Diseases and Pests of Ornamental Plants*, 3rd edn. The Ronald Press Company, New York.

Rishbeth, J. (1986) Some characteristics of English *Armillaria* species in culture. *Transactions of the British Mycological Society* 86, 213–218.

Ruzin, S. E. (1999) *Plant Microtechnique and Microscopy*. Oxford University Press, Oxford.

Sato, T. and Sato, S. (1982) Aeciospore surface structure of the Uredinales. *Transactions of the Mycological Society of Japan* 23, 51–63.

Schumann, G. and D'Arcy, C. (2010) *Essential Plant Pathology*. APS Press, St Paul, Minnesota.

Schwartz, H.F. and Mohan, S.K. (1996) *Compendium of Onion and Garlic Diseases*. APS Press, St Paul, Minnesota.

Scott, K.J. and Chakravorty, A.K. (1982) *The Rust Fungi*. Academic Press, London.

Shaw, D.S. (1988) The *Phytophthora* species. *Advances in Plant Pathology* 6, 27–51.

Sherf, A.F. and MacNab, A.A. (1986) *Vegetable Diseases and Their Control*, 2nd edn. John Wiley and Sons, New York.

Shin, H.D. and Choi, Y.J. (2006) *Peronosporaceae of Korea*. Plant Pathogens of Korea 12. National Institute of Agricultural Science and Technology, Suwon, Korea.

Simmonds, J.H. (1965) A study of species of *Colletotrichum* causing ripe fruit rots in Queensland. *Queensland Journal of Agricultural and Animal Sciences* 22, 437–459.

Skidmore, D.I. and Ingram, D.S. (1985) Conidial morphology and the specialization of *Bremia lactucae* Regel (Peronosporaceae) on hosts in the family Compositae. *Botanical Journal of the Linnean Society* 91, 503–522.

Smiley, R.W. (1983) *Compendium of Turfgrass Diseases.* American Phytopathological Society, St Paul, Minnesota.

Smith, B.J. and Black, L.L. (1990) Morphological, cultural, and pathogenic variation among *Colletotrichum* species isolated from strawberry. *Plant Disease* 74, 69–76.

Snowdon, A.L. (1990a) *A Colour Atlas of Post-harvest Diseases and Disorders of Fruits and Vegetables.* Volume 1: General Introduction and Fruits. Wolfe Scientific, London.

Snowdon, A.L. (1990b) *A Colour Atlas of Post-harvest Diseases and Disorders of Fruits and Vegetables.* Volume 2: Vegetables. Wolfe Scientific, London.

Spencer, D.M. (ed.) (1981) *The Downy Mildews.* Academic Press, London.

Stamps, D.J., Waterhouse, G.M., Newhook, F.J. and Hall, G.S. (1990) *Revised Tabular Key to the Species of Phytophthora.* Mycological Papers No. 162. CAB International Mycological Institute, Kew, UK.

Summerell, B.A. and Nelson, P.E. (2003) *Fusarium: Paul E. Nelson Memorial Symposium.* APS Press, St Paul, Minnesota.

Sutton, B.C. (1980) *The Coelomycetes. Fungi Imperfecti with Pycnidia, Acervuli and Stromata.* Commonwealth Mycological Institute, Kew, UK.

Sutton, B.C. (1992) The genus *Glomerella* and its anamorph *Colletotrichum.* In: Bailey, J.A. and Jeger, M.J. (eds) *Colletotrichum: Biology, Pathology and Control.* CAB International, Wallingford, UK, pp. 1–26.

Termorshuizen, A.J. and Swertz, C.A. (2011) *Dutch Rust Fungi/Roesten van Nederland.* AJ Termorshuizen (privately published).

Toussoun, T.A. and Nelson, P.E. (1968) *A Pictorial Guide to the Identification of Fusarium Species According to the Taxonomic System of Snyder and Hansen.* Pennsylvania State University Press, University Park, Pennsylvania.

UNECE (2014) *Guide to Seed Potato Diseases, Pests and Defects.* UNECE, Geneva, Switzerland.

Vargas, J.M. Jr (1994) *Management of Turfgrass Diseases.* Lewis Publishers, Boca Raton, Florida.

Wale, S., Platt, H.W. and Cattlin, N. (2008) *Diseases, Pests and Disorders or Potatoes. A Colour Handbook.* Manson Publishing, London.

Walker, J.C. (1969) *Plant Pathology*, 3rd edn. McGraw-Hill, New York.

Waller, J.M. (1992) *Colletotrichum* diseases of perennial and other cash crops. In: Bailey, J.A. and Jeger, M.J. (eds) *Colletotrichum: Biology, Pathology and Control.* CAB International, Wallingford, UK, pp. 131–142.

Waterhouse, G.M. (1963) *Key to the Species of Phytophthora de Bary.* Commonwealth Mycological Institute, Kew, UK.

Waterhouse, G.M. and Brothers, M.P. (1981) *The Taxonomy of Pseudoperonospora.* Mycological Papers No. 148, Commonwealth Mycological Institute, Kew, UK.

Waterhouse, G.M. and Waterston, J.M. (1966) *CMI Descriptions: 111. Phytophthora cactorum.* CMI Descriptions of Pathogenic Fungi and Bacteria. CABI, Wallingford, UK.

Weir, B.S., Johnston, P.R. and Damm, U. (2012) The *Colletotrichum gloeosporioides* species complex. *Studies in Mycology.* 73, 115–180.

Wiese, M.V. (1987) *Compendium of Wheat Diseases.* APS Press, St Paul, Minnesota.

Wilson, G.W. (1914) Studies in North American Peronosporales – VI. Notes on miscellaneous species. *Mycologia* 6, 192–210.

Wilson, M. and Henderson, D.M. (1966) *British Rust Fungi.* Cambridge University Press, Cambridge, UK.

Winston, J. (1999) *Describing Species.* Columbia University Press, New York.

Wormald, H. (1954) *The Brown Rot Disease of Fruit Trees.* Technical Bulletin No. 3. Ministry of Agriculture Fisheries and Food, London.

Wormald, H. (1955) *Diseases of Fruits and Hops.* Crosby Lockwood & Son Ltd., London.

Yerkes, W.D. and Shaw, C.G. (1959) Taxonomy of the *Peronospora* species on Cruciferae and Chenopodiaceae. *Phytopathology* 49, 499–507.

Zaracovitis, C. (1965) Attempts to identify powdery mildew fungi by conidial characters. *Transactions of the British Mycological Society* 48, 553–558.

7 Cultural Characterization

Charles R. Lane*
Fera Science Ltd, Sand Hutton, York, UK

7.1 Introduction

A favourable environment is one of the three key components for disease initiation and development and, ultimately, of the level of damage caused by any fungal plant pathogen. Deviation from optimal environmental conditions will influence both plant and fungal physiology and these, in turn, will affect disease severity. Environmental changes may alter nutrition, metabolism, growth and reproduction, and lead to the death of fungal cells; the changes may also manifest themselves in the appearance and activity of the organism and, from the plant's perspective, in the expression of certain genes involved in disease resistance or susceptibility.

Manipulation of the physical and chemical environment can be used to help differentiate fungal taxa for identification and to study how they might perform in natural environments; the knowledge so gained could be exploited for disease management. For example, Martín-García et al. (2015) investigated the influence of temperature on germination of *Quercus ilex* in soil infested with *Phytophthora cinnamomi, P. gonapodyides, P. psychrophila* and *P. quercina* to predict the impacts of climate change. Of these

environmental factors, temperature, water availability, relative humidity (RH), pH and light are the most widely studied physical factors, although others – such as aeration and pressure – may also be investigated. Chemical factors such as nutrients and fungicides may also be used to influence fungal physiology. Environmental changes may influence mycelial growth, number, type and longevity of spores formed, germination and infection; these effects may then affect the infection cycle and disease development. The simplest and most common way to monitor these effects is to study the radial growth rate (Protocol 7.1), biomass production (Protocol 7.2), propagule production (Protocol 7.3) and propagule germination (Protocol 7.4) characteristics of fungal pathogens. Quantification is still predominantly based on manual assessment of these morphometric measurements; however, more automated ways have been developed. For example, De Ligne *et al.* (2019) used image analysis to assess the impact of environmental conditions on the growth dynamics of two fungal plant pathogens. As imaging and processing become cheaper, then more automated techniques may become increasingly common.

* E-mail: charles.lane@fera.co.uk

DOI: 10.1079/9781800620575.0007

7.2   Effect of Physical Factors on Fungal Physiology

Temperature

Most fungal plant pathogens prefer warm temperate conditions and can be described as mesophiles (which grow at temperature ranges of 5–35°C, and with optima in the range of 20–30°C), but there are examples of fungi that are active in more extreme or stressful environments. These may be described as psychrotolerant (cold tolerant – maximum growth temperatures above 20°C, although they can grow at 10°C) or psychrophilic (cold loving – optimum growth temperatures that range up to 16°C, with maximum growth temperatures of about 20°C). However, there are a number of important plant pathogenic fungi of cereals and turf grasses – referred to as the 'snow moulds' – that thrive only in cool seasons or cold regions. These fungi are able to grow at the surface of frozen soil beneath snow cover that maintains temperatures at the soil surface of just above freezing, even when air temperatures can be at −20°C. Common diseases in this category include pink snow mould (caused by *Microdochium nivale*), speckled snow mould (*Typhula* species) and Sclerotinia snow mould (*Sclerotinia borealis*) (Wiese, 1977). Fungi may also be described as thermotolerant (heat tolerant – fungi that can grow when temperatures are below 20°C but can also grow at 40–50°C) or thermophilic (heat loving – fungi that can grow at temperatures ranging from above 20°C to above 50°C); these are fungi such as those used to produce compost which are able to exploit these more extreme ecological niches (Cooney and Emmerson, 1964).

Temperature may be used to help identify taxa based on differences in optimal radial growth rate (e.g. to separate pathogenic from saprophytic species of Trichoderma green mould in mushroom production; Samuels *et al.*, 2002), to study sporulation (e.g. in disease forecasting for potato late blight, caused by *Phytophthora infestans*; Hartill *et al.*, 1990) and determine thermal death points for disease management (e.g. hot water treatment of flower bulbs to control basal rots caused by fungal pathogens such as *Fusarium oxysporum*).

Water

The availability (not the amount) of water in the environment needs to be understood to study fungal growth and sporulation. For example, although ice is entirely made up of water, it is unavailable to most organisms (Zak and Wildman, 2004). Water availability is often measured as water potential, which is a summation of osmotic, matric and gravitational potentials, as described in the following equation:

$$\text{Water potential} = 8.314\, T \ln a_w / V_w \quad \text{(Eqn 7.1)}$$

where water potential is measured in MPa, T = temperature (°K), a_w = water activity (the ratio of the vapour pressure of water in a material to the vapour pressure of pure water at the same temperature) and V_w = partial molar volume of water.

Most fungi grow best at a water potential in the range of 0 to −1 MPa, but some fungi are able to grow at extremes of low water potential (<−30 MPa). The term water potential is commonly used in plant pathology, while in food spoilage studies the term water activity is used. Osmotic potential describes the influence of salts or sugars on water availability – fungi may be described as osmophilic or osmotolerant, depending on their adaptation to environments such as salt marshes or preserved foods such as salted meats or jams. In soils and growing media, water availability is primarily due to matric potential effects – fungi may be described as xerotolerant or xerophilic and found in such environments as semi-arid to arid ecosystems or dried foodstuffs. Manipulation of water availability is used in the preservation of foodstuffs (by the addition of salt or sugars) or in disease control, such as the practice of adding sodium chloride to the surfaces of mushroom beds – referred to as 'salting' – to control fungal diseases.

Relative humidity

The water content of the atmosphere is expressed in terms of RH, which is derived from the same ratio as water activity but is expressed as a percentage (e.g. a_w 0.9 = 90% RH). Moisture affects sporulation, longevity and particularly germination, and these can influence the number of infection cycles and disease severity. Most fungal pathogens of the aerial parts of plants require a film of water to infect hosts, e.g. the downy mildews. However, others such as the powdery mildews have overcome this requirement by producing relatively large spores with a high water content which are able to initiate infection on dry surfaces as long as there is high RH in the vicinity of the plant surface. Reduction of RH is commonly used in crop management to reduce fungal diseases by improving airflow – via means such as increasing ventilation by opening glasshouse vents, the use of circulation fans, increased plant spacing or canopy manipulation. A good example of this is the removal of foliage in vines to reduce infection by *Botrytis cinerea* – the cause of grey mould.

pH

Most fungi are acidophilic and grow well between pH 4 and pH 6, but many species can grow, although to a lesser extent, between pH 3 and pH 8 (Bachofen, 1986). The pH optima of most fungi lies within the pH range of most environments, typically between pH 4 and pH 9, although some environments, such as fruit juices or acidic soils, may be more extreme. For example, clubroot of crucifers caused by *Plasmodiophora brassicae* is most severe at about pH 5.7, but its development is arrested by increasing the pH to above 6.0; there is good disease control at pH 7.2 and the disease completely ceases at pH 7.8 and above (Alford, 2000). The addition of lime to raise the pH of soil used to grow cruciferous crops is a common agricultural practice. Soil pH may also influence nutrient availability for plants and may weaken them and make them more susceptible to diseases. Acidophilia may be used practically in the laboratory to develop isolation media such as cherry decoction agar to favour fungal as opposed to bacterial growth.

Light

This is frequently used to manipulate sporulation and inoculum production but can also affect secondary metabolite production to aid identification. Spectral quality and radiation

levels may vary considerably between different light sources, so care must be taken in choosing incubation conditions. The most used light sources are fluorescent tubes, which are preferred to tungsten bulbs as the latter may emit a considerable amount of heat. The two most common types of fluorescent tubes used are white 'daylight' tubes (visual spectrum 380–750 nm; sold as cool white, standard white, warm white, etc., which all vary slightly) and black light tubes (near UV; range 320–420 nm; maximum emission 360 nm) of varying wattages (20, 40, 80 W). Care must be taken to select the correct UV tubes as 'germicidal' tubes (254 nm) will have a deleterious effect on fungi. The performance of light sources may deteriorate with age, so they should be changed after several years, depending on their usage. There are numerous instances of the use of diurnal illumination (a combination of alternating black and white light) to manipulate spore production; for example, three different species of *Monilinia* found on stone and pome fruits may be separated based on the diurnal production of conidia under alternating black and white light (Lane, 2002). The influence of light has also been used commercially to manage fungal diseases in protected horticulture. The use of special UV-absorbing vinyl film that blocks the transmission of certain wavelengths of light has been used to control grey mould (*B. cinerea*) (West *et al.*, 2000). Pigment production in certain genera such as *Fusarium* and *Phoma* is a well-established method for assisting identification (Leslie and Summerell, 2006).

The determination of fungal growth characteristics in media of varying pH, water potential and relative humidity is described in Protocol 7.5.

7.3 Effects of Chemical Factors on Fungal Physiology

Fungal physiology may also be affected by the chemical environment that the pathogen inhabits. This can be artificially manipulated to get a better understanding of how the fungus concerned may interact with potential hosts and perform in the natural environment, and how to develop strategies that may help to control fungal pathogens. The ability to degrade various carbon or protein sources such as cellulose or lignin has been used to develop an understanding of the potential ecological niche that an organism may exploit, but knowledge of these characteristics has limited value for classification as they may change with different phases of growth (Pugh and Boddy, 1988). Different forms of the same enzyme (such as esterases and pectinases), which are referred to as isoenzymes or isozymes, have also been used to help differentiate taxa (Paterson and Bridge, 1994). However, these techniques will not be covered here as they have become less commonly used with the development of more sophisticated molecular assays for fungal identification.

The effect of chemicals on fungal plant pathogens is still widely studied, both in looking for new active ingredients and in monitoring the efficacy of existing products for disease management. The types of chemicals used in plant disease control are extensive, ranging from inorganic chemicals such as copper or sulfur compounds, to organic chemicals such as triazoles and strobilurins, and to more novel substances such as mineral oils and bicarbonates. Most fungicides are directly toxic to pathogens and act by inhibiting the synthesis of fungal cell wall components, damaging cell wall membranes, inactivating enzymes and coenzymes, or interfering with respiration. Inevitably, strains have developed that have overcome these fungicides, so the ability to monitor the sensitivity of fungal pathogens to them is a key part of long-term effective use of these pesticides (Protocol 7.6).

Protocol 7.1

Determination of radial growth rate

Method

1. Dispense sterile media (15–20 ml) into sterile Petri dishes and allow to cool.

2. Sterilize a cork borer (5–10 mm diameter) by flaming in alcohol and allow to cool; cut sufficient agar plugs from the colony margin of a culture of the pathogen under study, in addition to a few spare plugs in case they break up when removed.

3. Transfer agar plugs aseptically mycelium-side down using a sterile scalpel to the centre of each Petri dish. Consider three to five replicates per culture or treatment.

4. Label dishes with culture reference number, treatment and date, and draw x, y perpendicular axes on the bottom of the dish (see Fig. 7.1).

5. Incubate under appropriate conditions.

6. Record radial growth rate on axes at relevant time intervals based on expected growth rate of the fungus. If growth rate is unknown, monitor daily to begin with.

7. Calculate mean growth rates.

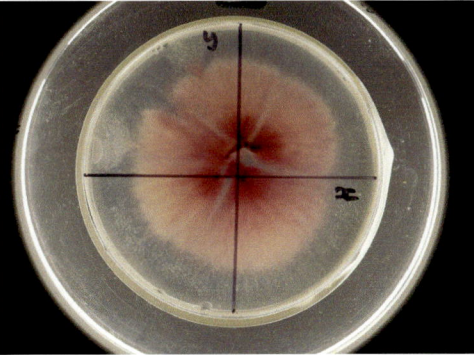

Fig. 7.1. Petri dish with x, y axes drawn on bottom of plate. (© UK Crown copyright – courtesy of Fera.)

Protocol 7.2

Determination of biomass production

> **Materials**
> - Erhlenmeyer flasks
> - Sterile media
> - Drying oven
> - Desiccator
> - Fine balance (to three decimal places)
> - Filter papers
> - Buchner funnel filtration system or similar

Method

1. Dispense liquid media into sterile Erhlenmeyer flasks (250 ml or similar) and allow to cool.

2. Sterilize a cork borer (5–10 mm diameter) by flaming in alcohol and allow to cool; cut sufficient agar plugs from the colony margin of a culture of the pathogen under study, in addition to a few spare plugs in case they break up when removed.

3. Transfer agar plugs with mycelium aseptically into flasks using a sterile scalpel. Consider three to five replicates per culture or treatment.

4. Label dishes with culture reference number, treatment, date.

5. Flasks may either be shaken (e.g. on a continuously rocking platform) or not shaken.

6. Incubate under appropriate conditions.

7. Label appropriately sized filter papers with a unique reference number using a soft pencil and place in a drying oven (60°C) for at least 2 days. Weigh.

8. Place oven-dried filter papers in a desiccator and allow to cool for at least 2 h.

9. Reweigh filter papers and record weight.

10. Pour contents of flask through Buchner funnel filtration system or similar using labelled filter papers to remove fungal growth.

11. Place filter paper in drying oven and repeat steps 7–9.

12. Calculate mean dry weight.

Protocol 7.3

Determination of propagule production

<table>
<tr><td>

Materials

- Cork borer
- Scalpel
- Vortex shaker
- Haemocytometer or eyepiece counting grid
- Compound microscope

</td></tr>
</table>

Method

1. Repeat steps 1–5 as described in Protocol 7.1, except that there is no need to label the x, y axes.

2. Aseptically remove a piece of agar of known size from the culture using a sterile cork borer or scalpel. The position that the piece of agar is removed from should be consistent, e.g. from the margin of the colony or at a midway point from the margin to the centre point; it is best not to include any of the original inoculum plug.

3. Place the agar plug in 1 ml sterile distilled water.

4. Shake vigorously (e.g. using a Vortex shaker) to produce a suspension and, using a haemocytometer or a compound microscope fitted with a counting grid in the eyepiece, determine the number of propagules present and calculate the propagules ml^{-1}.

Protocol 7.4

Assessment of propagule germination

Materials
- Muslin cloth
- Haemocytometer
- Compound microscope
- Inoculating loop
- Tap water agar plates
- Incubator

Method

1. Prepare a propagule suspension in sterile distilled water (as described in Protocol 7.3) and filter through sterile muslin to remove any agar.

2. Using a haemocytometer, adjust the propagule concentration to approximately 1×10^5 ml^{-1}.

3. Streak out propagule suspension with a sterile inoculating loop onto tap water agar (or similar low-nutrient agar).

4. Incubate for 24 h or longer, as necessary, under suitable conditions and observe propagules under a compound microscope.

5. Any propagule with a germ tuber longer than its width should be considered as germinated (you may also wish to record length, or branching pattern or presence of secondary spore production, etc.).

6. Results are commonly expressed as a percentage germination rate.

Protocol 7.5

Determination of growth characteristics over ranges of pH, water potential and relative humidity

Materials

- Filter sterilization equipment
- Laminar flow hood
- Sterile media
- pH meter
- Pipettes
- NaOH, HCl, NaCl, KCl, glycerol, polyethylene glycol (PEG 6000)
- Petri dishes, flasks or tightly sealed bottles
- Coverslips or fine nylon mesh

Methods

pH

1. Filter-sterilize (0.2 μm filter) suitable quantities of 1 mol l⁻¹ NaOH and HCl.
2. Dispense liquid media and determine pH of a representative sample.
3. Amend media with filter-sterilized NaOH/HCl to desired pH and record volume of acid/base added.

Discard this media.

4. Aseptically amend remaining media with known volumes of acid/base based on step 3.
5. Inoculate amended media as required and record biomass production (see Protocol 7.2).

Osmotic potential

1. Determine the osmotic potential of basal medium by psychrometry (reference) at desired temperature (e.g. corn meal agar at 25°C and osmotic potential −0.48 MPa; V-8 juice at 25°C and osmotic potential −0.22 MPa).
2. Aseptically amend with NaCl as described by Lang (1967) and shown in Table 7.1.
3. Inoculate amended media as required and record radial growth rate for biomass production (see Protocol 7.2) as necessary.

Matric potential

1. Aseptically amend media with appropriate quantity of PEG 6000 according to the equation described by Michel and Kaufmann (1973) (see Eqn 7.2 below) and as shown in Table 7.2.
2. Ensure medium is cool before pouring, mix well and seal dishes once poured.
3. Solid agar may not solidify below −1.5 MPa, so an absorbent 'float' may be required (Baudoin and Davis, 1987).
4. Inoculate amended media as required and record biomass production as necessary.

$$\text{Matric potential} = -\left(1.18\times10^{-2}\right)C - \left(1.18\times10^{-4}\right)C^2 + \left(2.67\times10^{-4}\right)CT + \quad \text{(Eqn 7.2)}$$
$$\left(8.39\times10^{-7}\right)C^2T$$

where matric potential is in MPa, T = temperature (°C) and C = concentration of PEG 6000 (g kg⁻¹ H₂O).

Table 7.1. Osmotic potential of V-8 broth and corn meal agar amended with NaCl.

Osmotic potential (−MPa)	V-8 broth amended with NaCl (g)	Osmotic potential (−MPa)	Corn meal agar amended with NaCl (g)
0.2	0	0.5	0
1.0	10.4	0.9	11.6
2.0	24.4	1.8	23.6
3.0	36.4	2.8	29.2
3.9	47.2	3.7	41.2
4.9	58.4	5.1	58.8
5.8	70.0	6.1	70.4

Relative humidity (RH)

1. Mix glycerol and water (Johnson, 1940) to obtain desired RH range (see Table 7.3).
2. Autoclave solutions and once cooled dispense into tightly sealed containers (Fig. 7.2).
3. Place an aliquot of spore suspension onto a suitable sterile platform, e.g. glass coverslip or sterile nylon mesh.

4. Place in laminar flow cabinet at room temperature to evaporate any water.
5. Suspend sterile platform in atmosphere generated in step 2 and incubate as necessary.
6. Remove sterile platform and place propagule side down on agar to determine germination as described in Protocol 7.4.

Table 7.2. Amount of polyethylene glycol (PEG 6000) added to corn meal agar to give 1000 g of amended media at different matric potentials at 25°C.

Matric potential (–MPa)	PEG 6000 (ml)
0.5	0
1.0	297
1.5	368
2.0	425
2.5	475
3.0	525
4.0	610
5.0	685

Table 7.3. Volume of glycerol added to distilled water to make 1000 ml solution at 25°C.

Relative humidity (%)	Glycerol (ml)
100	0
97	155.6
93	250
85	350
72	500
62	600
44	750
38	830
17	920
0	1000

Fig. 7.2. Incubation chamber for relative humidity studies. Mesh square, platform, screw-top bottle and tightly securing lid (from left to right). (© C.R. Lane.)

Protocol 7.6

Fungicide sensitivity testing

Materials
- Sterile Petri dishes/flasks
- Cork borer
- Incubator
- Laminar flow hood
- Pipettes
- Media
- Fungicides
- Ethanol
- Distilled water
- Sterile microtitre plates
- ELISA plate reader

Method

Poison plate method

1. Dissolve fungicide in sterile distilled water or other suitable solvent such as ethanol to obtain desired concentration.

2. Sterilize media and cool to near setting point ($c.50°C$) for solid media.

3. Mix fungicide and cooled media thoroughly and then dispense as required.

4. Inoculate media aseptically and incubate at optimal cultural conditions.

5. Assess growth rate, propagule production and propagule germination as described in Protocols 7.1, 7.3 and 7.4, as appropriate.

Assessment of fungal growth based on light absorbance

1. Choose a suitable liquid medium for fungal growth that is compatible with the fungicide to be tested (e.g. Czapek Dox, glucose peptone broth) and sterilize.

2. Make up fungicide solutions at 20 times concentration in ethanol.

3. Prepare a sterile spore suspension ($c.5 \times 10^5$ spores ml^{-1}) using the liquid medium.

4. Place 10 µl of the fungicide/ethanol solution into the bottom of the wells of a sterile microtitre plate.

5. Allow the ethanol to evaporate in an aseptic environment.

6. Add 200 µl of the nutrient media/spore suspension to each well.

7. Include negative controls of fungicide-amended media but without any inoculum.

8. Incubate at optimal conditions for approximately 6 days.

9. Record the absorbance of each well using an ELISA plate reader at 405 nm.

10. Compare the absorbance of inoculated and uninoculated wells.

References

Alford, D.V. (2000) *Pest and Disease Management Handbook*. Blackwell Science, Oxford.

Bachofen, R. (1986) Microorganisms in extreme environments. Introduction. *Experientia* 42, 1179–1182.

Baudoin, A.B.A.M. and Davis, L.L. (1987) Effects of osmotic potential and matric potential on radial growth of *Geotrichum candidum*. *Transactions of the British Mycological Society* 88, 323–328.

Cooney, D.G. and Emmerson, R. (1964) *Thermophilic Fungi. An Account of their Biology, Activities and Classification*. W.H. Freeman, San Francisco, California.

De Ligne, L., Vidal-Diez de Ulzurrun, G., Baetens, J.M., Van den Bulcke, J., Van Acker, J. and De Baets, B. (2019) Analysis of spatio-temporal fungal growth dynamics under different environmental conditions. *IMA Fungus* 10, 7.

Hartill, W.F.T., Young, K., Allan, D.J. and Henshall, W.R. (1990) Effects of temperature and leaf wetness on the potato late blight. *New Zealand Journal of Crop Horticultural Science* 18, 181–184.

Johnson, C.G. (1940) The maintenance of high atmospheric humidities for entomological work with glycerol–water mixtures. *Annual Review of Biology* 27, 295–299.

Lane, C.R. (2002) A synoptic key for differentiation of *Monilinia fructicola*, *M. fructigena* and *M. laxa*, based on examination of cultural characters. *Bulletin OEPP/EPPO Bulletin* 32, 489–493.

Lang, A.R.G. (1967) Osmotic coefficients and water potentials of sodium chloride solutions from 0 to 40°C. *Australian Journal of Chemistry* 22, 2017–2033.

Leslie, J.F. and Summerell, I. (2006) *The Fusarium Laboratory Manual*. Blackwell Publishing, Ames, Iowa.

Martín-García, J., Solla, A., Corcobado, T., Siasou, E. and Woodward, S. (2015) Influence of temperature on germination of *Quercus ilex* in *Phytophthora cinnamomi*, *P. gonapodyides*, *P. quercina* and *P. psychrophila* infested soils. *Forest Pathology* 45, 215–223.

Michel, B.E. and Kaufmann, M.R. (1973) The osmotic potential of polyethylene glycol 6000. *Plant Physiology* 51, 914–916.

Paterson, R.R.M. and Bridge, P.D. (1994) *Biochemical Techniques for Filamentous Fungi*. IMI Technical Series Volume 1. CAB International, Wallingford, UK.

Pugh, G.J.F. and Boddy, L. (1988) A view of disturbance and life strategies in fungi. *Proceedings of the Royal Society of Edinburgh* 94B, 3–11.

Samuels, G.J., Dodd, S.L., Gams, W., Castlebury, L.A. and Petrini, O. (2002) *Trichoderma* species associated with the green mold epidemic of commercially grown *Agaricus bisporus*. *Mycologia* 94, 146–170.

West, J.S., Pearson, S., Hadley, P., Wheldon, A.E., Davis, F.J., Gilbert, A. *et al*. (2000) Spectral filters for the control of *Botrytis cinerea*. *Annals of Applied Biology* 136, 115–120.

Wiese, M.V. (1977) *Compendium of Wheat Diseases*. APS Press, St Paul, Minnesota.

Zak, J.C. and Wildman, H.G. (2004) Fungi in stressful environments. In: Mueller, G.M., Bills, G.F. and Foster, M.S. (eds) *Biodiversity of Fungi: Inventory and Monitoring Methods*. Elsevier, Oxford, pp. 303–315.

8 Serological Techniques for Diagnosis

Christopher Thornton[1] and Charles R. Lane[2]*
[1]*Biosciences, Hatherly Building, University of Exeter, Prince of Wales Road, Exeter, UK;* [2]*Fera Science Ltd, Sand Hutton, York, UK*

8.1 Introduction

Hybridoma technology allows the production of monoclonal antibodies (mAbs) to fungi that are genus-specific (Thornton *et al.*, 2002) and species-specific (Thornton *et al.*, 1993). These can be used to develop highly specific and sensitive tests for the detection of plant pathogens in soil and *in planta* (Thornton *et al.*, 1993, 2002, 2004; Thornton and Talbot, 2006; Thornton, 2008a, 2009a), and for the diagnosis of human mycoses (Thornton, 2008b, 2009b, 2020). A number of species of fungi that were previously regarded as benign soil saprotrophs, plant pathogens or post-harvest storage pathogens (Thornton and Wills, 2015), such as *Aspergillus* spp., *Fusarium* spp., *Rhizopus oryzae* and *Scedosporium* spp., have emerged over recent years as the cause of life-threatening infections of immuno-compromised humans or of wild animals (Thornton, 2008b, 2009b, 2020; Al-Maqtoofi and Thornton, 2016; Dillon *et al.*, 2017; Davies and Thornton, 2022). This has led to the adaptation and exploitation of techniques previously developed for the detection of plant pathogens to the diagnosis of human and animal mycoses (Thornton, 2008b, 2009b, 2020; Al-Maqtoofi

and Thornton, 2016; Dillon *et al.*, 2017; Davies and Thornton, 2022). The protocols outlined in this chapter are, therefore, broadly applicable to both plant pathology and medical mycology.

In the agricultural and food hygiene sectors, antibody-based diagnostics have largely been superseded by the introduction of nucleic acid-based detection systems, which promise greater specificity and sensitivity. However, these systems do not always live up to expectations and are not necessarily the best solution. Furthermore, these techniques are mainly restricted to diagnostic laboratories equipped to perform such tests, as they require specialist equipment and expertise. While conventional enzyme immunoassays such as ELISA are similarly restrained by laboratory facilities, the development of lateral-flow devices (LFD), employing similar technology to that used in pregnancy test kits or COVID-19 home-testing kits, has revolutionized the use of antibody-based diagnostics, allowing translation of laboratory immunoassays into the field. The ease-of-use of lateral-flow devices, and the rapidity with which they provide results compared with conventional serological tests, has led to sustained interest in antibody-based diagnostics. Yet, the development of ELISA and

* E-mail: charles.lane@fera.co.uk

©CAB International 2023. *Fungal Plant Pathogens: Applied Techniques, 2nd Edition*
(eds C.R. Lane, P.A. Beales and K.J.D. Hughes)
DOI: 10.1079/9781800620575.0008

in-field or Point of Care (POC) test kits is reliant on the availability of suitable antibodies. The ethical concerns over antibody production in animals, timescales and cost of production (for a small economic market) have limited serological diagnostics for fungal plant pathogens. However, where antibodies exist, they still offer a very simple, robust way of in-field or POC surveillance and diagnosis. The mass market adoption and familiarity with COVID-19 LFD home test kits may well lead to a new interest in serological methods. The production of recombinant antibodies, that are generated *in vitro* using, for example, phage display technology, offers an alternative system of production with potentially increased reproducibility and control, decreased production time and animal-free systems. To date, the on-site detection of plant pathogens has been successfully deployed in the areas of plant quarantine, biological control and food production (Danks and Barker, 2000; Lane, 2006; Thornton, 2008a).

This chapter describes qualitative and quantitative immunoassays for the detection and visualization of fungi, with the supposition that mouse mAbs are available for the species of interest. For literature that describes the preparation of immunogens for mouse immunization, the generation of species-specific mAbs and the characterization of MAbs and their antigenic determinants, the reader is referred elsewhere (Thornton, 2001, 2009a).

8.2 Antibodies

Fungi consist of highly conserved antigens (invariably immuno-dominant glycoproteins) on their cell surfaces which preclude the use of cross-reactive rabbit or goat antisera for their detection. Immunization of rabbits or goats with fungal antigens elicits a humoral immune response that results in the production of antibody from multiple plasma B cell clones. Antiserum from these animals therefore consists of polyclonal antibodies targeted against specific and against conserved (non-specific) antigens or epitopes. Hybridoma technology allows the identification and isolation of individual plasma B cell clones secreting antibodies of predefined specificities. Immortalization of the specific B cells by fusion with myeloma cells (cancerous B cells)

allows the production of hybrid cells (hybridomas) that retain the characteristics both of the B cell (antibody production) and of the myeloma (indefinite growth *in vitro*). Using tissue culture, limitless supplies of highly specific mAbs can be produced *in vitro* for use in fungal-specific diagnostic assays. The assays described in this chapter relate to the use of mAbs that are generated as hybridoma tissue culture supernatants. They can be adapted for use with polyclonal antisera, and for this usage, the reader is guided to protocols described elsewhere (Thornton, 2001).

8.3 Enzyme-linked Immunosorbent Assay (ELISA)

For the purposes of this chapter, the enzyme-linked immunosorbent assay (ELISA) can be defined as a method for detecting fungal antigens that utilizes antibody-linked enzyme-substrate reactions. The enzymes are usually conjugated to antibodies (either to antibodies specific for the fungus – a direct ELISA, or to anti-immunoglobulin – an indirect ELISA). The amount of enzyme conjugate is determined from the turnover of an appropriate substrate. Enzymes that have substrates which yield easily detected coloured products are used, such as alkaline phosphatase (AP) and its substrate *p*-nitrophenolphosphate (PNP), which is converted to the yellow product *p*-nitrophenol. The amount of coloured product is quantified by determining the absorbance of light of an appropriate wavelength: in this case, light at 405 nm. For systems that use the enzyme horseradish peroxidase (HRP) and the substrate tetramethylbenzidine (TMB), the absorbance is measured at 450 nm. Absorbance of light by the product is directly related to the concentration of the specific antibody bound to antigen immobilized in the solid phase. This means that antigen concentrations can be quantified using standard curves that relate known amounts of antigen to absorbance (Thornton *et al.*, 2002). It also allows threshold values or cut-off points of absorbance to be established above which an unknown sample is deemed positive for the target fungus (Thornton, 2008b, 2009b).

Choice of the enzyme–substrate pair is an important consideration when devising ELISA tests for fungi. For the detection of fungi in soil extracts, HRP is the most suitable antibody reporter system. However, owing to the presence of endogenous plant peroxidases, HRP cannot be used for the *in planta* detection of fungi; here, it must be replaced with AP reporter systems. While AP has found widespread use in enzyme immunoassays, the kinetics of the enzyme–substrate reaction means that at low antigen concentrations the production of the colorimetric signal can take many hours to develop, thereby delaying the assay result. However, by linking AP to a commercially available assay amplification procedure that employs a unique enzyme cycling system (diaphorase and alcohol dehydrogenase), the signal from the primary enzyme AP is multiplied many times, and amplification factors of 100-fold can be achieved. The length of the colorimetric reaction is also reduced from several hours to 30 min, making it comparable to HRP-based immunoassays. The procedure outlined in Protocol 8.1 is a standard indirect ELISA method that has found widespread use in the detection and quantification of fungi in soil and *in planta*, and in monitoring levels of airborne fungal inocula (Kennedy *et al.*, 2000; Wakeham *et al.*, 2008).

8.4 Immunofluorescence (IF)

Immunofluorescence is a qualitative immunoassay that can be used for visualizing fungal propagules in which specific antibodies tagged with a fluorochrome are used to detect antigens that are either displayed on the surface of cells or secreted into extracellular matrices. Fluorochromes commonly used as reporter molecules are fluorescein isothiocyanate (FITC) and Texas Red (TR). These fluoresce under different wavelengths of ultraviolet light, so an epifluorescence microscope fitted with a range of excitation and emission filters is required. For FITC, 495 nm (excitation) and 525 nm (emission) filters are needed. The peak emission of TR is 615 nm. Other less commonly used fluorochromes are Alexa 430 (peak emission 541 nm) and Cascade Blue (peak emission 421 nm).

Immunofluorescence can provide useful information on the spatio-temporal expression of fungal antigens and their potential usefulness as diagnostic markers (Thornton *et al.*, 1993, 2002; Thornton and Talbot, 2006; Thornton,

2008b, 2009b). For example, fluorescence present at the tip of a hypha, but absent from dormant conidia, would indicate the production of an extracellular, constitutively expressed antigen. Such an antigen would provide an appropriate target for the detection of live infective biomass of a fungal plant pathogen, because secretion in fungi occurs at the tip of elongating hyphae (Thornton *et al.*, 2002; Thornton, 2008b; Ryder *et al.*, 2012). In IF, a sample is judged as positive for antigen–antibody binding if it fluoresces, and negative if it does not. However, the subjectivity of such an assay requires comparison with a negative control. In most cases, a negative control consisting of hybridoma tissue culture medium minus antibody is acceptable. In other cases, a more appropriate control might include comparative fluorescence with a closely related non-reactive species of the fungus. This is particularly suitable where species-specific MAbs are being tested.

The IF procedure outlined in Protocol 8.2 can be applied to various aspects of fungal biology – from the tracking of antigen production in the developing fungus (Thornton *et al.*, 2002; Thornton, 2008b) to the detection and enumeration of fungal propagules immobilized to glass slides using Burkard volumetric spore traps (Kennedy *et al.*, 2000; Wakeham *et al.*, 2008). For the visualization of fungi *in planta* using IF and immuno-enzymatic staining procedures, the reader should refer to Thornton and Talbot (2006).

8.5 Lateral-flow Device (LFD) Technology

The reliance of quantitative plate-based ELISA and qualitative slide-based IF procedures on cumbersome and expensive equipment restricts these tests to diagnostic laboratories where these assays are performed on a regular basis. However, the incorporation of highly specific mAbs into semi-quantitative LFDs has enabled the production of user-friendly tests for the detection of fungal pathogens in the field and at the time of inspection. Consequently, LFD assays have found broad applicability as on-site diagnostic tests for viral, bacterial, oomycete and fungal plant pathogens in agriculture (Danks and Barker,

2000; Thornton *et al.*, 2004; Lane, 2006; Thornton, 2008a) and as point-of-care diagnostics for human and animal mycoses (Thornton, 2008b, 2020; Dillon *et al.*, 2017; Davies and Thornton, 2022).

Several assays have now been developed for the detection of pathogens *in planta* (Lane, 2006; Celik *et al.*, 2009) and as saprotrophic contaminants in soil (Thornton *et al.*, 2004; Thornton, 2008b). Briefly, a specific mAb is immobilized to a defined capture zone on a porous nitrocellulose membrane, while the same mAb conjugated to colloidal gold particles serves as the detection reagent. Samples of solubilized antigens are added to a release pad containing the antibody–gold conjugate. The antibody–gold conjugate binds to the target antigen, passes along the porous membrane by capillary action and binds to the mAb immobilized in the capture zone. Once an antigen extract is prepared and applied to the LFD, the test result is recorded usually at 15 min or 30 min. Bound antigen–antibody–gold complex is seen as a coloured line (commonly red or blue, depending on the conjugate used) with an intensity that is proportional to the antigen concentration. Anti-mouse immunoglobulin immobilized to the membrane in a separate zone acts as an internal control. In the absence of the target antigen, no complex is formed in the zone containing solid-phase antibody, and a single internal control line is seen. In the presence of the target antigen, two lines are clearly visible.

As with IF tests, visual interpretation of LFD results is arguably open to bias due to operator subjectivity. For this reason, an LFD should always be read at the specified time after application of the sample to the device, as notified by the manufacturer. In situations where positive–negative threshold data are available from a quantitative ELISA that incorporates the same antibody–antigen interaction as the LFD, reference samples of known concentrations of the target antigen can be run in parallel to allow semi-quantitative calibration of the LFD device. Alternatively, a portable Cube reader can be used to determine optical densities of test lines (Kennedy and Wakeham, 2008; Davies and Thornton, 2022) to allow the establishment of LFD positive–negative thresholds for detection.

Antigens are extracted from plant material or from soil samples using simple extraction procedures that are equally appropriate for ELISA

diagnostics. Protocol 8.3 describes an antigen extraction procedure that was developed for the detection of *Trichoderma* spp. in composts that uses a genus-specific LFD (Thornton, 2008a). *Trichoderma* species are cosmopolitan saprotrophic fungi that are used in the biological control of plant diseases (Whipps, 1997) and as plant growth stimulants (Harman *et al.*, 2004). In addition, certain strains of *Trichoderma aggressivum* are noxious compost-borne pathogens of cultivated mushrooms (Seaby, 1987), while a number of thermotolerant species, most notably *Trichoderma longibrachiatum*, have emerged as medically important opportunistic pathogens of healthy and immunocompromised humans (Thornton and Wills, 2015). The LFD provides a simple, sensitive and specific onsite assay for the detection of *Trichoderma* spp. The mAb MF2 contained within the test is available commercially, as are mAbs to other fungal pathogens (https://www.iscadiagnostics.com; accessed 24 February 2023) (Fig. 8.1). The LFD detects strains of *T. aggressivum* that are aggressive competitors of the cultivated mushrooms *Agaricus bisporus* and *Pleurotus ostratus*, and so can be used to determine the presence of competitive *Trichoderma* strains in composts destined for use in mushroom cultivation. Once a contaminated sample is identified by LFD, the contaminant species can be isolated using selective isolation procedures and identified using PCR-based techniques such as sequencing of the internally transcribed spacer (ITS)1-5.8S-ITS2 rRNA (ribosomal RNA)-encoding region (Thornton *et al.*, 2002; Thornton, 2009b) or other such methods (see Chapter 9).

Protocol 8.4 describes an additional extraction process (the 'ball and bottle' technique) used for a number of fungal and oomycete plant pathogens, including *Phytophthora*, *Pythium* and *Rhizoctonia* species (see Fig. 8.3). This simple system is ideally suited to field testing of plant material and has been used successfully in the fight against the spread of *Phytophthora ramorum* (Lane, 2006). The LFD uses genus-specific antibodies, and so detects all *Phytophthora* spp. Consequently, once a *Phytophthora*-positive sample is identified using LFD, samples require further tests to identify the causal species. This can be achieved either by direct isolation of the organism from infected material and identification using its morphological characteristics, or by PCR using DNA eluted from the LFD device. The reduced cost, ease of use and speed of the diagnostic procedure has revolutionized the detection and monitoring of this important emerging pathogen.

Table 8.1 provides details of the solutions and reagents that are used in all three types of immunodiagnostic assays that are introduced above and are further described in Protocols 8.1–8.4.

Table 8.1. Solutions and reagents used in ELISA, immunofluorescence (IF) and lateral-flow device (LFD) immunodiagnostic assays.

Reagent	Composition
PTFE-coated multi-well slides	
Immunofluorescence fixative solution	Ethanol/chloroform/3% paraformaldehyde (6:3:1 by vol)
Mounting medium	
Nunc Immuno Maxisorp 96-well flat-bottomed microtitre plates	
Phosphate-buffered saline (pH 7.4)	137 m mol l^{-1} NaCl, 2.7 m mol l^{-1} KCl, 8 m mol l^{-1} Na_2HPO_4 and 1.5 m mol l^{-1} KH_2PO_4
Phosphate-buffered saline (pH 7.4) with Tween-20	PBS containing 0.05% polyoxyethylene (20) sorbitan monolaurate
Goat anti-mouse polyvalent (IgA, IgG, IgM) horseradish peroxidase conjugate	
Goat anti-mouse polyvalent (IgA, IgG, IgM) alkaline phosphatase conjugate	
Goat anti-mouse polyvalent (IgA, IgG, IgM) fluorescein isothiocyanate conjugate	
Tetramethylbenzidine (TMB) substrate for horseradish peroxidase	
p-Nitrophenylphosphate (PNP) substrate for alkaline phosphatase	
TMB substrate solution for ELISA	5 ml distilled water (dH_2O), 5 ml 0.2 mol l^{-1} sodium acetate, 195 μl 0.2 mol l^{-1} citric acid, 5 μl 30% H_2O_2 and 100 μl of a 10 mg ml^{-1} solution of TMB dissolved in dimethyl sulfoxide (DMSO)
Peroxidase substrate stop solution	3 M H_2SO_4
PNP substrate solution for ELISA	12% Diethanolamine (pH 9.8) containing 0.6 mg ml^{-1} PNP
AMPAK amplification system for alkaline phosphatase	
Bovine serum albumin, Cohn Fraction V	

Protocol 8.1

Indirect enzyme-linked immunosorbent assay (indirect ELISA)

> ### Materials
>
> - Monoclonal antibody as hybridoma tissue culture supernatant (TCS)
> - Distilled water (dH$_2$O)
> - Maxisorp 96-well microtitre plates
> - Microtitre plate reader fitted with 405 nm and 450 nm absorbance filters
> - Phosphate-buffered saline (pH 7.4) (PBS)
> - Phosphate-buffered saline with Tween-20 (pH 7.4) (PBST)
> - Goat anti-mouse polyvalent horseradish peroxidase (HRP) conjugate
> - Goat anti-mouse polyvalent alkaline phosphatase (AP) conjugate
> - Tetramethylbenzidine (TMB) substrate solution
> - Peroxidase stop solution
> - *p*-Nitrophenylphosphate (PNP) substrate solution
> - AMPAK alkaline phosphatase amplification system

The assay described here is based on the use of 50 µl samples, although other volumes such as 100 µl may also be used, in which case all values should be scaled up accordingly.

Method

1. Pipette 50 µl of antigen extract into the wells of microtitre plates. Tap the wells gently to ensure even distribution of solution over well surfaces. If only using a portion of the plate, it is recommended for some assays to fill at least an additional row or column of wells with buffer to ensure even incubation conditions. Place the plates in sealed plastic bags and incubate overnight at 4°C (or 2–4 h at 37°C) to allow immobilization of antigens.

2. Wash plates four times with PBST (5 min each washing time/step), once with PBS (5 min) and once with dH$_2$O (brief rinse). It is important to ensure that the edge of the plate is washed as well as the central wells, and some assays recommend washing the outside wells first – before washing the entire plate – to ensure an even wash.

3. Pipette 50 µl of hybridoma TCS into antigen-coated wells (concentration depends on the assay, and typically varies from neat to a 1:10 dilution). Positive and negative controls should be added as appropriate. Negative control wells should contain an equal volume of mAb-free tissue culture medium (TCM). Duplicate or triplicate wells for samples and controls should be prepared to improve the performance of the assay. Place the plates in sealed plastic bags and incubate at 23°C for 1 h. The incubation temperature may vary. Certain assays require a temperature of 37°C.

4. Remove mAb TCS or control TCM by inversion of plates and remove the fluid by flicking. Wash the wells three times with PBST (5 min each time).

5. Prepare 1000-fold dilutions of goat anti-mouse polyvalent HRP or anti-mouse AP conjugates by diluting in PBST. Add 50 µl to each well and incubate at 23°C (or 37°C) for 1 h in sealed plastic bags.

6. Wash the plates three times with PBST and once with PBS. Add 50 µl of TMB substrate to wells containing HRP conjugate, incubate plates for 30 min at 23°C (or 37°C) and add an equal volume of peroxidase substrate stop solution. Read absorbance of wells at 450 nm.

7. For wells containing AP-conjugated antibody, add 50 µl of PNP substrate solution to washed wells and monitor absorbance at 405 nm over time. Compared with HRP-TMB colorimetric reactions, AP-PNP reactions can take several hours to develop, depending on the concentration of the target antigen immobilized to the microtitre plate. The AMPAK amplification system can be used to increase the speed and sensitivity of the AP colorimetric reaction. Wash and diluent buffers provided by the manufacturer should be used throughout according to the instructions accompanying the kit.

8. The threshold value for detection of an antigen in ELISA is often established using the absorbance value obtained with the negative TCM control. Here, an absorbance value greater than that of the TCM control would be considered positive for the target antigen, but other assays typically use 2–3 times the mean negative control to establish the threshold. Alternative strategies can also be used for the setting of thresholds. For example, a more appropriate method for the establishment of a threshold in ELISA using antigen extracts from plant material might be the use of extracts from known uninfected material as the control. A leaf sample would be considered positive for a pathogen if its antigen extract had an absorbance value greater than that of the extract from known uninfected leaf material. There may also be a requirement for a minimum mean absorbance value to be reached by the positive controls.

Protocol 8.2

Indirect immunofluorescence (indirect IF)

Materials

- Monoclonal antibody as hybridoma tissue culture supernatant (TCS)
- Distilled water (dH$_2$O)
- PTFE-coated multi-well slides
- Phosphate-buffered saline (pH 7.4) (PBS)
- Goat anti-mouse polyvalent fluorescein isothiocyanate (FITC) conjugate
- Mounting medium
- Coverslips
- Epifluorescence microscope fitted with 495 nm (excitation) and 525 nm (emission) filters

Method

1. Sterilize the PTFE-coated multi-well slides by autoclaving at 121°C for 15 min.

2. Apply fungal spores to the wells in PBS or allow spores to germinate in the wells by incubation in an appropriate nutrient solution, for example 1% sterile filtered glucose solution (Thornton, 2008b, 2009b). Spores and germ tubes of many species of fungi adhere readily to glass slides through the exudation of extracellular matrices. This allows the fixation of fungal materials in the solvent-based solutions that are needed for preservation of antigens and subsequent visualization by IF.

An alternative strategy for the immobilization of material for IF utilizes PTFE slides embedded in agar (Thornton, 2001). Place a sterile PTFE slide in the base of a 9 cm Petri dish. Overlay the slide with molten agar and allow this to solidify. Remove the agar above each well using a sterile cork borer and scalpel. Inoculate the agar with the fungus and incubate under appropriate conditions. Once the glass in each well has been overgrown, carefully excise the slide from the agar and thoroughly air-dry at room temperature. Thereafter, the material should be fixed as described in step 3.

3. Remove the bathing solution from the slides by carefully touching the edge of the wells with absorbent paper. Allow the wells and adherent cells to air-dry fully and then fix the material by bathing the slides sequentially in immunofluorescence fixative solution for 3 min, in 95% methanol for 4 min and in dH$_2$O for 30 s.

4. Remove excess fluid and air-dry slides as described in step 1.

5. Apply 40 µl of mAb TCS to test wells, or mAb-free tissue culture medium (TCM) to control wells, and incubate at 23°C for 30 min. Several antibodies can be tested on a single multi-well slide, as the PTFE coating prevents the movement of antibody solutions between wells.

6. Wash the wells by gently bathing the slides four times in PBS (5 min each rinse).

7. Incubate wells with 40 µl of FITC conjugate diluted 1 in 40 in PBS. This step should be performed in the dark at 23°C for 30 min.

8. Wash the slides as described in step 6 and apply coverslips with mounting medium. The slides should be kept in a moist environment in the dark at 4°C until they are examined using an epifluorescence microscope.

Protocol 8.3

Lateral-flow device (LFD) extraction procedure for detection of antigens in soil samples

Materials

- Lateral-flow device (LFD)
- Phosphate-buffered saline (pH 7.4) (PBS)
- Phosphate-buffered saline with Tween-20 (pH 7.4) (PBST)
- Bovine serum albumin (BSA)
- 50 ml plastic Falcon tubes
- Graduated 1 ml disposable pipette

Method

1. Place 15 ml (approximately 6 g) of moistened soil into a 50 ml plastic tube and add to this 30 ml of PBS.

2. Shake the contents of the tube vigorously for 1 min and repeat this procedure at 10 min intervals for up to 1 h.

3. Allow the contents of the tube to settle for a further hour by placing the tube in an upright position.

4. While waiting for the contents to settle, prepare the LFD running buffer by making a 1% solution of BSA in PBST.

5. Carefully remove 1 ml of the PBS extraction solution from above the sediment using the graduated pipette. Add this to 30 ml of the running buffer in a new plastic tube to make the antigen extract. Mix thoroughly.

6. Place the LFD on a flat surface and apply three drops of the antigen extract (roughly equivalent to 100 μl) to the release port of the LFD (Fig. 8.1).

7. Allow the test to develop for 15 min. After this time, an intense red line will appear towards the 'C' of the viewing window (Fig. 8.2, left); this is the internal control that shows the test has worked. If a second line appears towards the 'T' of the viewing window (Fig. 8.2, right), this is a positive result and indicates that the fungus is present in the soil sample.

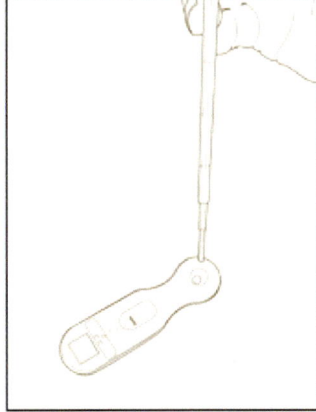

Fig 8.1. Lateral-flow device (LFD) test of soil extract, showing addition of 100 μl of soil extract to the LFD release port.

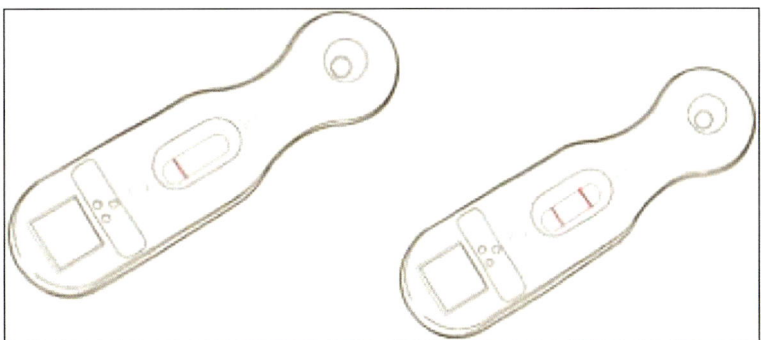

Fig 8.2. Results of lateral-flow device tests after 15 min: antigen-negative (control (C) line only) (left) and antigen-positive (control (C) and test (T) lines) (right).

Protocol 8.4

Lateral-flow device (LFD) extraction procedure for detection of antigens in plant material

Materials

- Extraction bottle
- Pipette
- Lateral-flow device (LFD)
- To prevent any cross-contamination it may be advisable to wear gloves when breaking up the tissue before placing it in the extraction bottle. However, if this is not possible, wipe or wash hands between uses. In practice, there is a low risk of carry-over between samples.

Methods

1. For softer plant tissue (e.g. leaves, flowers, fruits), take an approximately 1–2 cm square of necrotic tissue (Fig. 8.3a) or take several smaller pieces from a couple of spots (Fig. 8.3b).

2. Fold up the piece of tissue and crush it to help break down the sample (Fig. 8.3c).

3. For harder plant tissue (e.g. stems, bark), shave off five small (*c*.1 cm long) slivers of the outer tissue from a necrotic area (Fig. 8.3b, d) and crush if possible. Clean knife before reuse to prevent any carry-over.

4. Place the sample in the extraction bottle (Fig. 8.3e) and shake vigorously for 60 s (Fig. 8.3f).

5. With the pipette, draw up the extraction solution (there should be a slight green-to-brown coloration if the extraction has been successful) (Fig. 8.3g).

6. Place 2–3 drops in the sample well (Fig. 8.3h). If liquid does not start to flow across the membrane after 30 s (Fig. 8.3i), add one or two further drops until fluid flows across the membrane. Do not use more than 3–4 drops as this will flood the membrane, resulting in an invalid test.

7. Leave for 2–3 min.

8. Check to see if a blue line has appeared at 'C' (Fig. 8.3j, k).

9. Check to see if a blue line has appeared at 'T' (Fig. 8.3j, k).

10. You may check for lines up to 10 min, but results after this time should not be considered. Faint blue lines may appear after about 30 min or longer, but these should be disregarded.

11. Interpretation of results:

- No blue control line – test invalid. Retest with the same extract but a new LFD.
- Blue control line appears – test valid.
- Test line – no blue line within 10 min (Fig. 8.3k, upper LFD (A)) – sample negative for target organism.
- Test line – blue line within 10 min (Fig. 8.3k, lower LFD (B)) – sample positive for target organism.

However, if you believe the symptoms are typical for the target organism, either repeat the LFD ensuring good sample breakdown and vigorous shaking or, if you have any doubts, send a sample to a diagnostic laboratory for testing.

If the leaf is very soft (e.g. lilac) a green tinge may appear on the membrane, or if the tissue is very necrotic or woody a red brown tinge may appear.

(a)

Fig. 8.3. Testing procedure using a lateral-flow device (LFD) for plant leaf and stem plant material: (a) sample leaf; (b) sample stem; (c) crush leaf sample; (d) slice and crush stem sample; (e) place sample in bottle; (f) shake vigorously for 60 s; (g) check to see whether the extraction liquid is discoloured; (h) pipette 2–3 drops onto sample well in LFD; (i) pale blue fluid runs across membrane; (j) test kit after 30 s – faint control lines appearing at (C) arrowed, no test lines appearing at (T); (k) test kit after 3 min – test valid and sample negative (A), test valid and sample positive (B). (© UK Crown Copyright – courtesy of Fera.)

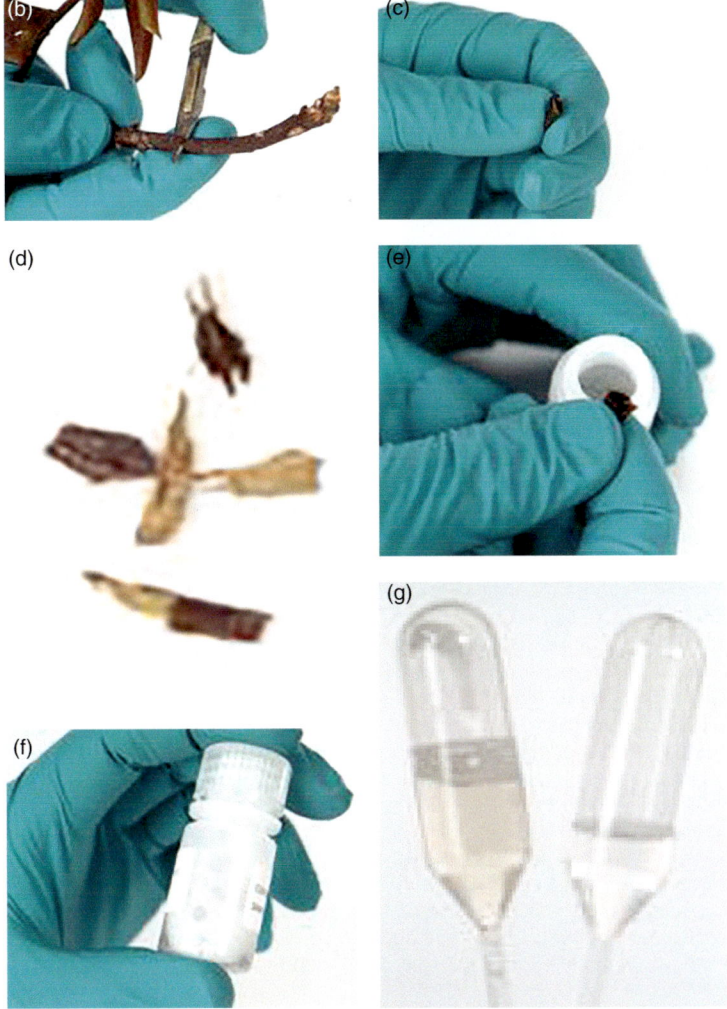

Fig. 8.3. Continued.

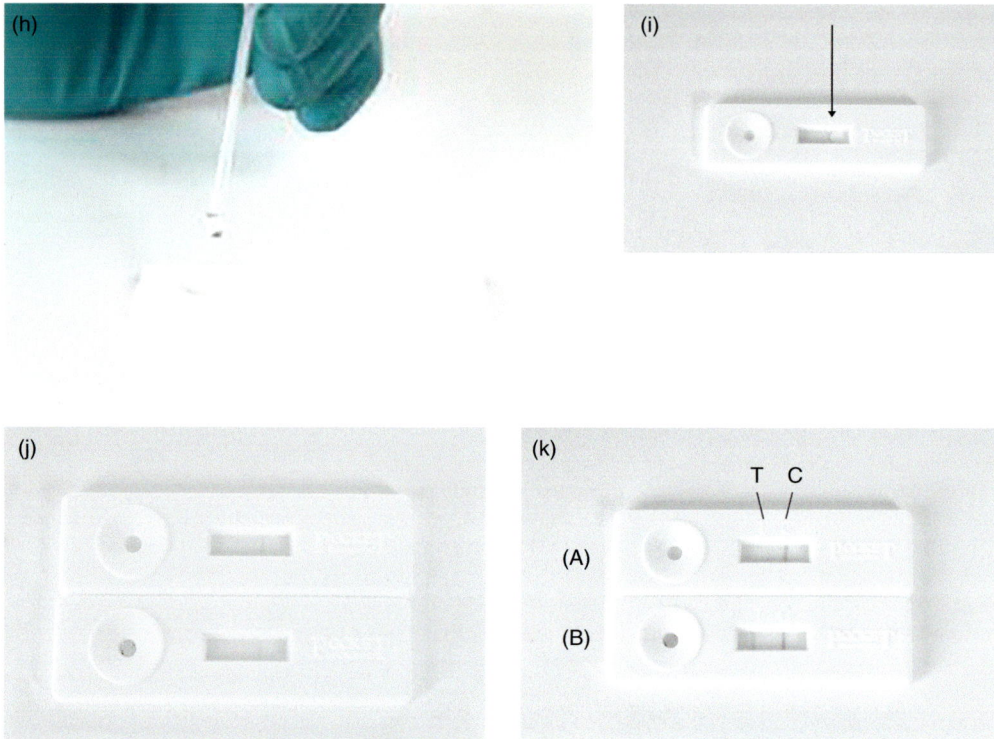

Fig. 8.3. Continued.

References

Al-Maqtoofi, M. and Thornton, C.R. (2016) Detection of human pathogenic *Fusarium* species in hospital and communal sink biofilms by using a highly specific monoclonal antibody. *Environmental Microbiology* 18, 3620–3634.

Celik, M., Kalpulov, T., Zutahy, Y., Ish-shalom, S., Lurie, S. and Lichter, A. (2009) Quantitative and qualitative analysis of *Botrytis* inoculated on table grapes by qPCR and antibodies. *Postharvest Biology and Technology* 52, 235–239.

Danks, C. and Barker, I. (2000) On-site detection of plant pathogens using lateral-flow devices. *Bulletin OEPP/EPPO Bulletin* 30, 421–426.

Davies, G.E. and Thornton, C.R (2022) Development of a monoclonal antibody and a serodiagnostic lateral-flow device specific to *Rhizopus arrhizus* (syn. *R. oryzae*), the principal global agent of mucormycosis in humans. *Journal of Fungi* 8, 756.

Dillon, M.J., Bowkett, A.E., Bungard, M.J., Beckman, K.M., O'Brien, M.F., Bates, K. *et al.* (2017) Tracking the amphibian pathogens *Batrachochytrium dendrobatidis* and *Batrachochytrium salamandrivorans* using a highly specific monoclonal antibody and lateral-flow technology. *Microbial Biotechnology* 10, 381–394.

Harman, G.E., Howell, C.R., Viterbo, A., Chet, I. and Lorito, M. (2004) *Trichoderma* species – opportunistic, avirulent plant symbionts. *Nature Reviews Microbiology* 2, 43–56.

Kennedy, R. and Wakeham, A.J. (2008) Development of detection systems for the sporangia of *Peronospora destructor*. *European Journal of Plant Pathology* 122, 147–155.

Kennedy, R., Wakeham, A.J., Byrne, K.G., Meyer, U.M. and Dewey, F.M. (2000) A new method to monitor airborne inoculum of the fungal plant pathogens *Mycosphaerella brassicicola* and *Botrytis cinerea*. *Applied and Environmental Microbiology* 66, 2996–3003.

Lane, C. (2006) Diagnosis of *Phytophthora ramorum*. *Outlooks on Pest Management* 17, 169–171.

Ryder, L.S., Harris, B.D., Soanes, D.M., Kershaw, M.J., Talbot, N.J. and Thornton, C.R. (2012) Saprotrophic competitiveness and biocontrol fitness of a genetically modified strain of the plant-growth-promoting fungus *Trichoderma hamatum* GD12. *Microbiology (Reading)* 158, 84–97.

Seaby, D.A. (1987) Infection of mushroom compost by *Trichoderma* species. *Mushroom Journal* 179, 355–361.

Thornton, C.R. (2001) Immunological methods for fungi. In: Talbot, N.J. (ed.) *Molecular and Cellular Biology of Filamentous Fungi: A Practical Approach*. Oxford University Press, Oxford, UK, pp. 227–257.

Thornton, C.R. (2008a) Tracking fungi in soil with monoclonal antibodies. *European Journal of Plant Pathology* 121, 347–353.

Thornton, C.R. (2008b) Development of an immunochromatographic lateral-flow device for rapid serodiagnosis of invasive aspergillosis. *Clinical and Vaccine Immunology* 15, 1095–1105.

Thornton, C.R. (2009a) Production of monoclonal antibodies to plant pathogens. In: Burns, R. (ed.) *Plant Pathology, Techniques and Protocols*. Humana Press, New York, pp. 63–74.

Thornton, C.R. (2009b) Tracking the emerging human pathogen *Pseudallescheria boydii* by using highly specific monoclonal antibodies. *Clinical and Vaccine Immunology* 16, 756–764.

Thornton, C.R. (2020) Detection of the 'big five' mold killers of humans: *Aspergillus*, *Fusarium*, *Lomentospora*, *Scedopsorium* and Mucormycetes. *Advances in Applied Microbiology* 110, 1–61.

Thornton, C.R. and Talbot, N.J. (2006) Immunofluorescence microscopy and immunogold EM for investigating fungal infections of plants. *Nature Protocols* 5, 2506–2511.

Thornton, C.R. and Wills, O.E. (2015) Immunodetection of fungal and oomycete pathogens: established and emerging threats to human health, animal welfare and global food security. *Critical Reviews in Microbiology* 41, 27–51.

Thornton, C.R., Dewey, F.M. and Gilligan, C.A. (1993) Development of monoclonal antibody-based immunological assays for the detection of live propagules of *Rhizoctonia solani* in soil. *Plant Pathology* 42, 763–773.

Thornton, C.R., Pitt, D., Wakley, G.E. and Talbot, N.J. (2002) Production of a monoclonal antibody specific to the genus *Trichoderma* and closely related fungi, and its use to detect *Trichoderma* spp. in naturally infested composts. *Microbiology* 148, 1263–1279.

Thornton, C.R., Groenhof, A.C., Forrest, R. and Lamotte, R. (2004) A one-step, immunochromatographic lateral flow device specific to *Rhizoctonia solani* and certain related species, and its use to detect and quantify *R. solani* in soil. *Phytopathology* 94, 280–288.

Wakeham, A.J., Kennedy, R., Byrne, K.G., Keane, G. and Dewey, F.M. (2008) Immunomonitoring air-borne fungal plant pathogens: *Mycosphaerella brassicicola*. *Bulletin EPPO/EPPO Bulletin* 30, 475–480.

Whipps, J.M. (1997) Developments in the biological control of soil-borne plant pathogens. *Advances in Botanical Research* 26, 1–34.

9 Nucleic Acid-based Techniques for Diagnosis

Jenny Tomlinson*, Ashleigh Elliott and Kinda Alraiss
Fera Science Ltd, Sand Hutton, York, UK

9.1 Introduction

Nucleic acid-based methods have the potential to provide accurate and timely diagnosis of fungal plant pathogens to better inform decisions relating to disease management and control (Martin *et al.*, 2000; McCartney *et al.*, 2003; Mumford *et al.*, 2006; Luchi *et al.*, 2020). Due to their sensitivity, specificity, speed and relative ease of use, such methods can also be useful tools in the investigation of infection processes, disease progression and epidemiology.

Nucleic acid-based methods can have advantages over both serological methods, which can lack sensitivity and specificity, and cultural methods, which can take several days (or longer) to complete (McCartney *et al.*, 2003). While the morphological identification of fungal pathogens requires specialized mycological expertise, nucleic acid-based techniques are generic, so the skills required are transferable to a very wide range of target pathogens. Nucleic acid-based methods can be applied to any life stage of a pathogen and have the potential to detect non-viable or non-culturable pathogens. The potential for automation also makes nucleic acid-based approaches particularly suitable for routine high-throughput testing.

In order to develop a nucleic acid-based detection method, information is required about the nucleic acid sequence of the target pathogen, as well as about organisms which are related to the target species or are likely to be encountered within the same test matrices. The majority of routinely used methods are based on the polymerase chain reaction (PCR), in which a region of DNA is amplified by DNA polymerase in a reaction mixture subjected to repeated cycles of heating and cooling (Henson and French, 1993). Real-time PCR assays (Schena *et al.*, 2004) are generally faster and more sensitive than conventional PCR followed by gel electrophoresis, but the higher equipment costs of real-time PCR may make conventional PCR a method of choice in some scenarios.

Nucleic acid-based detection methods generally comprise three stages: (i) extraction of nucleic acid from the sample; (ii) amplification of a target-specific region of nucleic acid; and (iii) detection of the amplification product either at the end of the reaction or as it accumulates in real time. This chapter considers nucleic acid extraction methods for plant samples and other matrices, conventional and real-time PCR and associated detection methods, and alternatives to PCR-based methods.

---

* E-mail: jenny.tomlinson@fera.co.uk

DOI: 10.1079/9781800620575.0009

9.2 DNA Extraction

The first step in any DNA-based test is generally to obtain a good-quality nucleic acid extract. In some cases, it is possible to extract amplifiable nucleic acid from mycelium or spores grown in culture by brief boiling, or even by adding a minute amount of material directly to an amplification reaction. However, in test plant material directly, and other matrices (such as soil, growing media, water), it is usually necessary first to extract nucleic acid from the sample. Samples which might be tested for fungal pathogens are extremely diverse, ranging from different plant tissues to fungal structures and matrices such as soil and water. A considerable challenge for the detection of pathogens in plant material and soil is the presence of substances that can be inhibitory to downstream tests such as PCR. For example, in plant material such substances include polysaccharides and polyphenolic compounds, particularly in woody or necrotic material (Demeke and Adams, 1992; Koonjul *et al.*, 1999) and humic acid from soil (Tsai and Olson, 1992). Extraction methods vary in terms of DNA purity and yield, with some methods being better suited to particular sample types or downstream tests. When selecting an extraction method for a particular application, additional factors that may be taken into account include speed, throughput, per-test costs and the equipment required.

Disruption of material

Regardless of the extraction method used, the initial step in the extraction of nucleic acid is disruption of the sample. The chosen method of disruption will depend primarily on the sample type (e.g. leaf, stem or woody plant tissue, or mycelium or spores from a culture), but also on the sample size and the number of samples to be tested. Soft tissues such as the leaves and petioles of some plants are easily disrupted, e.g. using a mortar and pestle or a hand-held roller.

Pre-freezing in liquid nitrogen can aid homogenization of some samples, and the use of liquid nitrogen also has the benefit of reducing the activity of endogenous nucleases in the sample, which can degrade nucleic acid at room temperature. Tissue homogenizers such as the HOMEX 6 (Bioreba, Reinach, Switzerland), Precellys® 24 (Bertin Instruments) and other mechanical grinders can be suitable for the disruption of samples of different sizes ranging from softer plant tissues to more challenging matrices (Fig. 9.1). The extent of disruption required for optimal extraction depends to some extent on the target pathogen. Complete homogenization is required for the optimal extraction of DNA from plant material infected with some pathogens, such as obligate parasites, but this may not be necessary for pathogens that are found on the surface of the host plant. Partially homogenizing plant material can have the advantage of reducing the release of inhibitors which could

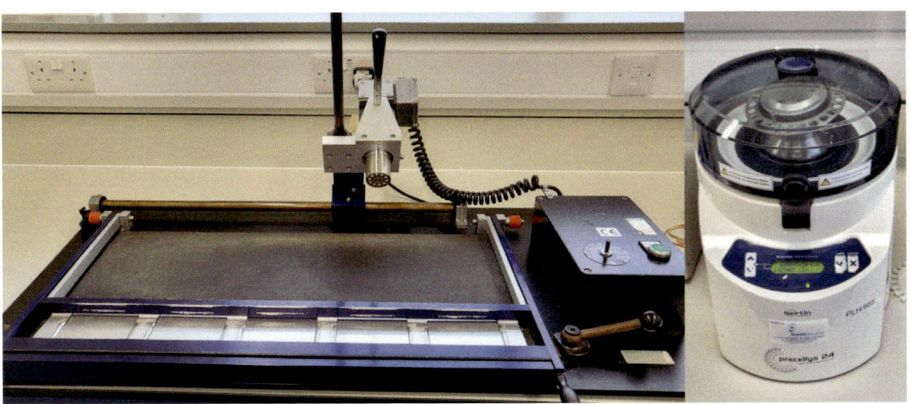

Fig. 9.1. HOMEX and Precellys instruments for sample homogenization. (Image Courtesy Fera-Science Limited © Copyright Fera-Science Limited.)

adversely affect subsequent tests. Different fungal structures may also require different methods for optimal disruption, e.g. mycelium is more easily disrupted than thick-walled spores or sclerotia.

Manual extraction methods

Traditional methods for the extraction of nucleic acid from plant material use buffers prepared in-house and require relatively basic laboratory equipment (e.g. a centrifuge or water bath) and plasticware. These methods can be somewhat laborious when large numbers of samples are processed at one time, but are suitable for processing small numbers of samples and can have considerable cost advantages over the use of commercially available extraction kits. Manual methods can also easily be scaled up to allow testing of bulked or pooled samples, which can be useful when testing asymptomatic material (Suarez *et al.*, 2005). High yields of high-quality nucleic acid can be obtained by methods using an extraction buffer containing the cationic surfactant cetyltrimethyl ammonium bromide (CTAB) followed by chloroform extraction (Lodhi *et al.*, 1994). Such a method is described in Protocol 9.1. These methods are often considered to be a 'gold standard' for extraction from a wide range of plant matrices, and if in doubt are a useful starting point for extraction from an unfamiliar matrix. However, manual extraction methods are generally too laborious for high-throughput testing and the number of manipulations involved can increase the risk of cross-contamination and errors leading to incorrect diagnostic results.

Commercially available extraction kits

In addition to traditional manual extraction methods, various kits for nucleic acid extraction from plant material are commercially available. The use of these kits is generally less time consuming than manual methods, allowing more samples to be processed at one time, but methods using kits are often less flexible in terms of sample size. The nucleic acid yield may be lower than for 'gold standard' manual methods but is also

likely to be highly consistent for a given sample type. Most commercially available kits are based on the process of binding nucleic acids to silica in the presence of chaotropic (disrupting) salts (Boom *et al.*, 1990), with the two most common formats being spin columns and magnetic beads. Nucleic acid is bound to either a silica membrane or magnetic silica particles, allowing contaminants (proteins, polysaccharides, etc.) to be washed away, followed by release of the nucleic acid into an elution buffer. Methods using spin columns generally result in relatively high yields of high-quality nucleic acid, and can be completed quite quickly, but involve numerous centrifugation steps, although automation is also possible using an instrument such as the QIAcube (Qiagen). Magnetic bead-based methods (Berensmeier, 2006) are also suitable for automation using platforms such as the KingFisher™ (ThermoFisher Scientific, Waltham, Massachusetts) (Fig. 9.2), BioSprint (Qiagen, Hilden, Germany) and Maxwell® (Promega, Madison, Wisconsin) systems, allowing higher throughput than more laborious methods. A disadvantage of many automated methods is the tendency to require relatively small sample sizes, with an increase in cost and a reduction in throughput for larger format extractions.

9.3 Conventional PCR

Following extraction of nucleic acid from a sample, the next step is the amplification of a target-specific region of nucleic acid. Amplification methods based on PCR have been developed for the detection of a wide range of plant pathogens (Henson and French, 1993). A thermostable DNA polymerase is used to amplify the region of DNA between two target-specific primers, such that the presence of the expected amplification product indicates that the target organism was present in the test sample. Amplification by PCR requires repeated cycles of heating and cooling: the reaction is heated to 95°C to denature the double-stranded target, after which the temperature is reduced to allow annealing of the primers to the target DNA, followed by extension of the primers by DNA polymerase (Fig. 9.3).

In some cases, nested PCR, in which a second pair of primers is designed within the region

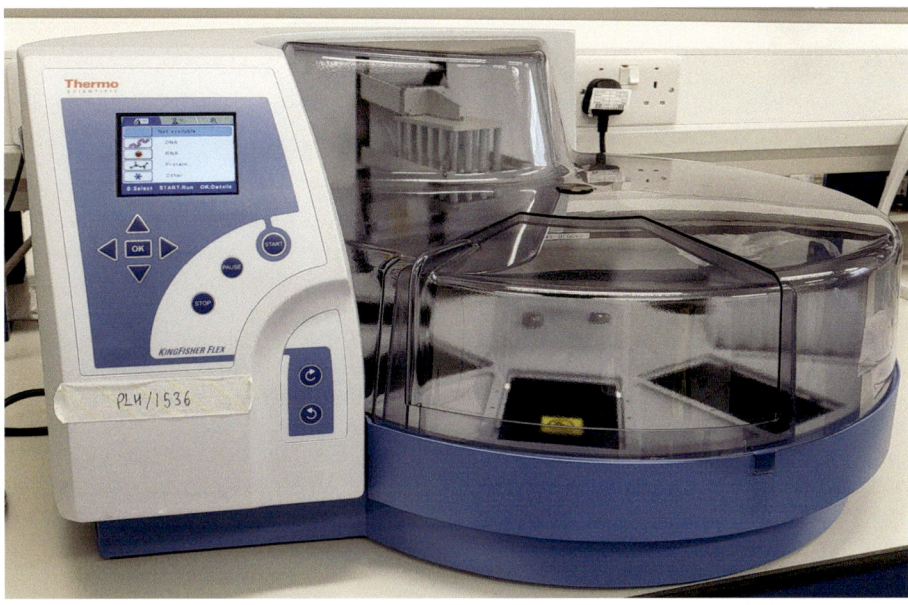

Fig. 9.2. KingFisher Flex instrument for automated handling of magnetic beads for nucleic acid extraction. (Image Courtesy Fera-Science Limited © Copyright Fera-Science Limited.)

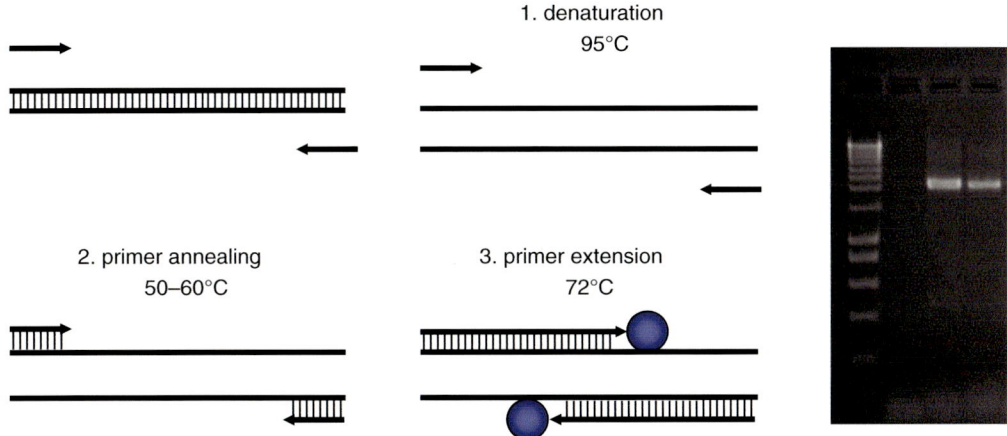

Fig. 9.3. Polymerase chain reaction (PCR). Left: A target-specific region of DNA is amplified by denaturation of the double-stranded template (1); annealing of target-specific primers (arrows) (2); and extension of the primers by Taq DNA polymerase (3). Right: PCR products visualized by agarose gel electrophoresis (lane 1: DNA ladder; lane 2: negative reaction; lanes 3 and 4: positive reactions). (Image Courtesy Fera-Science Limited © Copyright Fera-Science Limited.)

amplified by the first pair of primers, is used to increase sensitivity. The use of a DNA polymerase which is inactive until incubation at 95°C (a 'hot-start' polymerase) prevents the formation of unintended products caused by non-specific primer binding at room temperature.

Assay development

In order to design a PCR-based assay, it is necessary to identify a suitable sequence which is present in the target species but not in related non-target organisms or other organisms likely to be found in the same host material. As well as being dependent on the availability and quality of relevant sequence data in public databases such as GenBank® (Benson *et al.*, 2009) or generated in-house, the development of PCR-based detection methods also relies heavily on the availability of suitable reference material or cultures for assay validation. For any assay, it is also necessary to consider the taxonomic level to which identification is required – from genus (or higher) to subspecies, which will depend on the context in which the assay is to be used. Clear and well-understood taxonomic concepts enable a rational approach to design, while the availability of definitively identified reference material enables validation of an assay in terms of analytical sensitivity (i.e. the limit of detection) and specificity (whether the assay detects all target organisms and no non-target organisms) before it is used to test unknown samples.

The target sequence for a species-specific PCR assay should ideally be conserved within the species but be sufficiently variable between species to allow discrimination between target and non-target organisms. Regions of the ribosomal RNA (rRNA) genes (rDNA) (in particular, the internal transcribed spacers, ITS1 and ITS2, and the intergenic spacer, IGS) have been used as targets for the detection of a range of fungal pathogens (Schena *et al.*, 2004; Ma and Michailides, 2007). These targets have the benefit of occurring in multiple copies, thereby increasing assay sensitivity. In some cases, however, there are insufficient differences in these regions to allow discrimination between closely related species. Other conserved genes have been sequenced for a range of fungal species and used as targets for assay design, including β-tubulin, elicitin, laccase, elongation factor and various others (Schena *et al.*, 2004; Ioos *et al.*, 2006; Schena and Cooke, 2006; Ma and Michailides, 2007).

In the absence of relevant sequence data, short arbitrary primers can be used to attempt to amplify target-specific fragments referred to as RAPD (random amplified polymorphic DNA) markers (see Chapter 10) which can be used as the basis for species-specific primer design (McCartney *et al.*, 2003; Schena *et al.*, 2004; Ma and Michailides, 2007). This approach can be laborious and the identification of a suitable target region is not guaranteed, but it can allow the design of assays for pathogens for which little or

no sequence information is currently available. More recently, comparative genomic approaches have been used to identify species-specific markers for the development of diagnostic assays (e.g. Malapi-Wight *et al.*, 2016; Tang *et al.*, 2017).

Software is available for the design of primers for PCR and real-time PCR, both commercially (e.g. Primer Express™ (Applied Biosystems, Foster City, California); Visual OMP™ (DNA Software Inc., Ann Arbor, Michigan)) and free of charge (e.g. Primer3™; Rozen and Skaletsky, 2000). Primers are selected on melting temperature, guanine/cytosine (GC) content and other predicted physical characteristics (Dieffenbach *et al.*, 1993; SantaLucia, 2007). When designing primers an alignment of target and non-target sequences, positioning mismatches at the 3′ end of the forward and reverse primers maximizes specificity.

Practical considerations for the use of PCR-based methods

In order to interpret the results of PCR-based diagnostic tests, it is necessary to consider factors which could cause incorrect results (both false positives and false negatives) and take measures to avoid them. PCR can be prone to the inhibitory effects of substances such as polyphenolics and polysaccharides in plant material (see Section 9.2), resulting in failed amplification, so it is preferable to use high-quality DNA extracts. The effects of inhibition can be reduced by diluting extracts before adding them to the PCR reaction, or by using additives such as BSA (bovine serum albumin), PVP (polyvinylpyrrolidone) or DMSO (dimethyl sulfoxide) in the PCR reaction mix (Bickley and Hopkins, 1999). Conversely, due to the high sensitivity of PCR-based methods, it is also essential that precautions are taken to avoid false-positive results caused by cross-contamination. This is particularly true for nested PCR methods, where the amplification product of a first round of PCR is added to a second PCR reaction. Nucleic acid extraction, PCR reaction set-up and post-PCR analysis should be carried out in spatially separated areas (ideally in different rooms) using designated pipettors for each step. PCR reagents should always be stored separately from DNA extracts, preferably in small aliquots to avoid contamination of reagent stocks resulting from frequent handling. PCR reactions should be set up in a laminar flow cabinet, which should be regularly decontaminated by UV irradiation or the use of chemical decontaminants.

The use of appropriate controls is essential for the interpretation of PCR results: positive controls and no-template controls indicate that the reactions have been set up correctly in the absence of contamination. Healthy control samples and extraction blanks (containing no sample but subjected to the nucleic acid extraction process) are also valuable in identifying any contamination introduced during extraction. It is advisable to carry out PCR reactions in duplicate to reduce the risk of pipetting errors leading to incorrect results.

Following amplification, the products of conventional PCR are commonly analysed by agarose gel electrophoresis (Fig. 9.3) in which DNA fragments can be separated by size as they migrate through an agarose gel in response to an applied voltage. The size of the amplification product can be compared to a DNA ladder (a mixture of DNA fragments of known size) to ensure that it is the expected length. In this way, specific products can be distinguished from non-specific amplification artefacts such as primer dimers (non-specific products generated by interaction between primers or secondary structure within a primer), which are typically smaller in size and visible as a diffuse band. Amplification products can be further analysed, for example by sequencing or performing restriction digests in order to confirm that the intended product has been amplified or to gain more information about the sample.

Materials and methods for conventional PCR are described in Protocol 9.2.

9.4   Real-time PCR

In conventional PCR-based testing, amplification products are detected at the end of amplification. However, it is also possible to monitor PCR reactions in 'real-time' (i.e. during thermal cycling) using fluorescence detection, and such methods have advantages over conventional PCR. As no post-amplification steps are required, real-time

PCR results can be obtained more quickly, and because the amplification products are typically shorter than conventional PCR products, more rapid thermal-cycling conditions can be used. Real-time PCR assays are generally more sensitive than conventional PCR, and the lack of post-amplification steps also makes real-time PCR less susceptible to contamination. These advantages have enabled real-time PCR assays to become established for the routine detection of a broader range of plant pathogens than conventional PCR (Schena *et al.*, 2004; Mumford *et al.*, 2006). Numerous platforms are available for performing real-time PCR (Fig. 9.4). Reactions are typically run in 96- or 384-well plates, allowing high-throughput testing. Instruments differ somewhat in terms of their multiplexing capability and the fluorescent dyes with which they are compatible. The two main approaches to real-time monitoring of PCR use either non-specific intercalating dyes (such as SYBR™ Green) or sequence-specific probes (Wong and Medrano, 2005).

Intercalating dyes

Dyes such as SYBR Green which fluoresce on intercalation between the strands of double-stranded DNA can be used to monitor the generation of PCR products in real time. As well as detecting specific products, this method will detect any non-specific products or amplification artefacts, so it is generally necessary to perform a 'melt curve' analysis at the end of amplification. Double-stranded amplification products melt at a characteristic temperature determined by their length and base composition; at this temperature, the amplification product becomes single stranded, resulting in a large decrease in fluorescence. Monitoring SYBR Green fluorescence as the temperature is slowly increased allows the melting temperature of the amplification product(s) to be determined and enables specific amplification products to be distinguished from any non-specific products or artefacts.

Probe-based real-time PCR

The other strategy for real-time monitoring of PCR is the use of sequence-specific probes.

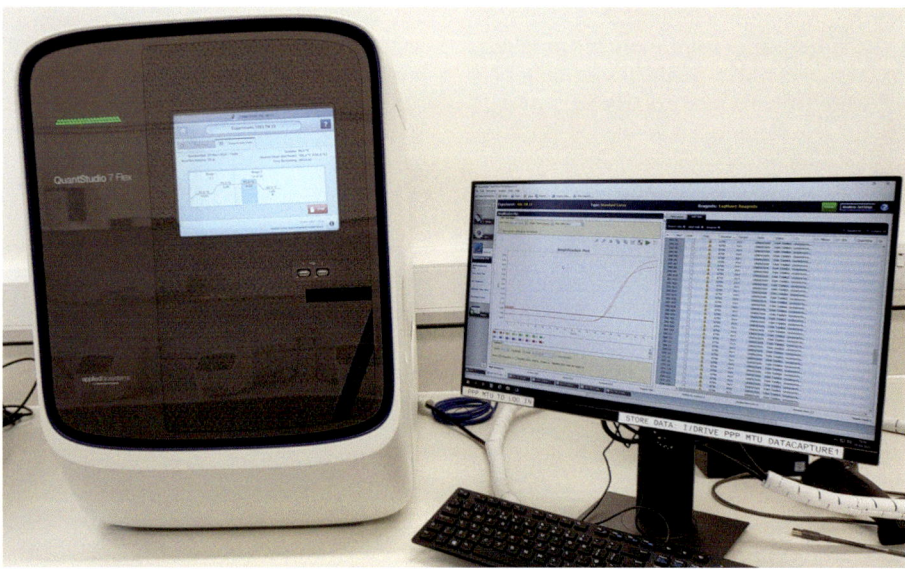

Fig. 9.4. QuantStudio real-time PCR platform. (Image Courtesy Fera-Science Limited © Copyright Fera-Science Limited.)

TaqMan™ PCR (Holland *et al.*, 1991; Lee *et al.*, 1993) is the most widely used probe-based real-time PCR chemistry. A target-specific probe is added to the PCR reaction which is labelled at the 5′ end with a fluorescent reporter and at the 3′ end with a quencher. In the probe's intact state, reporter fluorescence is quenched by the quencher by fluorescence resonance energy transfer (FRET) (Wong and Medrano, 2005). The probe is designed to bind downstream from one of the primers (typically the forward primer); the primer is extended by the action of *Taq* polymerase, and when the downstream probe is encountered, the 5′–3′ exonuclease activity of *Taq* polymerase cleaves the probe, separating the reporter from the quencher. The separation of the reporter and quencher results in a fluorescent signal, which can be detected in real time (Fig. 9.5). Materials and methods for TaqMan real-time PCR are described in Protocol 9.3.

Because binding of the probe to the target is required for a fluorescent signal to be generated, probe-based real-time PCR is generally more specific than SYBR Green PCR. The fluorescence signal is also not affected by amplification artefacts, although the generation of large amounts of unintended products (such as primer dimers) will still adversely affect the efficiency of amplification.

Other probe formats (reviewed by Wong and Medrano, 2005), include molecular beacons (Tyagi and Kramer, 1996) and Scorpions® probes (Whitcombe *et al.*, 1999), both of which have been used for the detection of fungal plant pathogens (Bates and Taylor, 2001; Bonants *et al.*, 2004).

Multiplex real-time PCR

Real-time PCR probes can be labelled with fluorescent dyes that emit fluorescence at different wavelengths, allowing more than one assay to be run in the same tube (multiplex amplification) by using differently labelled probes for each assay. A useful strategy is the multiplex amplification of both pathogen and host DNA in the same reaction (Weller *et al.*, 2000; Winton *et al.*, 2002). The use of an assay designed to detect the host plant (either a specific host or a range of plant species) assists in the interpretation of negative results by indicating whether an extract contained amplifiable DNA and whether amplification was affected by inhibition. In an alternative approach, the level of inhibition in a nucleic acid extract can be assessed by monitoring the amplification of a synthetic

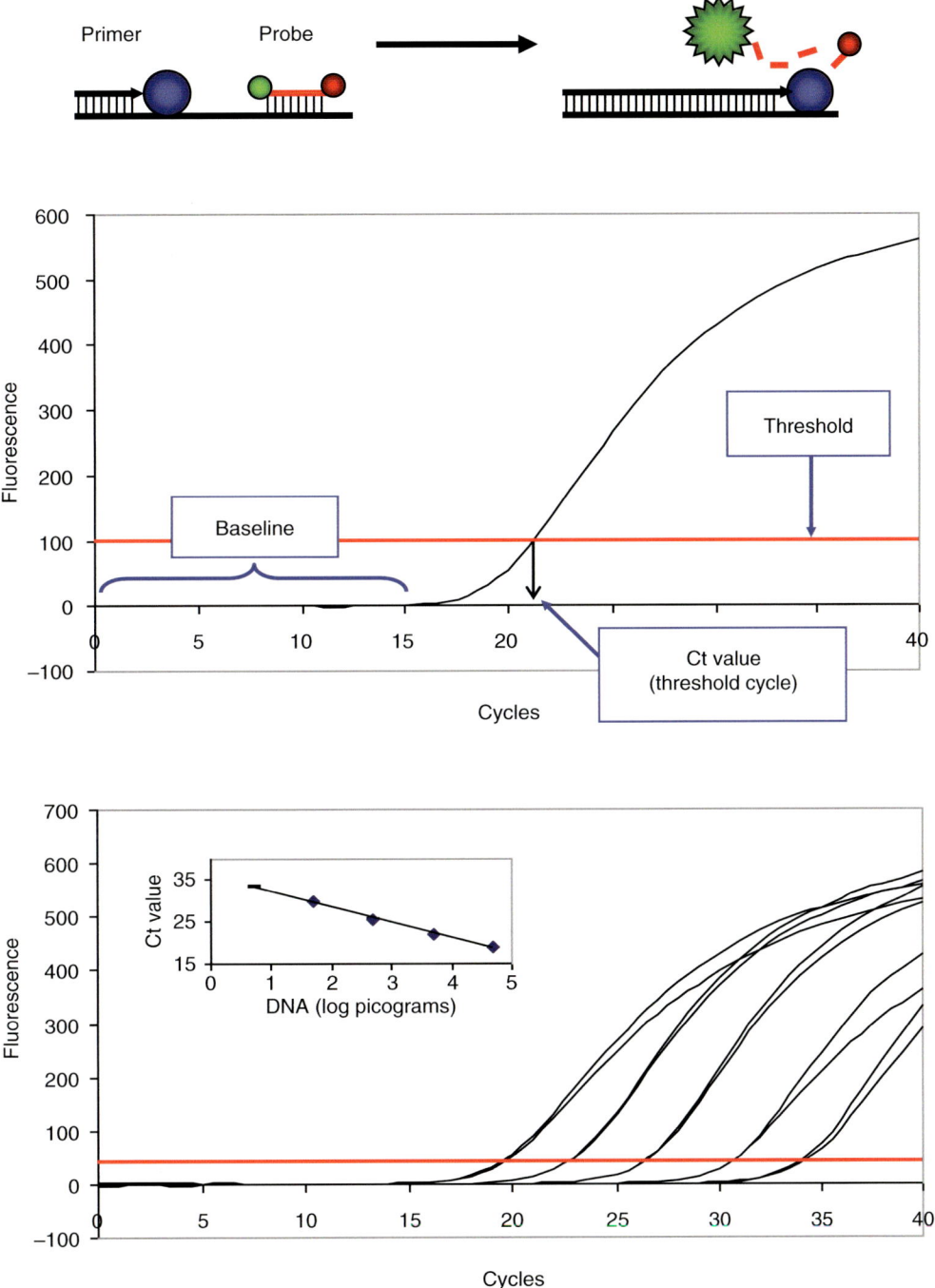

Fig. 9.5. TaqMan real-time PCR. Top: The TaqMan probe is cleaved by *Taq* DNA polymerase, which separates the reporter from the quencher, resulting in an increase in fluorescence. Middle: Real-time PCR amplification plot showing the cycle threshold (Ct) value (cycle at which fluorescence exceeds the threshold). Bottom: Ct value is proportional to the logarithm of the number of copies of the target sequence in the real-time PCR reaction. (Image Courtesy Fera-Science Limited © Copyright Fera-Science Limited.)

exogenous control template in multiplex with the target assay (Kox *et al.*, 2005).

Interpretation of real-time PCR results

Real-time PCR results are generally interpreted in terms of cycle threshold (Ct) values – the PCR cycle at which fluorescence exceeds a threshold value. The Ct value reflects the starting amount of DNA in the reaction, such that a lower Ct value indicates the presence of more copies of the target sequence (Fig. 9.5). If amplification is 100% efficient, a 2-fold difference in concentration results in a 1-cycle difference in Ct value and a 10-fold difference in concentration results in an approximately 3.3 cycle difference. Most real-time PCR assays are run for 40 cycles of heating and cooling. Real-time PCR assays can be extremely sensitive; the limit of detection of some assays can approach a single copy of the target sequence, making it particularly important to take precautions to avoid contamination during nucleic acid extraction and reaction set-up. As high Ct values can indicate the presence of extremely small quantities of target pathogen, it can be prudent in some circumstances to use a conservative cut-off value (around 35–37 cycles), above which samples should be retested (using the same or a different method), to avoid false-positive results caused by contamination.

Quantification

A significant advantage of real-time PCR over conventional PCR methods is the ability to detect pathogens quantitatively (Schena *et al.*, 2004). The Ct value is proportional to the logarithm of the number of copies of the target sequence present at the start of the reaction. Therefore, the amount of template in a sample can be inferred from the Ct value for that sample by reference to a standard curve of the Ct values for a dilution series of known amounts of target (Fig. 9.5). Quantification can be absolute (in terms of copy number, amount of DNA, number of spores, etc.), or relative to a reference sample (for a review of approaches to quantification methods, see Wong and Medrano, 2005). Quantitative

detection can be useful in applications such as the assessment of inoculum levels in soil or other matrices (e.g. Cullen *et al.*, 2002) or the study of disease processes (e.g. Suarez *et al.*, 2005).

Digital PCR is an alternative approach to quantification (Morley, 2014) in which the PCR reaction is partitioned into thousands of nano-litre-scale reactions. Amplification in each partition is detected at the end of the reaction, indicated by a fluorescent signal, and the frequency of positive reactions allows absolute quantification of the target. Where the reactions are partitioned into droplets in an oil emulsion, this is referred to as droplet digital PCR (ddPCR). As detection of amplification in the nanolitre-scale reactions takes place as an end-point measurement, quantification by digital PCR has the potential to be less affected by inhibitors than quantification by real-time PCR (Rački *et al.*, 2014).

9.5   Isothermal Amplification

PCR-based methods allow rapid, sensitive and specific detection of plant pathogens. However, thermal cycling equipment, and especially real-time PCR equipment which performs concurrent thermal cycling and fluorescence monitoring, is relatively complex and expensive, so the routine use of real-time PCR is generally restricted to larger laboratories. In contrast, isothermal amplification methods, which do not require thermal cycling, can be carried out using only basic equipment such as a water bath or heated block for incubation at a single temperature. Various approaches have been developed for amplification of nucleic acids without thermal cycling, including methods based on transcription, strand displacement, rolling-circle amplification and loop formation; these are reviewed by Gill and Ghaemi (2008). Two isothermal amplification methods which have been applied to the detection of plant pathogens are loop-mediated isothermal amplification (LAMP) and recombinase polymerase amplification (RPA).

LAMP is a DNA amplification method that uses the formation of single-stranded loop structures in the amplification product to allow primers to bind without cycles of thermal denaturation and annealing (Notomi *et al.*, 2000; Nagamine *et al.*, 2001, 2002). An advantage of LAMP is that a very large amount of double-stranded DNA is generated in the reaction, allowing a range of end-point detection methods to be used. For example, intercalating dye can be added at the end of the reaction to produce a colour change that is visible to the naked eye (Tomlinson *et al.*, 2007). Alternatively, the large amount of magnesium pyrophosphate generated as a by-product of polymerization can be observed with the naked eye as a white precipitate, or measured using a simple turbidimeter (Mori *et al.*, 2001, 2004). LAMP can also be carried out using an intercalating dye for real-time fluorescence detection. Performing real-time LAMP on a portable platform such as the Genie® instrument (OptiGene, UK) allows testing to be carried out in non-laboratory conditions. LAMP reactions are typically completed in under 30 min.

RPA uses recombinase to denature a double-stranded DNA template to allow primer binding and subsequent amplification without the need for thermal cycling (Piepenburg *et al.*, 2006). RPA has a relatively low reaction temperature and short reaction time and is compatible with probe-based detection approaches that allow detection in real-time via a fluorescent signal or using lateral-flow devices to detect labels incorporated into the amplification product.

Both LAMP and RPA show increased tolerance of substances which can be inhibitory to PCR, allowing relatively crude extracts to be tested and removing the need for complicated DNA extraction processes. Together with the short reaction times and lack of thermal cycling equipment, this makes these methods suitable for use outside the laboratory, e.g. at disease outbreak sites or in under-resourced settings (Tomlinson and Boonham, 2008; Donoso and Valenzuela, 2018).

Protocol 9.1

CTAB-type DNA extraction method (adapted from Lodhi *et al.*, 1994)

Materials

- Equipment suitable for disruption/homogenization of sample[1]
- CTAB buffer[2]
- Chloroform
- GITC buffer
- 5 mol l[-1] NaCl
- Isopropanol
- 70% Ethanol
- Molecular-biology-grade water
- Pipettors
- Filter tips for pipettors
- Benchtop centrifuge
- Microcentrifuge tubes
- Water bath or heated block
- −20°C freezer
- Fume hood for handling chloroform

Method

1. Disrupt samples[1] (e.g. plant material, mycelium) in 10 volumes of CTAB extraction buffer[2] (i.e. 1 ml buffer per 100 mg sample) using a suitable method. Transfer approximately 1 ml of homogenate to a microcentrifuge tube, then incubate at 65°C for 15–30 min.

2. Centrifuge at approximately $13,000 \times g$ for 5 min.

3. Carefully remove 700 µl of the supernatant without disturbing the pellet and transfer to a fresh tube.

4. Add 200 µl of chloroform and mix to emulsion by inverting the tube.

5. Centrifuge at approximately $10,000 \times g$ for 5–10 min.

6. Carefully remove 500 µl of the upper (aqueous) layer and transfer to a fresh tube.

7. At this stage the sample can be transferred to a magnetic particle processor (proceed to step 8) or continued manually (proceed to step 10).

8. The sample is mixed with 500 µl isopropanol and 50 µl magnetic beads and transferred to a magnetic particle processor, following the manufacturer's instructions.

9. The magnetic beads are moved through one wash of 1 ml GITC buffer[3] and two washes of 1 ml 70% ethanol and DNA eluted in 200 µl of molecular grade water or TE buffer[4] (proceed to step 16).

10. Add 250 µl 5 mol l[-1] NaCl and 500 µl ice-cold isopropanol and mix well.

11. Incubate at −20°C for 30 min.

12. Spin for 10 min at approximately $13,000 \times g$.

13. Remove the supernatant, then wash the pellet by adding 400 µl 70% ethanol and centrifuge at approximately $13,000 \times g$ for 4 min.

14. Decant off the ethanol and dry the pellet to remove residual ethanol.

15. Resuspend the pellet in 100 µl of molecular-biology-grade water or TE buffer[4].

16. Store the resuspended DNA at −20°C.

Notes

[1] Methods for disrupting/homogenizing samples include the use of a mortar and pestle (with or without pre-freezing of the sample), a bead beater or a flat-bed grinder such as the Bioreba HOMEX 6. The method should be selected based on the nature of the sample and the number of samples (see Section 9.2).

[2] CTAB buffer: 2% CTAB (hexadecyltrimethylammonium bromide, synonym cetyltrimethyl ammonium bromide); 100 mmol l[-1] Tris-HCl, pH 8.0; 20 mmol l[-1] EDTA; 1.4 mol l[-1] NaCl; 1.0% Na_2SO_3; and 2.0% PVP (polyvinylpyrrolidone)-40. Stock buffer (containing the first four reagents) can be autoclaved and stored at room temperature. PVP and Na_2SO_3 should be added fresh just before use.

[3] GITC buffer (5.25 M Guanidinium thiocyanate, 50 mM Tris-HCl pH 6.4, 20 mM EDTA, 1.3% Triton X-100).

[4] TE buffer: 10 mmol l[-1] Tris-HCl (pH 8.0), 1 mmol l[-1] EDTA.

Protocol 9.2

Conventional PCR

Table 9.1. PCR components.

Component	Final concentration
PCR buffer	1×
MgCl$_2$	1.5 mmol l^{-1}
dNTPs	0.2 mmol l^{-1} each (dATP, dCTP, dGTP and dTTP)
Forward primer[1]	500 nmol l^{-1}
Reverse primer[1]	500 nmol l^{-1}
Taq polymerase	0.025 U µl^{-1}
Water	To make up final reaction volume of 24 µl

Method

1. Reactions should be set up in a laminar flow cabinet to prevent contamination; DNA and RNA should not be handled in the cabinet.

2. Set up reactions containing the following components (see Table 9.1). Make up a master mix for the number of reactions to be run plus at least four extra reactions to allow for pipetting inaccuracies.

Observe the precautions described in Section 9.3 for the prevention of contamination of the PCR reagents.

3. Vortex briefly, then pipette 24 µl aliquots of master mix into 0.2 ml reaction tubes (alternatively use strips of reaction tubes or 48- or 96-well plates, as appropriate).

4. Add 1 µl sample DNA extract, or water for no-template control reactions, to give a total reaction volume of 25 µl. Do not perform this step in the cabinet used for setting up reactions.

5. Transfer the plate to a suitable thermal cycling instrument, and run with the following cycling conditions: 95°C for 1–10 min,[2] followed by 30–40 three-step cycles of 95°C for 30 s, 60°C for 60 s[3] and 72°C for 60 s.[4]

6. Cycling conditions should be optimized for the assay being used (melting temperature of primers, length of amplicon).

7. Analyse products by agarose gel electrophoresis on a 1–2% agarose gel. For a basic protocol for agarose gel electrophoresis, see Sambrook and Russell (2006).

Notes

[1] Primer concentrations should be optimized for new primer pairs. Typical final concentrations range from 100 nmol l^{-1} to 1 µmol l^{-1}.

[2] The duration of the initial denaturation period depends on the brand of *Taq* polymerase being used.

[3] The annealing temperature should be optimized for new primer pairs. Typical annealing temperatures range from 50°C to 60°C.

[4] The duration of the extension step can be extended for longer products (approximately 1 min kb^{-1}).

Protocol 9.3

TaqMan™ real-time PCR using a QuantStudio™ 12K Flex Real-Time PCR System instrument

---

Materials

- *Taq* polymerase and PCR buffer (usually supplied with *Taq* polymerase)
- MgCl$_2$, dNTPs (deoxyribonucleotide triphosphates; supplied with *Taq* polymerase, or purchased separately)
- Alternatively, a commercial real-time PCR master mix for use with probes
- Target-specific primers (forward and reverse)
- Target-specific dual-labelled TaqMan probe
- Molecular-biology-grade water
- Pipettors – separate sets for setting up reactions and adding DNA to reactions
- Filter tips for pipettors
- Vortexer
- Laminar flow cabinet for setting up reactions

Method

1. Reactions should be set up in a laminar flow cabinet to prevent contamination; DNA and RNA should not be handled in the cabinet.

2. Set up reactions containing the following components (see Table 9.2). Make up a master mix for the number of reactions to be run plus at least four extra reactions to allow for pipetting inaccuracies.

Observe the precautions described in Section 9.3 for the prevention of contamination of the TaqMan real-time PCR reagents.

3. Vortex briefly, then pipette 24 μl aliquots of master mix into the wells of a 96-well plate.

4. Add 1 μl sample DNA extract, or water for no-template control reactions, to give a total reaction volume of 25 μl. Do not perform this step in the cabinet used for setting up reactions.

Table 9.2. TaqMan real-time PCR components.

Component	Final concentration
PCR buffer	1×
MgCl$_2$	5.5 mmol l^{-1}
dNTPs	0.2 mmol l^{-1} each (dATP, dCTP, dGTP and dTTP)
Forward primer[1]	300 nmol l^{-1}
Reverse primer[1]	300 nmol l^{-1}
TaqMan probe[1,2]	100 nmol l^{-1}
Taq polymerase	0.025 U μl^{-1}
Water	To make up final reaction volume of 24 μl

5. Transfer the plate to the QuantStudio 12K Flex Real-Time PCR System instrument (or other suitable real-time platform, following manufacturer's instructions), and run with the following cycling conditions: 95°C for 10 min,[3] followed by 40 two-step cycles of 95°C for 15 s and 60°C for 60 s.[4]

6. Results can be analysed primarily in terms of Ct values (the cycle at which normalized fluorescence exceeds a threshold value). Raw spectra and multi-component data should be examined to confirm integrity of reagents, and for troubleshooting purposes.

Notes

[1] Primer and probe concentrations can be optimized for new assays; however, these concentrations give good results for most assays.

[2] Labelled at the 5′ end with a fluorescent reporter (typically FAM, 6-carboxy-fluorescein) and at the 3′ end with a quencher (typically TAMRA – tetramethyl-rhodamin, or Black Hole Quencher 1).

[3] The optimal duration of the initial denaturation period depends on the brand of *Taq* polymerase being used.

[4] These cycling conditions are suitable for use with the majority of TaqMan assays.

References

Bates, J.A. and Taylor, E.J.A. (2001) Scorpion ARMS primers for SNP real-time PCR detection and quantification of *Pyrenophora teres*. *Molecular Plant Pathology* 2, 275–280.

Benson, D.A., Karsch-Mizrachi, I., Lipman, D.J., Ostell, J. and Sayers, E.W. (2009) GenBank. *Nucleic Acids Research* 37, Database issue, D26–D31.

Berensmeier, S. (2006) Magnetic particles for the separation and purification of nucleic acids. *Applied Microbiology and Biotechnology* 73, 495–504.

Bickley, J. and Hopkins, D. (1999) Inhibitors and enhancers of PCR. In: Saunders, G.C. and Parkes, H.C. (eds) *Analytical Molecular Biology: Quality and Validation*. Royal Society of Chemistry, Cambridge, UK, pp. 81–102.

Bonants, P.J.M., van Gent-Pelzer, M.P.E., Hooftman, R., Cooke, D.E.L., Guy, D.C. and Duncan, J.M. (2004) A combination of baiting and different PCR formats, including measurement of real-time quantitative fluorescence, for the detection of *Phytophthora fragariae* in strawberry plants. *European Journal of Plant Pathology* 110, 689–702.

Boom, R., Sol, C.J.A., Salimans, M.M.M., Jansen, C.L., Wertheim-van Dillen, P.M.E. and van der Noordaa, J. (1990) Rapid and simple method for purification of nucleic acids. *Journal of Clinical Microbiology* 28, 495–503.

Cullen, D.W., Lees, A.K., Toth, I.K. and Duncan, J.M. (2002) Detection of *Colletotrichum coccodes* from soil and potato tubers by conventional and quantitative real-time PCR. *Plant Pathology* 51, 281–292.

Demeke, T. and Adams, R.P. (1992) The effects of plant polysaccharides and buffer additives on PCR. *BioTechniques* 12, 332–334.

Dieffenbach, C.W., Lowe, T.M.J. and Dveksler, G.S. (1993) General concepts for PCR primer design. *Genome Research* 3, S30–S37.

Donoso, A. and Valenzuela, S. (2018) In-field molecular diagnosis of plant pathogens: recent trends and future perspectives. *Plant Pathology* 67, 1451–1461.

Gill, P. and Ghaemi, A. (2008) Nucleic acid isothermal amplification technologies – a review. *Nucleosides, Nucleotides and Nucleic Acids* 27, 224–243.

Henson, J.M. and French, R. (1993) The polymerase chain reaction and plant disease diagnosis. *Annual Review of Phytopathology* 31, 81–109.

Holland, P.M., Abramson, R.D., Watson, R. and Gelfand, D.H. (1991) Detection of specific polymerase chain reaction product by utilizing the 5′ to 3′ exonuclease activity of *Thermus aquaticus* DNA polymerase. *Proceedings of the National Academy of Sciences, USA* 88, 7276–7280.

Ioos, R., Laugustin, L., Schenck, N., Rose, S., Husson, C. and Frey, P. (2006) Usefulness of single copy genes containing introns in *Phytophthora* for the development of detection tools for the regulated species *P. ramorum* and *P. fragariae*. *European Journal of Plant Pathology* 116, 171–176.

Koonjul, P.K., Brandt, W.F., Farrant, J.M. and Lindsey, G.G. (1999) Inclusion of polyvinylpyrrolidone in the polymerase chain reaction reverses the inhibitory effects of polyphenolic contamination of RNA. *Nucleic Acids Research* 27, 915–916.

Kox, L.F.F., Boxman, I.L.A., Jansen, C.C.C. and Roenhorst, J.W. (2005) Reliability of nucleic acid amplification techniques: modified target RNA as exogenous internal standard for a real-time RT-PCR for potato spindle tuber pospiviroid. *Bulletin OEPP/EPPO Bulletin* 35, 117–124.

Lee, L.G., Connell, C.R. and Bloch, W. (1993) Allelic discrimination by nick-translation PCR with fluorogenic probes. *Nucleic Acids Research* 21, 3761–3766.

Lodhi, M.A., Ye, G.N., Weeden, N.F. and Reisch, B.I. (1994) A simple and efficient method for DNA extraction from grapevine cultivars and *Vitis* species. *Plant Molecular Biology Reporter* 12, 6–13.

Luchi, N., Ioos, R. and Santini, A. (2020) Fast and reliable molecular methods to detect fungal pathogens in woody plants. *Applied Microbiology and Biotechnology* 104, 2453–2468.

Ma, Z. and Michailides, T.J. (2007) Approaches for eliminating PCR inhibitors and designing PCR primers for the detection of phytopathogenic fungi. *Crop Protection* 26, 145–161.

Malapi-Wight, M., Demers, J.E., Veltri, D., Marra, R.E. and Crouch, J. (2016) LAMP detection assays for boxwood blight pathogens: a comparative genomics approach. *Scientific Reports* 6, 26140.

Martin, R.R., James, D. and Lévesque, C.A. (2000) Impacts of molecular diagnostic technologies on plant disease management. *Annual Review of Phytopathology* 31, 207–239.

McCartney, H.A., Foster, S.J., Fraaije, B.A. and Ward, E. (2003) Molecular diagnostics for fungal plant pathogens. *Pest Management Science* 59, 129–142.

Mori, Y., Nagamine, K., Tomita, N. and Notomi, T. (2001) Detection of loop-mediated isothermal amplification reaction by turbidity derived from magnesium pyrophosphate formation. *Biochemical and Biophysical Research Communications* 289, 150–154.

Mori, Y., Kitao, M., Tomita, N. and Notomi, T. (2004) Real-time turbidimetry of LAMP reaction for quantifying template DNA. *Journal of Biochemical and Biophysical Methods* 59, 145–157.

Morley, A.A. (2014) Digital PCR: A brief history. *Biomolecular Detection and Quantification* 1, 1–2.

Mumford, R.A., Tomlinson, J., Barker, I. and Boonham, N. (2006) Advances in molecular phytodiagnostics – new solutions for old problems. *European Journal of Plant Pathology* 116, 1–19.

Nagamine, K., Watanabe, K., Ohtsuka, K., Hase, T. and Notomi, T. (2001) Loop-mediated isothermal amplification reaction using a nondenatured template. *Clinical Chemistry* 47, 1742–1743.

Nagamine, K., Hase, T. and Notomi, T. (2002) Accelerated reaction by loop-mediated isothermal amplification using loop primers. *Molecular and Cellular Probes* 16, 223–229.

Notomi, T., Okayama, H., Masubuchi, H., Yonekawa, T., Watanabe, K. *et al.* (2000) Loop-mediated isothermal amplification of DNA. *Nucleic Acids Research* 28, e63.

Piepenburg, O., Williams, C.H., Stemple, D.L. and Armes, N.A. (2006) DNA detection using recombination proteins. *PLoS Biology* 4, 1115–1121.

Rački, N., Dreo, T., Gutierrez-Aguirre, I., Blejec, A. and Ravnikar, M. (2014) Reverse transcriptase droplet digital PCR shows high resilience to PCR inhibitors from plant, soil and water samples. *Plant Methods* 10, 42.

Rozen, S. and Skaletsky, H.J. (2000) Primer3 on the WWW for general users and for biologist programmers. In: Krawetz, S. and Misener, S. (eds) *Bioinformatics Methods and Protocols: Methods in Molecular Biology*. Humana Press, Totowa, New Jersey, pp. 365–386.

Sambrook, J. and Russell, D.W. (2006) Agarose gel electrophoresis. *Cold Spring Harbour Protocols* doi:10.1101/pdb.prot4020.

SantaLucia, J. Jr (2007) Physical principles and Visual-OMP software for optimal PCR design. In: Yuryev, A. (ed.) *Methods in Molecular Biology: PCR Primer Design*. Humana Press, Totowa, New Jersey, pp. 3–33.

Schena, L. and Cooke, D.E.L. (2006) Assessing the potential of regions of the nuclear and mitochondrial genome to develop a 'molecular toolbox' for the detection and characterization of *Phytophthora* species. *Journal of Microbiological Methods* 67, 70–85.

Schena, L., Nigro, F., Ippolito, A. and Gallitelli, D. (2004) Real-time quantitative PCR: a new technology to detect and study phytopathogenic and antagonistic fungi. *European Journal of Plant Pathology* 110, 893–908.

Suarez, M.B., Walsh, K., Boonham, N., O'Neill, T., Pearson, S. and Barker, I. (2005) Development of real-time PCR (TaqMan) assays for the detection and quantification of *Botrytis cinerea* in planta. *Plant Physiology and Biochemistry* 43, 890–899.

Tang, J., Zheng, L., Jia, Q., Liu, H., Hsiang, T. and Huang, J. (2017) PCR markers derived from comparative genomics for detection and identification of the rice pathogen *Ustilaginoidea virens* in plant tissue. *Plant Disease* 101, 1515–1521.

Tomlinson, J. and Boonham, N. (2008) Potential of LAMP for detection of plant pathogens. *CAB Reviews: Perspectives in Agriculture, Veterinary Science, Nutrition and Natural Resources* 3, 066.

Tomlinson, J., Barker, I. and Boonham, N. (2007) Faster, simpler, more specific methods for improved molecular detection of *Phytophthora ramorum* in the field. *Applied and Environmental Microbiology* 73, 4040–4047.

Tsai, Y.-L. and Olson, B.H. (1992) Detection of low numbers of bacterial cells in soils and sediments by polymerase chain reaction. *Applied and Environmental Microbiology* 58, 754–757.

Tyagi, S. and Kramer, F.R. (1996) Molecular beacons: probes that fluoresce upon hybridisation. *Nature Biotechnology* 14, 303–308.

Weller, S.A., Elphinstone, J.G., Smith, N.C., Boonham, N. and Stead, D.E. (2000) Detection of *Ralstonia solanacearum* strains with a quantitative, multiplex, real-time, fluorogenic PCR (TaqMan) assay. *Applied and Environmental Microbiology* 66, 2853–2858.

Whitcombe, D., Theaker, J., Guy, S.P., Brown, T. and Little, S. (1999) Detection of PCR products using self-probing amplicons and fluorescence. *Nature Biotechnology* 17, 804–807.

Winton, L.M., Stone, J.K., Watrud, L.S. and Hansen, E.M. (2002) Simultaneous one-tube quantification of host and pathogen DNA with real-time polymerase chain reaction. *Phytopathology* 92, 112–116.

Wong, M.L. and Medrano, J.F. (2005) Real-time PCR for mRNA quantitation. *BioTechniques* 39, 75–85.

10 DNA Fingerprinting Techniques

David Cooke*

The James Hutton Institute, Invergowrie, Dundee, Scotland, UK

10.1 Introduction

The paucity of phenotypic variation in many fungal plant pathogens drives the need for other fingerprinting techniques that underpin initial pathogen identification and the tracking of genetic variation in populations. Although they have been available for at least 30 years, it is perhaps in the last 15 years that advanced molecular techniques, with a particular focus on variation at the level of DNA sequence, have become more commonplace. Increased access to equipment and reagents for PCR and more affordable DNA sequencing services allows both fine-scale fingerprinting and identification by comparisons against ever-expanding international DNA sequence databases. It is not all plain-sailing, however; a challenge for today's scientists lies in the correct interpretation of database 'matches', sifting high-quality, robust data from that which is erroneous. Initiatives such as the 'International Barcode of Life' (https://ibol.org/; accessed 22 February 2023) seek to standardize methods and harness these data more effectively. Advances such as high-throughput sequencing technologies offer exciting opportunities to collect molecular data that improve our understanding of the evolution and biology of fungal pathogens.

The terms genotyping and DNA fingerprinting may be used interchangeably.

Integrated disease management is aided by the study of pathogen characteristics (or phenotypes) such as aggressiveness or virulence on current crop cultivars, response to fungicide-active ingredients or mating type. However, a lack of resolution of these phenotypic traits in fungi often leaves many questions unanswered. For example, if an important fungicide fails to suppress a pathogen, it is important to know more about the genetic basis of that failure. Understanding whether it was due to a single mutation in a specific localized population or to a wider problem occurring simultaneously in many populations would influence the response and management strategy. There has thus been a major shift in recent years towards the examination of variation in the underlying genetic make-up of an organism, i.e. its genotype, rather than its phenotype. Scientific advances in the methods of DNA-based fingerprinting are allowing the discrimination of individuals based on minor, and therefore often relatively recently evolved, differences in DNA sequence. Such progress has revolutionized the study of fungal diversity and advanced our knowledge of the population genetics and phylogeography of fungal and other microbial pathogens.

* E-mail: david.cooke@hutton.ac.uk

©CAB International 2023. *Fungal Plant Pathogens: Applied Techniques, 2nd Edition*
(eds C.R. Lane, P.A. Beales and K.J.D. Hughes)
DOI: 10.1079/9781800620575.0010

Historically, a range of methods has been used to examine the biochemical or genetic diversity of pathogen populations. Early methods looked for differences in total protein extracts separated by molecular weight using polyacrylamide gel electrophoresis (PAGE). These were displaced by methods in which specific stains were used to discriminate allelic variants of particular proteins. Such allozyme analysis was widely adopted but has been replaced by DNA-based fingerprinting. The first DNA-based assays were those based on restriction fragment length polymorphisms (RFLPs), which involved the digestion of relatively large amounts of total fungal DNA with restriction enzymes that cleave the DNA strand at specific base-pair combinations. These digested DNA fragments were size-separated by agarose gel electrophoresis and transferred to a nylon membrane which was subsequently exposed to a labelled fragment of DNA (or probe). This probe bound (or hybridized) to complementary DNA fragments on the membrane, and the hybridization was visualized as a banding pattern or 'fingerprint' that revealed genetic differences among the fungal isolates being tested. Such methods were effective but lengthy, technically demanding and could be difficult to interpret compared with other methods.

PCR amplification has yielded a new generation of methods that have revolutionized fungal DNA fingerprinting and identification. These fit into two broad categories that examine genetic variation in different ways. First, there are methods that amplify many fragments of unknown DNA sequence that are sorted by size and visualized as a banding pattern. These include randomly amplified polymorphic DNA (RAPD) and amplified fragment length polymorphism (AFLP). Although AFLPs were improved by accurate product sizing on capillary sequencers (Fig. 10.1), both methods have been largely superseded by a second category that uses highly specific PCR primer pairs to amplify multiple DNA fragments at known locations, termed loci, in the genome. These are further characterized by accurate sizing on a capillary sequencer or by DNA sequencing to detect allelic variation at these loci. Simple sequence repeat (SSR) analysis identifies alleles of differing length (Fig. 10.2, Section 10.2) and sequencing detects single nucleotide polymorphisms (SNPs, Section 10.3) in DNA sequence across fragments of the genome. The sequencing approach is used to identify a specific DNA barcode (Section 10.4). A third hybrid approach, termed genotyping by sequencing (GBS), uses restriction enzymes to digest genomic

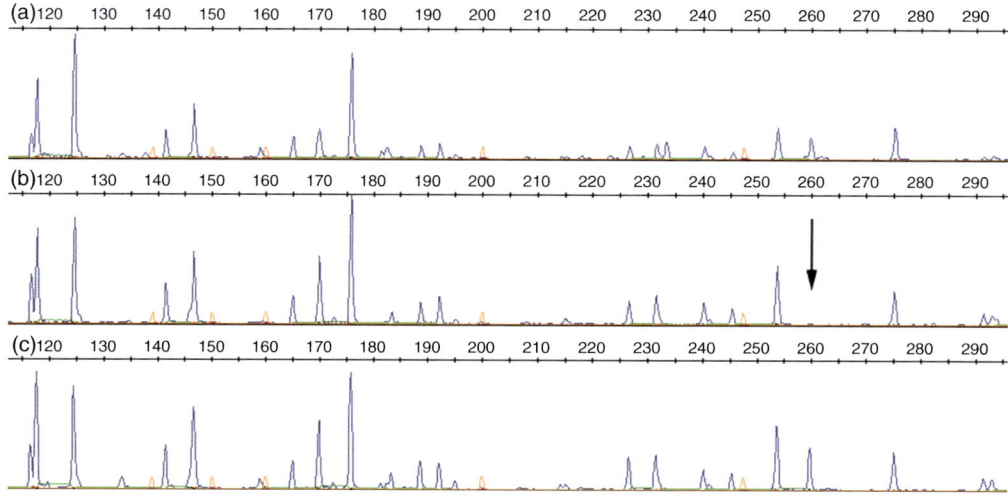

Fig. 10.1. FAM (fluorescent reporter 6-carboxy-fluorescein)-labelled AFLP (amplified fragment length polymorphism) peak profiles of three isolates (a–c) of a clonal population of *Phytophthora rubi* amplified with the E11 and M15 selective primers. The majority of peaks are present in all three isolates, but a deletion of a 260 bp peak is marked with an arrow in isolate b. The smaller orange peaks are molecular size markers (139–250 bp) labelled with the fluorescent dye LIZ. (Content is the author's own annotated screen shot from GeneMapper™ v.5.0 (Applied Biosystems, Foster City, California).)

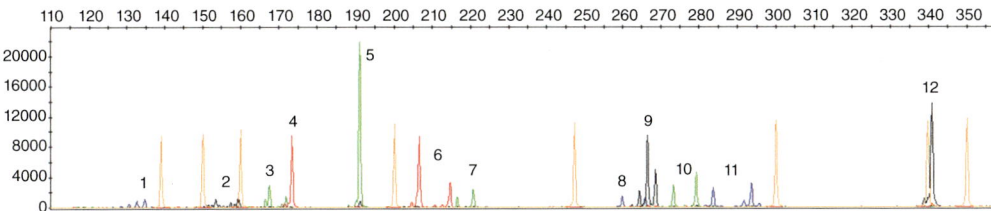

Fig. 10.2. Peaks generated from multiplex amplification of 12 SSR (simple sequence repeat) loci of an isolate of *P. infestans*. Each of the 12 loci is labelled with one of four different fluorescent dyes (NED, PET, FAM or VIC) and run on an Applied Biosystems 3730 DNA Analyzer. Some loci are homozygous with one peak representing a single allele (e.g. 4, 5 and 12) while others are heterozygous with two peaks, indicating two alleles (e.g. 6, 10 and 11). In some cases, stutter peaks are apparent (1). The uniform orange peaks are the LIZ-labelled molecular size markers (139–350 bp). (Content is the author's own annotated screen shot from GeneMapper™ v.5.0 (Applied Biosystems, Foster City, California).)

DNA followed by amplification and sequencing of a specific sub-set of the fragments using high-throughput methods. In all cases, an individual fungal isolate is characterized to determine its haplotype or multi-locus genotype (MLG) which is then compared to that of other individuals in the population.

This chapter will provide an update to a review by Cooke and Lees (2004) and focus on SSRs, SNP discovery and DNA barcoding methods, which are currently the most commonly applied methods in fungal DNA fingerprinting. High-throughput and massively parallel DNA sequencing technology is revolutionizing the field. Several platforms, such as Illumina's sequencing-by-synthesis, are now available that can sequence hundreds of millions of short (50–500 bp) DNA fragments in a single run. Other powerful options are the nanopore sequencing platforms, such as those of Oxford Nanopore, that read the sequence of single DNA strands as they pass through engineered protein pores. Sequence error rates have fallen which now makes this technology appropriate for fingerprinting (Kahvejian *et al.*, 2008; Wibberg *et al.*, 2021). While not yet available to most research teams, technical advances are bringing down the costs and increasing the application of such technology. The recent publication of the whole genome sequences of over 1000 isolates of yeast (Peter *et al.*, 2018), for example, offers the ultimate in base pair by base pair fungal DNA fingerprinting. Such phylogenomic studies allow extremely detailed sequence comparisons of isolates from which the mechanisms and processes that drive fungal diversity can be explored.

Another burgeoning high-throughput sequencing application is the field of metabarcoding. Total DNA is extracted from a particular environmental sample and the phylogenetic diversity of a subsection of microbial life in this so-called 'microbiome' is screened via high-throughput DNA sequencing of a specific barcode locus (e.g. Simon and Daniel, 2009). Such environmental DNA (eDNA) analysis is distinct from fingerprinting methods described above that detect within-species genetic variation by instead fingerprinting between-species DNA sequence variation across a community. The methods are culture independent and thus particularly powerful for identifying microbes that have, to date, proved impossible to culture or are, as yet, unknown to science. Many studies have focused specifically on fungal diversity, but this approach has great potential for probing both the biodiversity and functional diversity of many microbial communities (Stat *et al.*, 2017). There are, however, many challenges in the efficient and accurate interpretation of such large volumes of sequence data that require the ongoing development of sophisticated bioinformatic tools and databases.

10.2 SSRs (Simple Sequence Repeats)

SSRs (also known as microsatellites or short tandem repeats, STRs) are sections of DNA sequence in which short combinations of 1–6 nucleotides are tandemly repeated. The sequence

AGAGAGAGAGAGAGAGAG, for example, represents eight AG repeats. Repeat length varies dramatically from a few to hundreds or thousands but, in practice, regions with 4–30 repeats are normally used as genetic markers for fingerprinting (Lees *et al.*, 2006; Li *et al.*, 2013). SSRs are prone to slippage-based errors during DNA replication, which leads to differing numbers of the repeat and thus to length variation. This length variation at each specific locus can be scored by PCR amplification of the small (100–500 bp) fragment using primer pairs based on the flanking sequence, followed by very accurate sizing of the PCR product, previously on PAGE gels, but generally now via capillary electrophoresis systems. Since its first use in the late 1980s, this method has become one of the most widely used means of fingerprinting (Ellegren, 2004); it is, for example, the basis of most criminal forensic DNA profiling. The mutation rate of SSRs is sufficiently high to generate differences among individuals within a population, but also is stable enough to track allelic forms at appropriate timescales. Of the many repeat combinations, it is dinucleotide (e.g. GTGTGTGTGT or GT_5) and trinucleotide repeats (e.g. CAGCAG-CAGCAG or CAG_4) that are commonly used in microbial fingerprinting.

Unlike the multi-locus methods such as AFLP markers, fewer SSR markers per isolate are examined. Typically, 5–15 polymorphic SSR loci are identified, the alleles at each are scored and these data are combined to generate a fingerprint or MLG for that specific isolate. Each locus is clearly defined and, in the case of diploid or polyploid organisms, both alleles are discriminated (i.e. they are codominant markers). The assays are easily scored and objective, and an exchange of protocols and reference isolates allows different laboratories to generate standardized data sets that can be compared meaningfully. An additional advantage is that the PCR primers for each locus are generally species specific, which means that the pathogen DNA can be amplified directly from infected plant material without the need to isolate the pathogen into pure culture. Analysis of historic or herbarium samples is also possible, and pathogen DNA pressed directly from an active lesion onto FTA® cards (Whatman, Kent, UK) has proved a very effective way of fingerprinting *Phytophthora infestans* isolates.

A disadvantage of SSRs is the long lead-in time needed for their discovery and optimization before their application. If sequence data are available, bioinformatic tools can be used to 'mine' for SSRs and a large selection of candidate markers screened against a panel of isolates from the fungal population. Low-cost and high-throughput sequencing technologies have reduced the time required to generate appropriate sequence data. Once sufficient polymorphic markers are discovered, the parameters of the PCR assay must be optimized to yield peaks that are easy to score. Some markers are prone to form 'stutter' products owing to slippage during DNA replication by *Taq* polymerase (e.g. marker 1 in Fig. 10.2) but this may be reduced by the addition of short oligonucleotide extensions to the 5′ end of the unlabelled primer (so-called 'pigtails'; Brownstein *et al.*, 1996).

A final step in the optimization is to reduce costs and increase throughput by using multiplex PCR. The mixing of multiple primer pairs in a single reaction can yield unpredictable artefact amplification, so a careful stepwise process of combining increasing numbers of primers to amplify reference isolates is recommended. The primer concentration and the amount of PCR product used on the gel/capillary system needs to be adjusted to standardize the peak heights because PCR amplification efficiency varies between primer sets. Commercial kits with PCR reagents optimized for multiplex PCR are available and may simplify the process. The procedure presented in Protocol 10.1 (Li *et al.*, 2013) is that used for the multiplex amplification of 12 SSR loci in *P. infestans* by EuroBlight (The Potato Late Blight Network for Europe: www.euroblight. net; accessed 22 February 2023).

10.3   SNP (Single Nucleotide Polymorphism) Analysis

SNPs (pronounced 'snips') are base-pair differences in DNA sequence that occur due to substitutions, insertions or deletions (Kwok, 2003). Such mutations are the main source of genetic variation in any genome and SNPs occur at frequencies of, for example, 1.1 kb^{-1} and 2.8 kb^{-1} of sequence in humans and yeast, respectively. The identification of SNPs within fungal isolates and the scoring of any variability within the population is thus a practical fingerprinting method. The scale of SNP analysis varies dramatically: from the tracking of a specific single base-pair mutation in a gene that conditions fungicide resistance (Torriani *et al.*, 2009) in a fungal pathogen to studies examining associations between human SNP variation and disease in which many millions of SNPs are scored. SNPs may be selected within non-coding DNA, where the outcome of a change is likely to be selectively neutral. Conversely, a SNP that encodes a change in an amino acid in a functional gene may be subject to positive or negative selection within the population. The ability to target specific functional mutations in this manner is an advantage of SNPs and, in combination with neutral (e.g. non-coding) SNPs, they can be powerful markers for fingerprinting fungal populations.

The first step is to identify sufficient genetically unlinked SNPs, which, in the absence of available sequence data, is a daunting and lengthy process. Also, as the variation involves single base-pair changes that do not alter the PCR product length, SNPs are more technically challenging to discriminate than SSRs. While multiple alleles of different lengths may be scored at a single SSR locus (i.e. they are multi-allelic), in SNPs each marker is bi-allelic (e.g. a cytosine (C) or thymine (T) residue) and less information is therefore gained per marker scored. In practice then, many more SNP than SSR loci must be identified and scored for an equivalent level of genotypic resolution (Brumfield *et al.*, 2003).

Factors such as these have tended to reduce the application of SNPs as a tool for fingerprinting fungal populations and limited publications in this field. Several SNP assays are available, but all have some limitations and there is no single 'recommended' assay. Historically, assays were based on PCR primers with specificity against the actual SNP nucleotide, whereas others amplified a product that includes the SNP and both flanking regions. This relied on differences in restriction digestion, DNA melting profiles or conformation polymorphisms between variant DNA strands to discriminate each allele. The ongoing fall in DNA sequencing costs is driving the replacement of such methods with direct amplification and parallel sequencing approaches in which short-fragment sequence data are screened for SNPs and used directly for genetic analysis.

High-throughput sequencing is an extension of prior multi-locus sequence typing (MLST) and identifies multiple sequence haplotypes in an individual fungal sample. The choice of target sequence depends on the objective of the study. For example, a study of *P. infestans* specifically targeted pathogen effector genes using a DNA bait-based enrichment process to extract the target genes for sequencing (e.g. Thilliez *et al.*, 2018). More commonly a selection of target regions at random is used as it requires no prior knowledge of the organism's genome. Genotyping-by-sequencing (GBS) is a widely used approach that uses DNA restriction enzymes to cut total genomic DNA followed by PCR amplification of the resulting cleavage site-defined short fragments and high-throughput sequencing to identify SNP variation in the resultant haplotypes (Poland *et al.*, 2012). These are increasingly common approaches for studies in which the objective is a broader view of population structure. Such sequence data are used in the reconstruction of gene genealogies or in phylogeographic analysis within fungal species using novel analysis tools. Improved sequencing chemistry, reduced costs and new analysis tools are increasing the applicability of this approach. It has, for example, been applied to the study of variability and mating patterns of vascular wilt pathogens (Milgroom *et al.*, 2014). There are many GBS profiling methods and it is beyond the scope of this chapter to present a single detailed protocol.

10.4 Barcoding and Metabarcoding

The assays discussed above are used to genetically fingerprint isolates within a fungal species. However, in many cases, it may be a challenge to identify the fungal species in the first place. There will always be a role for experienced mycologists, but there is increasing recognition of the power of DNA-based methods of species identification. Although the term 'DNA barcoding' was only used in 2003, the generic PCR primers currently used to amplify fungal ribosomal RNA (rRNA) genes (rDNA) were published almost 20 years ago (White *et al.*, 1990). Such methods were adopted widely from the mid-1990s onwards with the advent of automated sequencing machines.

An ever-expanding database of DNA sequences of key target genes such as the rDNA spacer regions and the mitochondrial cytochrome oxidase (COX) has resulted in powerful species identification systems. PCR amplification of the specific target gene from the fungal isolate is followed by sequencing and a similarity search against a database (e.g. BLAST) that returns 'hits' ranked according to their sequence similarity. In many cases, the sequences for a particular species are linked to a peer-reviewed publication and a voucher specimen, which provides a form of 'quality control'. However, unpublished material, uncorrected errors and taxonomic revisions are reasons to consider database search results as a guide rather than as providing a definitive answer. Account must be taken of other issues, such as intraspecific variation or the existence of distinct but closely related species with identical sequences. The Barcode of Life Data Systems (BOLD) provides a more focused single database of reference barcodes (Ratnasingham and Hebert, 2007). There has been much discussion on the gene, or genes, most suited for DNA barcoding (Chase and Fay, 2009). For simplicity, a single barcode gene would be ideal, but no single region appears to suit all forms of life and the consensus is now that more than one region is needed for fungi as both the CO1 (COX gene 1; Seifert *et al.*, 2007) and the ITS (internal transcribed spacer) regions have limitations in some groups. Difficulties include the absence of a clear 'barcode gap' between species in some taxa, the resolution of hybrid species and challenges with sequence alignments (e.g. in the case of the ITS region).

Barcoding individual isolates is useful, but the method is increasingly being used for characterizing all the species in an eDNA (environmental DNA) sample from a community. This is possible due to a parallel PCR amplification and high-throughput sequencing method called metabarcoding. A typical scenario would be to examine the fungal population diversity at a range of scales from a specific ecosystem via targeted sample of, for example, roots, soil or water. The eDNA extracted from each sub-sample is subject to amplification with a generic pair of PCR primers that amplify a species-specific barcode at the relevant taxonomic resolution (genus, family, etc.). These PCR products, often comprising multiple distinct barcodes, are then individually labelled with a short DNA sequence tag that is used to identify them after downstream sequencing. All the samples are then pooled to a single sequencing library that is sequenced to yield millions of short sequences. These are aligned and deconvoluted to generate a collection of specific barcodes for each eDNA sample. A computational pipeline is then required for the crucial quality control and subsequent fungal taxon identification stages. A brief outline of the stages for metabarcoding using the Illumina MiSeq system is provided in Protocol 10.2.

Protocol 10.1

Multiplex simple sequence repeat (SSR) analysis using fluorescent detection

Materials

DNA extracted from fungal isolates

- Amplification of SSR loci is relatively robust and insensitive to differences in DNA quality and quantity. Several different DNA extraction protocols have been successfully applied at the James Hutton Institute (JHI), including a rapid NaOH-based method (Wang *et al.*, 1993), commercial kits (e.g. Gentra® Puregene® DNA extraction kit, Qiagen Cat. No. 158567) and more complex phenol-chloroform based methods (Raeder and Broda, 1985). Extraction directly from infected host material using small (25 mm²) sections from the edge of fresh or dried lesions has proved successful, as has the use of FTA® cards (FTA Classic Card, Whatman WB120205) using the manufacturer's protocols for Plant DNA extraction. Fluorescent SSR detection is very sensitive and typically 10 ng of fungal DNA is sufficient for amplification.

PCR primers

- Forward and reverse primers for each SSR locus are required. One primer of each pair (normally the forward primer) must be labelled with a fluorescent dye. The dye used depends on the fluorescent detection system on the instrument used to size the PCR product (e.g. for Applied Biosystems™ machines FAM, NED, PET and VIC are used). Typically, 100 µM stock solutions are used and small aliquots of 10 µM prepared to minimize freeze–thaw cycles for the stocks. Fluorescent primers must also be protected from daylight to prevent degradation. In this protocol, the following 12 SSR loci are amplified in a single multiplex reaction: FAM (D13, SSR4, SSR8), NED (Pi02, PiG11, SSR11), PET (Pi4B, SSR2) and VIC (Pi04, Pi63, Pi70, SSR6). Primer concentrations are altered to standardize peak heights to approximately 10,000 (arbitrary units). See *Phytophthora infestans* protocols available from EuroBlight (www.euroblight.net; accessed 22 February 2023).

PCR reagents

- The protocol below is based on specialized reagents optimized for multiplex SSRs (Type-it Microsatellite Kit, Qiagen, Cat. No. 206243). Reagents for denaturation of the PCR product and a within-sample size marker are also required: Applied Biosystems Hi-Di™ Formamide (Cat No. 4401457) and Applied Biosystems GeneScan™ 500 LIZ™ dye Size Standard (Cat No. 4322682), both available from ThermoFisher.

Equipment

- PCR machine, pipettes and associated plastic-ware.
- Applied Biosystems (ABI) 3730 DNA Analyzer or equivalent capillary sequencer equipment.

Method

1. PCR amplification of SSR loci:
   - Set up reactions containing the components shown in Table 10.1. Make up a master mix for the number of reactions required (generally 48 or 96) plus 10% extra to allow for pipetting inaccuracies.
   - Observe the precautions described in Chapter 9 for the prevention of contamination of the PCR reagents.
   - Dispense 11.5 µl aliquots of master mix into 0.2 ml tubes in 96-well plates.
   - Add 1 µl sample DNA extract, or water for no-template control reactions, to give a total reaction volume of 12.5 µl. For FTA

Table 10.1. Master mix for PCR amplification of SSR (simple sequence repeat) loci.

Component	Final concentration
2× Qiagen Type-IT Multiplex PCR Mix	6.25 µl
Primer mix (24 primers, each pair at final concentrations ranging from 0.3 to 0.03 µM)	0.44 µl
HPLC water	4.81 µl
Total volume	11.5 µl

cards, place the dry disk in the well and add 1 µl HPLC water more per sample to the master mix.

- Transfer the plate to a thermal cycling instrument, and run with the following cycling conditions: 95°C for 5 min, 28 three-step cycles[1] of 58°C[2] for 90 s, 72°C for 20 s and 95°C for 30 s, followed by a single 30 min step at 60°C.

2. Running samples on capillary sequencer:
   - Optimize PCR product dilution. Generally, a 1:20 dilution is required for the next stage, but this will vary depending on the amplification efficiency.
   - Prepare the LIZ size standard and Hi-Di Formamide mix using 6 µl and 10.4 µl, respectively, per reaction. Dispense 10.2 µl of this mix into each well of a 96-well ABI 3730 plate (ABGene™ Thermo-Fast™ 96, non-skirted Cat. No. AB-0600) and add 0.6 µl of the diluted PCR product.

- After a short pulse on a centrifuge, seal plates and run on the ABI 3730. For this protocol, a 48 capillary system with 36 cm capillaries, POP 7 polymer and ABI 10X TBE (made up to 1× for use) is used, with electrokinetic injection for 10 s under ABI's default conditions. Detection is based on the settings for dye set G5 (which detects FAM, VIC, NED, PET and LIZ).

3. Process the results:
   - ABI's GeneMapper™ 5.0 is used to inspect and standardize the peaks into named bins.

Notes

[1] Cycle number is increased to 33 for FTA card samples.
[2] The annealing temperature should be optimized for new primer pairs. Typical annealing temperatures range from 50°C to 60°C.

Protocol 10.2

Metabarcoding

Materials

DNA extracted from substrates

- DNA from single microbial cultures is extracted. Since PCR is relatively insensitive to differences in DNA quality and quantity, a range of extraction procedures that release PCR-quality DNA can be applied. NaOH-based methods are rapid and simple (see Protocol 10.1).
- eDNA may be extracted from substrates such as water filters, soil or plant roots using a range of commercial DNA extraction and purification kits.

Reagents, primers and adapters

- Suitable generic primers for CO1 (cytochrome oxidase (COX) gene 1) or ITS (internal transcribed spacer) genes. Primers for rDNA amplification from White *et al*. (1990) are appropriate for most fungi. However, the ITS5 rather than ITS1 primer should be used for ITS amplification owing to problems of self-complementarity in the latter.
- Specific PCR primers for groups of organisms should be used. Examples include a set for *Phytophthora* and related downy mildew species (Scibetta *et al*., 2012), or fungal primers appropriate for different taxonomic groups (Bellemain *et al*., 2010).
- PCR reagents including a high-fidelity DNA polymerase to reduce PCR sequence error rates in downstream output.
- Metabarcoding reagents, including PCR fragment clean-up kits, DNA quantification assays such as PicoGreen, Illumina index kits and Flow Cells.

Equipment

- PCR machine, pipettes and associated plastic-ware.
- Sequencing platform such as Applied Biosystems 3730 capillary sequencer for low-throughput Sanger sequencing.
- High-throughput sequencing platform such as the Illumina MiSeq.

Method

The method below provides a highly abbreviated overview of the key stages of the Illumina metabarcoding protocol for the MiSeq sequencing system. Readers are directed to the detailed Illumina protocol (see Resources, below) and general guidance documentation (Pawlowski *et al.*, 2020).

1. PCR amplification of target barcode region from individual eDNA samples.

Amplification using primers suitable for your taxa of interest should be conducted using standard PCR conditions. Primers should generate a product of approximately 400 bp and need to be extended with Illumina adapters

that are required for later indexing steps. Using a high-fidelity polymerase is recommended to improve the amplification efficiency and reduce error rates. Nested PCR may be required to obtain the required sensitivity but does increase the risk of cross-contamination. Extreme care is needed to avoid PCR contamination through steps such as using separate laboratories for pre- and post-PCR work, using small aliquots of reagents and multiple negative control reactions per batch. Samples that amplify successfully can be moved to step two.

2. Preparation of the fragment library involving:

a. PCR cleanup

Use AMPure XP beads and magnetic separation to purify the PCR product of each sample of free primers and primer dimer products. Care is needed to ensure no carry-over of beads into the next stage.

b. Second stage index PCR

Eight cycles of PCR are run to add a unique oligonucleotide tag to every amplicon in each sample. This enables sample discrimination after sequencing (deconvolution). The use of proprietary Nextera® XT index primers from Illumina is recommended.

c. PCR cleanup two

Use AMPure XP beads and magnetic separation as in step 2a and check the sample amplification using standard gel electrophoresis.

3. Library quantification, pooling and normalization.

The first stage is to check and then normalize the concentrations of each amplification product using fluorometric assays with double-stranded DNA binding dyes (for example, PicoGreen kit from ThermoFisher). Once normalized, the individually indexed samples can be pooled into a single library. The number of samples that can be pooled depends on the depth of sequencing required per sample.

4. Library denaturing and sample loading.

Before preparing for sequencing, the pooled and diluted library is denatured with NaOH, further diluted with a hybridization buffer and then heat denatured. Depending on the expected sequence complexity of the library, it should be spiked with differing amounts of a PhiX control sequence. There are a range of Illumina kits for the MiSeq system and the choice depends, in part, on the length of the barcode and the depth of sequencing coverage required.

5. Post-PCR computational analysis.

After many cycles of sequencing, 1–30 million barcode reads are available. The Illumina protocols run a preliminary quality check and deconvolution to link each sequence to the original sample code. The data comprise many paired reads from each sample that need be processed to identify matches in the data to a reference database of the taxa of interest. The key steps of this processing pipeline are (1) pair the overlapping reads and trim away primer sequences, (2) classification of the sequences against a prepared reference database of known species or comparisons to external databases using BLAST and (3) report the output of the classifications per sample in human-readable format, ideally integrating associated sample metadata. THAPBI PICT is an example of a pipeline developed for *Phytophthora* but also applicable to other taxa (see Resources, below).

References

Bellemain, E., Carlsen, T., Brochmann, C., Coissac, E., Taberlet, P. and Kauserud, H. (2010) ITS as an environmental DNA barcode for fungi: an *in silico* approach reveals potential PCR biases. *BMC Microbiology* 10, 189.

Brownstein, M.J., Carpten, J.D. and Smith, J.R. (1996) Modulation of non-templated nucleotide addition by tag DNA polymerase: primer modifications that facilitate genotyping. *BioTechniques* 20, 1004–1010.

Brumfield, R.T., Beerli, P., Nickerson, D.A. and Edwards, S.V. (2003) The utility of single nucleotide polymorphisms in inferences of population history. *Trends in Ecology and Evolution* 18, 249–256.

Chase, M.W. and Fay, M.F. (2009) Barcoding of plants and fungi. *Science* 325, 682–683.

Cooke, D.E.L. and Lees, A.K. (2004) Markers, old and new, for examining *Phytophthora infestans* diversity. *Plant Pathology* 53, 692–704.

Ellegren, H. (2004) Microsatellites: simple sequences with complex evolution. *Nature Reviews Genetics* 5, 435–445.

Kahvejian, A., Quackenbush, J. and Thompson, J.F. (2008) What would you do if you could sequence everything? *Nature Biotechnology* 26, 1125–1133.

Kwok, P.Y. (2003) *Single Nucleotide Polymorphisms – Methods and Protocols*. Humana Press, Totowa, New Jersey.

Lees, A.K., Wattier, R., Sullivan, L., Williams, N.A. and Cooke, D.E.L. (2006) Novel microsatellite markers for the analysis of *Phytophthora infestans* populations. *Plant Pathology* 55, 311–319.

Li, Y., Cooke, D.E., Jacobsen, E. and van der Lee, T. (2013) Efficient multiplex simple sequence repeat genotyping of the oomycete plant pathogen *Phytophthora infestans*. *Journal of Microbiological Methods* 92, 316–322.

Milgroom, M.G., Jimenez-Gasco, M. del M., Olivares Garcia, C., Drott, M.T. and Jimenez-Diaz, R.M. (2014) Recombination between clonal lineages of the asexual fungus *Verticillium dahliae* detected by genotyping by sequencing. *PLoS One* 9(9), e106740.

Pawlowski, J., Apothéloz-Perret-Gentil, L., Mächler, E. and Altermatt, F. (2020) *Environmental DNA Applications in Biomonitoring and Bioassessment of Aquatic Ecosystems. Guidelines.* Environmental Studies no. 2010. Federal Office for the Environment, Bern, Switzerland.

Peter, J., De Chiara, M., Friedrich, A., Yue, J.X., Pflieger, D. *et al.* (2018) Genome evolution across 1,011 *Saccharomyces cerevisiae* isolates. *Nature* 556, 339–344.

Poland, J.A., Brown, P.J., Sorrells, M.E. and Jannink, J.-L. (2012) Development of high-density genetic maps for barley and wheat using a novel two-enzyme genotyping-by-sequencing approach. *PLoS One* 7: e32253.

Raeder, U. and Broda, P. (1985) Rapid preparation of DNA from filamentous fungi. *Letters in Applied Microbiology* 1, 17–20.

Ratnasingham, S. and Hebert, P.D.N. (2007) BOLD: The Barcode of Life Data System (www.barcodinglife.org). *Molecular Ecology Notes* 7, 355–364.

Scibetta, S., Schena, L., Chimento, A., Cacciola, S.O. and Cooke, D.E.L. (2012) A molecular method to assess *Phytophthora* diversity in environmental samples. *Journal of Microbiological Methods* 88, 356–368.

Seifert, K.A., Samson, R.A., deWaard, J.R., Houbraken, J., Lévesque, C.A. *et al.* (2007) Prospects for fungus identification using CO1 DNA barcodes, with *Penicillium* as a test case. *Proceedings of the National Academy of Sciences, USA* 104, 3901–3906.

Simon, C. and Daniel, R. (2009) Achievements and new knowledge unravelled by metagenomic approaches. *Applied Microbiology and Biotechnology* 85, 254–276.

Stat, M., Huggett, M.J., Bernasconi, R., DiBattista, J.D., Berry, T.E. *et al.* (2017) Ecosystem biomonitoring with eDNA: metabarcoding across the tree of life in a tropical marine environment. *Scientific Reports* 7, 12240.

Thilliez, G.J.A., Armstrong, M.R., Lim, T.Y., Baker, K., Jouet, A. *et al.* (2018) Pathogen enrichment sequencing (PenSeq) enables population genomic studies in oomycetes. *New Phytologist* 221, 1634–1648.

Torriani, S.F.F., Brunner, P.C., McDonald, B.A. and Sierotzki, H. (2009) QoI resistance emerged independently at least 4 times in European populations of *Mycosphaerella graminicola*. *Pest Management Science* 65, 155–162.

Wang, H., Qi, M. and Cutler, A.J. (1993) A simple method of preparing plant samples for PCR. *Nucleic Acids Research* 21, 4153–4154.

White, T.J., Bruns, T.D., Lee, S.B. and Taylor, J.W. (1990) Amplification and direct sequencing of fungal ribosomal RNA genes for phylogenetics. In: Innis, N., Gelfand, D., Sninsky, J. and White, T. (eds) *PCR – Protocols and Applications – A Laboratory Manual.* Academic Press, New York, pp. 315–322.

Wibberg, D., Stadler, M., Lambert, C., Bunk, B., Spröer, C. *et al.* (2020) High quality genome sequences of thirteen Hypoxylaceae (Ascomycota) strengthen the phylogenetic family backbone and enable the discovery of new taxa. *Fungal Diversity* 106, 7–28.

Resources

16S Metagenomic sequencing library preparation. Protocol from Illumina. Part #15044223 Rev. B. https://support.illumina.com/documents/documentation/chemistry_documentation/16s/16s-metagenomic-library-prep-guide-15044223-b.pdf (accessed 8 April 2022).

BLAST finds regions of similarity between biological sequences: BLAST Assembled RefSeq Genomes. Available at: https://blast.ncbi.nlm.nih.gov/ (accessed 8 April 2022).

EuroBlight: A Potato Late Blight Network For Europe. Available at: www.euroblight.net (accessed 8 April 2022).

THAPBI *Phytophthora* ITS1 Classifier Tool (PICT) developed by Peter J. Cock https://thapbi-pict.readthedocs.io/en/latest/index.html (accessed 8 April 2022).

11 Maintenance and Storage of Fungal Plant Pathogens

Matthew J. Ryan*, Anthony Kermode and David Smith
CABI Bioservices, Bakeham Lane, Egham, Surrey, UK

11.1 Introduction

The long-term preservation and stability of plant pathogens is essential for continuity and reproducibility in research into their properties, interactions and evolutionary change. Having representative strains for further study is essential not only for their pathology, but they also may have potential use as reference strains, research tools, the producers of compounds, fuel and food, and be the basis for teaching and education in microbiology. These organisms are often placed in the back of a refrigerator or kept in the incubator until they are needed again. If research is carried out using deteriorated, contaminated or incorrect strains, large investments in time, human resources and finance can be lost. Therefore, these organisms must be maintained without change to ensure reproducibility and sustainability. It is important that plant pathogens are maintained using methodologies that will not induce physiological or genetic change. For the growth of fungi, nutrient sources (growth media) are one of the most important factors, as are sporulation, temperature, light, aeration, pH and water activity.

Plant pathologists often establish their own laboratory collections to ensure that their key strains are maintained for future use. However, it is culture collections, positioned as custodians of *ex-situ* genetic resources, that have the crucial role of providing the authenticated biological material on which high-quality research is based (Ryan *et al.*, 2019). Authenticated fungi are essential for the quality of programmes such as the 'Darwin Tree of Life' project (https://www.darwintreeoflife.org/; accessed 28 February 2023). Collections are repositories for *ex-type* strains, strains linked to plant disease records and strains as part of patent deposits. They are providers of safe and confidential services that store key organisms for research and industry. Microorganisms cited in scientific papers should be deposited in culture collections to be available for confirmation of results and for further study (Stackebrandt *et al.*, 2014). Collections range from small private collections through to large service collections, and they have widely differing policies and holdings. Collections of organisms are normally linked to their use in operations related to the activities of the parent organization (Smith *et al.*, 2013).

The primary objective of preserving and storing an organism is to maintain it in a viable state, without morphological, physiological or genetic change, until it is required for future use. Ideally, complete viability and stability should be achieved, especially for important reference,

---

* E-mail: m.ryan@cabi.org

©CAB International 2023. *Fungal Plant Pathogens: Applied Techniques, 2nd Edition*
(eds C.R. Lane, P.A. Beales and K.J.D. Hughes)
DOI:10.1079/9781800620575.0011

research and industrial isolates. It is important that reproducible and quality-controlled procedures are adopted and culture collections have utilized various guidance and International Standards to ensure reproducibility in the strains they provide (Martin *et al.*, 2015). However, working, teaching or research collections may have to consider additional factors such as simplicity, availability and cost.

It is also worth bearing in mind that there is a move from addressing single axenic organisms towards the study, harnessing and conservation of microbial communities, the microbiomes of humans, plants, animals and the environment. This adds an additional level of complexity to the preservation and supply of microbial communities in which plant pathogens exist and function (Ryan *et al.*, 2020).

11.2 Isolation and Growth of Fungi

Generally, fungi grow best on media formulated from the natural materials from which they were originally isolated. The growth requirements for fungi may vary from strain to strain, although cultures of the same species and genera tend to grow best on similar media. The geographical and climatic areas from which a fungus was isolated can give an indication of suitable growth conditions; for example, those from hot deserts and the tropics may prefer high growth temperatures. Despite being able to grow fungi, many are yet to be cultured, although recent high-throughput culturomic approaches are proving successful (Sarhan *et al.*, 2019). These techniques are based on the diversified and multiple combinations of various growth media, culturing conditions, atmospheres and times of incubation. In the medical field, where culture-dependent methods allow the identification of only 75 (27%) of the 278 reported fungal species in human gut microbiota, the use of culturomics has led to significant numbers of new species being discovered (Hamad *et al.*, 2016, 2017).

Once culture conditions are found, cultures are usually maintained on agar slopes in culture (Universal) bottles. However, some fungi deteriorate when kept on the same nutrient source for prolonged periods, so the medium should be alternated from time to time. Most laboratories prefer not to keep large stocks of different media and the majority of isolates can be maintained on a relatively small range, depending on the specialization of the collection. Research has shown that cultures grow better on freshly prepared media, especially natural media made from a vegetable base (Smith and Onions, 1994; Smith *et al.*, 2001). The introduction of pieces of tissue, such as rice, grain, seeds, leaves, wheat straw or dung, often produces good sporulation.

Some examples of the specific growth requirements of fungi (Smith and Onions, 1994) are given below:

- Mucorales grow well on malt agar (MA) and will not grow in Czapek agar (CZ) as they lack the enzymes to digest sucrose.
- Many fungi thrive on potato dextrose agar (PDA), but this can be too rich, and encourage growth of mycelium with ultimate loss of sporulation; a period on potato carrot agar (PCA), a starvation medium, may encourage sporulation.
- Wood-inhabiting fungi and dematiaceous fungi often sporulate better on cornmeal agar (CMA) and oat agar (OA), both of which have less easily digestible carbohydrate.
- Cellulose-destroying fungi and spoilage fungi, such as *Trichoderma*, *Chaetomium* and *Stachybotrys*, retain their ability to produce cellulase when grown on a weak medium such as tap water agar (TWA) or PCA with a piece of sterile filter paper, wheat straw or lupin stem placed on the agar surface.
- All sorts of vegetable decoctions are possible, and apart from the advantages of standardization, it is reasonable to use what is readily available, e.g. yam media might be preferable to potato media in the tropics.
- For *Fusarium* species, synthetic nutrient agar (SNA, also known as Speziellar Nährstoffarmer agar) is suitable for good spore production without encouraging excessive mycelial growth; the recipe used for SNA by these authors is slightly different to that reported in Appendix 1, in that it uses 20 g of Agar No. 3 (Oxoid) and the pH is adjusted to 6.5 (with HCl and NaOH). PSA (potato sucrose agar) is suitable for the expression of

pigment production in *Fusarium*; this can be an important diagnostic character within some species.

Environmental factors

While different fungi have specific growth requirements where media are concerned, other factors that affect their growth are equally important. These are outlined below and discussed in Chapter 7.

Temperature

Most filamentous fungi are mesophilic, growing at temperatures within the range of 10–40°C, with optimum temperatures between 20°C and 35°C. Some species (e.g. *Aspergillus fumigatus*, *Talaromyces avellaneus*) are thermotolerant and will grow at higher temperatures, although they are still capable of growth within the 20–25°C range. A small number of species (e.g. *Chaetomium thermophilum*, *Penicillium dupontii* and *Thermoascus aurantiacus*) are thermophilic and will grow and sporulate at 50°C or higher, but fail to grow below 20°C, and have optimum growth between 40°C and 50°C. A few fungi (e.g. *Hypocrea psychrophila*) are psychrophilic and are unable to grow above 20°C, while many others (e.g. a wide range of *Fusarium* and *Penicillium* species) are psychrotolerant and are able to grow both at freezing point and at mesophilic temperatures.

Aeration

Nearly all fungi are aerobic and when cultured in tubes or bottles obtain sufficient oxygen through cotton wool plugs or loose bottle caps. A few aquatic hyphomycetes require additional aeration. In this case, air is bubbled through liquid culture media to enable normal growth and sporulation to occur.

pH

Most common fungi grow well over the range pH 3–7, although some can grow at pH 2 and below (e.g. *Moniliella acetoabutans*, *Aspergillus niger*, *Penicillium funiculosum*).

Water activity

All organisms need water for growth, but the amount required varies widely. Although the majority of filamentous fungi require high levels of available water, a few are able to grow at low water activity (e.g. *Eurotium* spp., *Xeromyces bisporus*). Fungi isolated from saline environments will only grow well on media containing high concentrations of salt (halophiles).

Light

Many species grow well in the dark, but others prefer daylight and some sporulate better under near-ultraviolet light. Most leaf- and stem-inhabiting fungi are light-sensitive and require light stimulation for sporulation. Cultures can be grown in transparent glass-fronted or illuminated incubators. However, some fungi are diurnal and require the transition from periods of light to dark to initiate sporulation.

Induction of sporulation by near UV light

Some fungi require irradiation by near UV light ('black' light) for production of sporulating structures. The fungi are grown in plastic Petri dishes or plastic Universal bottles for 3–4 days before irradiation (Pyrex® glass is not permeable to near UV). The edges of the dishes are sealed with adhesive PVC tape to prevent rapid drying and mite infestation (Smith and Onions 1994; Waller *et al.*, 1997).

The illuminators comprise:

- Three 1.22 m fluorescent lamps 130 mm apart.
- A black light tube held in the centre.
- Cool white tubes supported on each side.

The lamps are controlled by a time switch that gives a 12 h on/off cycle. The plastic Petri dishes, or the bottles, are supported on a shelf 320 mm below the light source and are illuminated until sporulation is induced. It should be noted that placing fungi under UV light may induce irreversible genetic damage, so any culture so placed should be destroyed after taxonomic evaluation. It is therefore advisable to keep stocks that are not subjected to UV light.

11.3 Preservation Regimes

There are several methods and techniques available for the storage and preservation of fungi. However, not all of them are suitable for each fungus, as each microorganism will differ in structure and form. It is important to identify the most appropriate preservation technique to keep microbial cultures alive, stable and free from contamination. Logistical factors, such as the availability of resources and an institution's overall infrastructure, have to be considered before cultures are preserved. Careful consideration should be given as to why an organism needs to be preserved and the duration of storage. A very important strain may be better preserved in a public service culture collection; a list of these is available at the World Data Centre for Microorganisms (http://www.wdcm.org/; accessed 28 February 2023). Some methodologies may be cheap in terms of consumables but can require frequent maintenance in the short term (e.g. continuous culture regimes, water) or medium term (oil, silica gel and soil storage), thus adding to labour costs and increasing the risk of change and of contamination. Cryopreservation in liquid nitrogen and freeze-drying cultures reduces maintenance but requires apparatus which is expensive.

Preservation techniques range from continuous growth methods to methods that reduce growth rates to the ideal situation in which metabolism is suspended. These methods can be divided into three groups: continuous growth techniques, drying and suspension of metabolism.

Continuous growth techniques involve frequent transfer from depleted to fresh nutrient sources. The need for frequent subculture (serial subculture, described in Protocol 11.1) can be delayed either by growing cultures on media with reduced carbon, nitrogen or other nutrients, or by storing cultures in a refrigerator, in a freezer (at −10°C to −20°C), under a layer of paraffin oil or in water (Fig. 11.1).

Drying of the resting stage (e.g. spores, cysts or sclerotia) of an organism can be achieved by air-drying in or above silica gel, in soil or in sand. Desiccation has been used successfully for the preservation of many microorganisms as the removal of water suspends the metabolism of the cell. Fungal spores have a lower water content than vegetative hyphae and can withstand desiccation, reviving when water becomes available. Drying can be carried out using many techniques. Air-drying is achieved by passing dry air over the culture or spores; this speeds the drying process by evaporation. Air-drying has been successful for some *Aspergillus* and *Penicillium* species (Smith and Onions, 1994). Drying can also be achieved using soil, silica gel or other desiccants. Silica gel storage (see Protocol 11.3) was first employed at CABI in 1970, and organisms stored at this time are periodically tested to determine ultimate shelf life (Sharma and Smith, 1999). Soil storage was initiated in the late 1960s, and over 900 *Fusarium* and related genera are stored by this method. Storage in sterile soil or sand is described in Protocol 11.5.

The suspension of metabolism normally involves reducing the water content available to cells by dehydration, e.g. by freeze-drying (Protocol 11.6) or cryopreservation (Protocol 11.7). Freeze-drying (lyophilization) is the sublimation of ice from frozen material at reduced pressure; it requires storage in an inert atmosphere either under vacuum or at atmospheric pressure in an inert gas. Lyophilization was first extensively used with fungal cultures by Raper and Alexander (1945). The methods and machinery have developed over the years to produce a reliable and successful preservation technique for sporulating microfungi and most bacteria (Ryan and Smith, 2007). Stability and long storage periods are the main advantages of freeze-drying, though the expense of the modern and quite complex machinery can be a deterrent. Cryopreservation generally implies storage at temperatures of around −70°C and below that impede chemical reactions (Fig. 11.2). This can be achieved in mechanical deep freezers (some can reach temperatures of −150°C) or in/above liquid nitrogen (Smith and Ryan, 2012). Achieving an adequate suspension of metabolism to a point where no physical or chemical reaction can occur requires storage at temperatures below −139°C (Morris, 1981). The rate of cooling and the protection from ice damage are critical to enable the preservation of the more delicate or recalcitrant fungi. Various approaches have proven successful at CABI, including encapsulation in alginate beads (Benson *et al.*, 2018) and controlled cooling for optimal survival (Ryan *et al.*, 2014). Vitrification (rapid

Fig. 11.1. Culture storage under mineral oil.

cooling of liquid medium in the absence of ice crystal formation) has also proven useful in the cryopreservation of the difficult-to-preserve fungi (Benson, 2008; Benson *et al.*, 2018).

11.4   Selection of Preservation Regime

Not all microorganisms will tolerate preservation by every method available (Ryan *et al.*, 2000). Indeed, for some it may be difficult to preserve them at all. These are classed as 'recalcitrant' organisms. Knowledge of an organism's life cycle and structure is important, as often only the stages that result in the production of spores or the resting structure can be preserved. The taxonomic position of a fungus should be considered before it is preserved, and knowledge of the identity of a fungus can greatly help when considering how to preserve it. However, one must bear in mind that surviving a technique often depends on the strain, as some strains are more robust or express different chemical characteristics than others.

Oomycota are best stored in liquid nitrogen using a cooling rate of *c.*−10°C min^{-1} and a cryoprotectant of 10% w/v glycerol or DMSO (dimethyl sulfoxide). Some strains do not survive freezing. Storage under mineral oil (described in Protocol 11.2) may be satisfactory for periods up to 6 months (three strains of *Phytophthora* have survived for 41 years at CABI). Cultures can also be kept viable in water storage and transferred every 2 years. The water storage of mycelial plugs is described in Protocol 11.4.

Liquid nitrogen storage is recommended for members of the zygomycetes, such as *Mucor*, *Rhizopus* and similar genera. Most isolates can even be freeze-dried. Not all survive dehydration, particularly in silica gel; for example, *Coemansia*, *Martensiomyces*, *Condiobolus*, *Entomophthora*, *Piptocephalis* and *Syzygites*. Of these genera, only *Piptocephalis* and *Coemansia* strains have been freeze-dried successfully at CABI.

Most ascomycetes which grow in culture can be freeze-dried or cryopreserved in liquid nitrogen. Conidial ascomyceteous fungi are relatively easy to preserve. *Aspergillus*, *Penicillium* and *Paecilomyces* species store for 6 months to 2 years on agar slants in a freezer at −20°C.

Basidiomycetes generally grow only as mycelium in continual culture. They can only be preserved by serial transfer on agar with or without oil or stored in liquid nitrogen. Those fungi

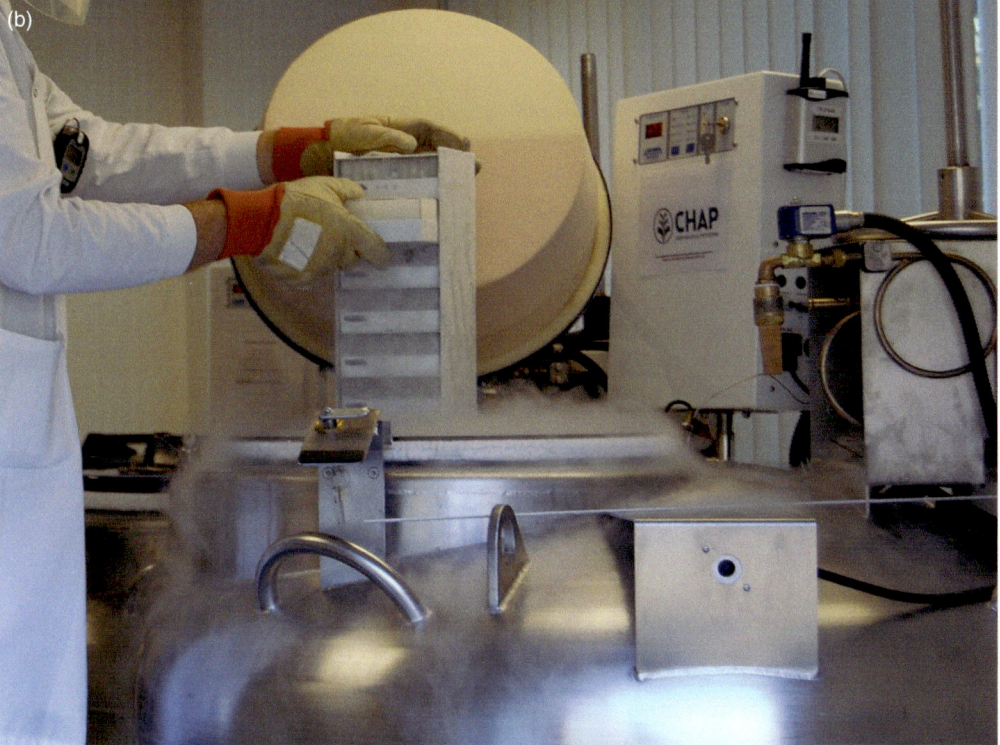

Fig. 11.2. Cryopreservation at ultra-low temperatures: (a) cryo-storage tanks and (b) recovery of ampoules from storage.

that produce thick-walled hyphae can be freeze-dried, but viability is usually low. Basidiospores harvested from fungi growing in their natural environment can usually be freeze-dried. Those genera and species that do not sporulate well in culture survive dehydration techniques poorly. However, most of these have survived long-term storage in liquid nitrogen. Rusts do not normally grow in culture but living collections can be maintained in good condition in liquid nitrogen on the host or as harvested spores. The Ustilaginales produce very disappointing cultures and survive best in liquid nitrogen, though it is possible to keep them by other means. If the spores can be harvested successfully, some survive freeze-drying quite well.

Single-celled vegetative yeasts survive freeze-drying well, although the best technique is in or above liquid nitrogen. Basidiomycete yeasts generally survive cryopreservation best but can also be freeze-dried. Most species survive silica gel storage, but recovery is usually very low.

Microbiome samples (representing all microbes whether harmful or not) that require the preservation and stability of all microbial content are best cryopreserved using specifically designed controlled cooling protocols (Cafa *et al.*, 2021).

11.5 Assessment of the Success of Preservation

The viability, purity and stability of strains must be assessed before and after preservation, and during storage. The assessments should be compared with the original results before the culture is made available outside the collection. The organism may be sent to the depositor for confirmation of properties. Viability is usually assessed by the percentage recovery of propagules, followed by a growth test. The culture should be grown on the most suitable medium, or host, to give optimum growth.

Cultures should be monitored to ensure that characteristics have not altered. Careful microscopic examination must be carried out to ensure that the culture is not mixed. It may be necessary to grow the culture under special conditions to determine whether there are contaminants present. Cultures from single-cell isolations can be prepared to give a better chance of growing

a pure culture, although this carries a risk of unintentional selection. To determine whether the cells remain stable during storage pre- and post-preservation, comparisons should be made. Morphology can be checked by comparison with organism descriptions or photomicrographs of the culture. Pathogenicity testing is not always possible, and licences are often needed for infecting hosts, along with appropriate containment and quarantine facilities (see Chapter 12). Genetic profiles can be created using a variety of molecular tools, e.g. sequencing and assay of properties (Chapter 10); biochemical properties can also be checked where appropriate, for instance by using specific carbon sources for the presence of enzymes. It is essential to record as much information and data as possible concerning the organism's properties before storage so that this information can be compared after preservation and storage. Additionally, all observations on the recovery of strains from preservation methods and storage are equally important. Important data elements are:

- organism descriptions;
- photomicrographs;
- metabolic profiles, e.g. HPLC and MALDI-TOF mass spectrometry profiles; and
- sequence data.

11.6 Storage of DNA

Collections may store DNA for their own use or make them available to researchers via a DNA bank. Collections often have very different strategies for the storage of DNA. The quality of DNA is often influenced by the stringency of the extraction method. The length of storage is often determined by the buffer in which the DNA is suspended and the temperature at which it is stored. Some DNA preparations are stored in commercially supplied buffer formulations, but alternatives such as TE buffer (Tris EDTA) or water are often used.

Storage of DNA at ultra-low temperatures is often recommended. Indeed, many DNA banks advocate storage at −80°C; e.g. the DNA Bank at the Royal Botanic Gardens at Kew, UK, which contains over 36,000 accessions, and the German DNA bank network. However, other methodologies allow for storage at around −20°C,

5°C or room temperature. For example, at ambient temperature the use of the water-free GenTegra™ matrix (from GENVAULT®) protects DNA from hydrolysis and oxidation. Whatman FTA® paper can be used as a means of storing DNA (on micro, mini and classic cards).

Various methodologies can be used to assess the quality of DNA, such as PCR fingerprinting methods (AFLP, amplified fragment length polymorphism; ISSR, inter simple sequence repeat, etc.), as described in Chapters 9 and 10.

11.7 Legislation, Safe Handling and Distribution

The collection, isolation, handling, maintenance and distribution of microorganisms are controlled by law at the national, regional and international levels (Kurtböke *et al.*, 2019) (see Chapter 12). Microbiologists and the providers of living biological resources must be aware of such legislation and operate within it. Legislation, regulations and guidelines will have variations in different countries. Safety regulations, for example the UK Health and Safety Executive's regulations on the Control of Substances Hazardous to Health (COSHH – https://www.hse.gov.uk/coshh/; accessed 28 February 2023), require that every employer makes a suitable and sufficient assessment of the risks to health and safety to which any person, whether employed by them or not, may be exposed (HSE, 2009). In Europe, the EC Directive 2000/54/EC on Biological Agents at work (https://osha.europa.eu/en/legislation/directives/exposure-to-biological-agents/77; accessed 28 February 2023) sets down the European regulations on the distribution of microorganisms outside the workplace and also extends to employer's duties to protect others (https://www.hse.gov.uk/biosafety/law.htm; accessed 28 February 2023).

the World Health Organization (WHO), US Department of Health and Human Services, Centers for Disease Control and Prevention (CDC), UK Advisory Group on Dangerous Pathogens (ACDP, an expert committee of the UK Department of Health and Social Care), European Federation of Biotechnology (EFB) and European Community (EC). In Europe, the EC Directive (2000/54/EC) on Biological Agents sets a common baseline, which has been strengthened and expanded in many of the individual member states. The definition and minimum handling procedures of pathogenic organisms are set by appropriate authorities in each country and are often the same or similar for all EC countries. In the UK, the ACDP lists four hazard groups (1–4) with corresponding containment levels. In line with this, microorganisms are normally classified into four risk groups depending on their potential to cause human disease and likelihood of spread, and the availability of effective prophylaxis or treatment (Advisory Committee on Dangerous Pathogens, 2021).

- *Risk Group 1*: a biological agent that is most unlikely to cause human disease.
- *Risk Group 2*: a biological agent that may cause human disease and which might be a hazard to laboratory workers but is unlikely to spread in the community. Laboratory exposure rarely produces infection and effective prophylaxis or treatment is available.
- *Risk Group 3*: a biological agent that may cause severe human disease and present a serious hazard to laboratory workers. It may present a risk of spread in the community but there is usually effective prophylaxis or treatment.
- *Risk Group 4*: a biological agent that causes severe human disease and is a serious hazard to laboratory workers. It may present a high risk of spread in the community and there is usually no effective prophylaxis or treatment.

Classification of microorganisms based on hazard

There are various classification systems available. These include the definitions for classification by

Quarantine regulations

Before the provision of cultures of non-indigenous pathogens, consideration must be given to plant health requirements (Chapter 12).

Convention on Biological Diversity

The Convention on Biological Diversity (CBD) requires that microbiologists seek prior informed consent from the country from which they wish to collect organisms. They will be required to agree terms on which benefits will be shared should they accrue from the use of these organisms. The benefit sharing may include monetary elements, but may also include information, technology transfer and training. A collection must ensure transparency in retaining the link between the country of origin and the end user of the genetic resources. Biological materials must be received and supplied within the spirit of the CBD ensuring material transfer agreements are in place. A collection must maintain contact and follow recommendations of its national CBD Contact Point and National Focal Point. Information on country status with regard to the CBD and the Nagoya Protocol can be found at the Access and Benefit Sharing Clearing House (https://absch.cbd.int/; accessed 28 February 2023). Information on how culture collections and microbial domain Biological Resource Centres (BRC) comply are published in various papers and best practice (Smith *et al.*, 2017b; Verkley *et al.*, 2020).

Regulations governing shipping of cultures

There are national and international regulations governing mailing and shipping of organisms (Smith and Ryan, 2019). Many national postal authorities permit the receipt and dispatch of non-hazardous biological substances, but most place restrictions on hazardous substances. For information, see the *UPU Convention Manual* (UPU, 2019), Section V Prohibitions and customs matters. Normally, the risk group definitions of the WHO are used to categorize materials. Long-distance or international transport is often by air, and then the IATA (International Air Transport Association) Dangerous Goods Regulations (DGR) are invoked. These require that shippers of microorganisms of Risk Groups 2, 3 or 4 must be trained every 2 years by IATA-certified and IATA-approved instructors if cultures are sent by air transport.

The transport of infectious substances (which are classified as Dangerous Goods, Class 6, Division 6.2, UN numbers 2814, UN 2900 and 3373), or of such genetically modified organisms as require to be transported under Class 9, UN 3245, requires specific packaging and shippers' declaration forms, which accompany the package in duplicate (Fig. 11.3). Definitions of the UN numbers cited here can be found in IATA's Dangerous Goods Regulations Manual, which is updated annually (latest version: DGR, 2022). The UN Model regulations have moved away from using WHO risk group definitions and defined two categories of Biological Substance: Category B (UN 3373) requires Packing Instruction 650, whereas microorganisms capable of causing disability, or life-threatening or fatal disease (UN 2814 or UN2900) are defined as Category A and require the stricter PI 602 Packing Instruction. Category A includes principally Risk Group 3 and 4, and Category B includes the majority of Risk Group 2 microorganisms. Further up-to-date information is available on IATA's website (http://www.iata.org/Pages/default.aspx; accessed 28 February 2023).

A collection must ensure that the staff responsible for the distribution of infectious substances have a current IATA shippers training certificate and ensure that organisms are packed and shipped in accordance with IATA requirements, if applicable. Non-infectious microorganisms may be sent by (air) mail, according to UPU requirements.

Additionally, of specific relevance is that microorganisms may be 'dual use' and particularly plant pathogens could carry a biosecurity risk as a result (Smith *et al.*, 2017a). Microorganisms in the wrong hands could be used in direct or indirect acts against humans, livestock, crops, food, water infrastructure and other economically valuable entities. Biosecurity is a major concern for several countries and regulations have been introduced, for example the Anti-terrorism, Crime and Security Act 2001 in the UK (https://www.legislation.gov.uk/ukpga/2001/24/contents; accessed 28 February 2023), in its schedule 5, pathogens and toxins, lists some plant pathogenic bacteria. Several fungi not in this list could be misused so access to them should be controlled. Collections should include their own specialist knowledge when undertaking assessments.

Fig. 11.3. Packaging of dangerous fungal pathogens must adhere to UN 2814, UN 2900, UN 3245 or UN 3373 (Image © CABI.)

Protocol 11.1

Serial subculture
(Smith and Onions, 1994)

Fungal transfer by subculture can be potentially disadvantageous, as frequent subculturing could result in contamination from other microorganisms, such as bacteria or airborne spores of other fungal species. It can also result in selection of genetic variants if atypical areas of growth are selected. The choice of medium is an important factor, as some fungi are notoriously difficult to culture (e.g. many mycorrhizal fungi). Most fungi will survive on malt agar (MA) or potato carrot agar (PCA) (Smith and Onions, 1994), but others require more specialized media. The precautions mentioned in the main text should ensure that, as far as possible, fungi maintain the characteristics exhibited on isolation from nature and do not mutate or show selection. It is a feature of the opportunistic nature of fungi to easily adapt to the environment. The use of cold storage can slow the rate of metabolism and thus increase the intervals between subculture. Storage at 4–7°C in a refrigerator, or cold room, can extend the transfer interval to 4–6 months from the average period of 2–4 months. Storage in a deep freeze (−7°C to −24°C) will allow many fungi to survive 4–5 years between transfers.

Materials

- Scalpel
- Fresh culture media (pre-poured in Petri dishes or Universal bottles)
- Class 2 microbiological safety cabinet
- Bunsen burner
- Alcohol for flaming
- Parafilm tape or similar

Method

1. Using a sterile flamed scalpel, cut a small block of agar/fungus from the actively growing area of the fungal culture.

2. Transfer the agar/fungus to fresh media (either to a Petri dish or Universal bottle).

3. Seal Petri dish with laboratory film and incubate under optimal growth conditions to establish growth. For cultures established in Universal bottles, ensure that the lid is not fully closed to allow for gaseous exchange.

4. Monitor daily to ensure that the cultures remain free of contamination and do not show signs of strain drift.

5. Once Petri dish or Universal bottle slope is covered with fungus, store at 4°C until required.

Protocol 11.2

Preservation of agar cultures under mineral oil (Smith and Onions, 1994)

This method is generally only used for yeasts and filamentous fungi but can be applied successfully to bacteria. It involves covering cultures with mineral oil to prevent dehydration and to slow down metabolic activity and growth through reduced oxygen tension. Preservation under oil is recommended for storage of organisms in laboratories with limited resources and facilities.

Mycelial forms such as *Pythium* and *Phytophthora* and mycelial basidiomycetes survive well using this method. It is particularly useful in the tropics as it prevents the cultures from drying out and protects against mite infestation. The method does not require expensive apparatus or chemicals and is easy to use. The disadvantage is that subculturing from oil stocks is messy and offers a potential for selection and strain drift; it requires culturing onto agar medium in a Petri dish to allow the fungus to grow away from the oil and then subculturing again onto a fresh dish or into new bottles.

Materials

- Medicinal-quality light liquid paraffin, specific gravity 0.830–0.890 (autoclaved twice at 121°C for 20 min on consecutive days)
- Sterile solid growth medium in 30 ml Universal bottles set at a 30° slope, using the most appropriate growth medium for the organism
- Metal segmented trays (375 × 175 mm, divided into 25 × 25 mm squares to take 60 × 30 ml Universal bottles (Wesbart))
- Sterile inoculating needle, loop or cork borer
- Class 2 microbiological safety cabinet
- Bunsen burner
- Alcohol for flaming

Method

1. Inoculate at least two Universal bottles for each strain to be maintained.

2. Label one culture as a reserve stock, the other(s) as working stock.

3. Incubate at optimum growth temperature until the organism has reached maturity.

4. Add 8–10 ml of sterile (twice autoclaved) liquid paraffin to cover the slope to a maximum depth of 10 mm over its highest point.

5. Store the oiled cultures, with the screw caps loose, in metal-divided racks at 15–20°C.

Recovery

1. Remove a portion of the working stock culture using a sterile needle or loop.

2. Drain as much oil as possible from the inoculum.

3. Inoculate fresh growth medium (it is often best to either inoculate a slope or an agar plate at an angle so that the adhering oil can drain and the organism can grow up and away from the oil at the point of inoculation). The reserve stock culture is used only when re-preservation becomes necessary – when all the inoculum has been removed, when it is contaminated or when the shelf-life expiry date set for the organism has been reached.

4. Once the colony has grown away from the oil, subculture again onto fresh medium.

Protocol 11.3

Preservation of cultures using silica gel (Smith and Onions, 1994)

Silica gel storage has several advantages: it is cheap, simple and does not require expensive apparatus. Cultures are relatively stable and allow a wide range of sporulating fungi (including representatives of the basidiomycetes) to be successfully preserved. Infestation by mites is unlikely, as they cannot survive the dry conditions that are encountered. Repeated inocula can be removed from a single bottle. Survival is often up to 25 years, according to species. The method is suitable for organisms that survive freeze-drying, including mycelial forms that produce sclerotia or chlamydospores.

Materials

- Coarse non-indicator silica gel
- Sterile 5% (w/v) solution of non-fat skimmed milk cooled to 5°C
- Sterile Pasteur pipettes
- 30 ml glass Universal bottles (a minimum of two per culture)
- 100 mm deep waterproof tray filled to a depth of 30 mm with water
- Refrigerator
- −20°C freezer
- Airtight storage boxes
- Indicator silica gel
- Class 2 microbiological safety cabinet
- Bunsen burner
- Alcohol for flaming

Method

1. One-third fill glass 30 ml Universal bottles with coarse non-indicator silica gel and sterilize by dry heat (180°C for 2–3 h).

2. Place bottles in a tray of water up to the top level of the silica gel and place the tray in a −20°C freezer (nominal); leave overnight to freeze.

3. Prepare a spore suspension in the cooled 5% (w/v) skimmed milk (at 5°C).

4. Add approximately 1 ml of the spore suspension, using a Pasteur pipette, to at least two bottles of the silica gel while they remain in the frozen water.

5. After 20 min or longer, when the bottles can be easily moved, remove them from the ice and agitate them to disperse the suspension.

6. Label one bottle as 'reserve stock' and the other(s) as 'working stock'.

7. Incubate the bottles at 25°C until the silica gel crystals readily separate when shaken; this may take 1 or 2 weeks.

8. Screw the bottle caps down tightly and store at 4°C in an airtight container; include an open Universal bottle containing indicator silica gel to absorb water.

Recovery

1. Sprinkle a few crystals from the working stock onto a suitable growth medium and incubate under appropriate growth conditions.

2. If the organism fails to grow, attempt again, this time streaking a silica gel crystal over the agar to dislodge the cells and discarding the crystal before incubation.

Protocol 11.4

Water storage of mycelial plugs (Smith and Onions, 1994)

The method of water storage is relatively simple and involves the storage of mycelial plugs (cut from healthy cultures grown in Petri dishes) in ampoules or small bottles of sterile distilled water. The advantages of storage in water are the low cost and easy application. However, the length of storage is often limited, and isolates may be subject to strain drift. Some fungi will not survive even short periods submerged.

Materials

- Sterile distilled water (10 ml in 30 ml Universal bottles, at least two per culture)
- Mature cultures on agar media in Petri dishes
- Metal segmented trays (375 × 175 mm, divided into 25 × 25 mm squares to take 60 × 30 ml Universal bottles (Wesbart))
- Sterile inoculating needle, loop or cork borer
- Class 2 microbiological safety cabinet
- Bunsen burner
- Alcohol for flaming

Method

1. Cut 6 mm^3 agar blocks through the colony of the organism, generally from the growing edge. Sporulating or non-filamentous organisms can be harvested without the agar and simply suspended in water. Recover the organism by placing a small amount of the suspension onto suitable growth medium.

2. Transfer 20–30 agar blocks to 10 ml of sterile distilled water in two or more 30 ml Universal bottles.

3. Label one bottle as reserve stock and the other(s) as working stock.

4. Screw the caps of the Universal bottles tightly and store in the culture store at 15–20°C.

Recovery

1. Remove an agar block from the working stock, place it colony-side down on a suitable growth medium and incubate under optimum growth conditions. Use the reserve stock when re-preservation is necessary.

Protocol 11.5

Storage in sterile soil or sand (Smith and Onions, 1994)

Preservation in sterile sandy loam soil may be one of the most practical and cost-efficient ways to preserve filamentous sporulating micro-organisms. The method is simple and involves inoculating sterile sand or soil with a spore suspension of the fungus. Other advantages of the method include good viability of cultures for up to 10 years, a reduced chance of mite infection and the option of obtaining repeated inocula from the same source. This method of storage is very successful with *Fusarium* species (Gordon, 1952; Booth, 1971). Therefore, soil storage should be used in preference to oil storage for the preservation of both *Fusarium* species and other fungi that show variation under oil. There are few disadvantages to the method, but it is not suitable for many fungi and variation may occur after storage.

Materials

- Garden soil or sand
- Sterile distilled water
- Sterile Pasteur pipettes
- 30 ml glass Universal bottles (a minimum of two per culture)
- Refrigerator
- Class 2 microbiological safety cabinet
- Bunsen burner
- Alcohol for flaming

Method

1. One-third fill 30 ml Universal bottles with garden soil or sand and autoclave on two consecutive days at 121°C for 20 min.

2. Prepare a suspension of the fungus, either mycelium or spores, or both, in sterile distilled water.

3. Add approximately 1 ml of the suspension, using a Pasteur pipette, to at least two bottles of sterile soil for each strain.

4. Label one bottle as 'reserve stock' and the other(s) as 'working stock'.

5. Incubate at 20–25°C for 5–10 days depending on the rate of growth of the fungus being stored. This initial growth period allows the fungus to use the available moisture and gradually to become dormant.

6. Screw the bottle caps down tightly and store in a refrigerator (4–7°C) or a cool room (15–20°C).

Recovery

1. Sprinkle particles of soil/sand from the working stock onto a suitable growth medium and incubate under appropriate growth conditions.

Protocol 11.6

Freeze-drying (Smith and Onions, 1994; Ryan and Smith, 2007)

Freeze-drying techniques for fungi can employ a two-stage centrifugal freeze-drying process or single phase with sealing in a shelf freeze-drier. Heat-sealable ampoules or vials are required as the maintenance of a vacuum is imperative for long-term survival. Injection vials are generally unsuitable for long-term storage due to the degradation of seals and loss of vacuum. Freeze-drying of sporulating fungi such as the anamorphic stages of ascomycetes and basidiomycetes is routinely undertaken; however, the technique is not so suitable for the oomycetes and other non-sporulating cultures. Research has also been carried out to establish whether lyophilized hyphae can be revitalized successfully after preservation. In most cases this has met with little success, but hyphae from *Claviceps* spp. (Pertot *et al.*, 1977), from a limited range of basidiomycetes (Bazzigher, 1962) and from some arbuscular mycorrhizal fungi (Tommerup and Kidby, 1979) have been revitalized successfully. Investigations by Tan *et al.* (1991a, 1991b) gave mixed results. Some cultures did not survive at all and others showed only limited viability. White rot basidiomycetes were revitalized successfully, although enzyme activity varied after lyophilization (Voyron *et al.*, 2009a).

Success with freeze-drying varies between isolates of the same species. In general, those fungi that grow and sporulate well in culture survive the process, while weak or deteriorated isolates tend to fail. It may be misleading to state categorically that one particular species will not survive freeze-drying. Also, the young vegetative hyphae of fungi do not usually survive freeze-drying. Sterile ascomata, chlamydospores, sclerotia and, in some cases, stroma and resting mycelium, have survived. However, it is generally only the spores (e.g. conidia, ascospores and basidiospores) that survive (Smith *et al.*, 2001; Ryan and Smith, 2007).

The apparatus for freeze-drying is very expensive to buy and maintain. Experienced staff are required to operate the system and for preparation of the material for preservation.

Materials for the spin (centrifugal) freeze-dry method

- For fungi: sterile 10% (w/v) skimmed milk +5% (w/v) inositol (autoclaved at 114°C for 10 min); different cell types may require different protective additives
- For bacteria: sterile 14% (w/v) sucrose +14% (w/v) peptone (autoclaved at 121°C for 15 min)
- Freeze-drier with spin-freeze and manifold accessories
- 0.5 ml (nominal capacity) neutral glass ampoules (Adelphi Healthcare Packaging Ltd) heat-sterilized (180°C for 2–3 h) and labelled with the strain number of the organism to be freeze-dried, and the batch date and date of the freeze-drying
- Sterile caps (heat-sterilized at 180°C for 2–3 h) fitted to the 0.5 ml ampoules either individually or in batches of ampoules of the same organism
- Sterile Pasteur pipettes
- Sterile non-absorbent cotton wool
- Air/gas glass constricting torch
- Gas glass-sealing torch
- Heat-resistant mat
- Phosphorus pentoxide

For revival of strains:

- Glass cutter in support handle
- Sterile distilled water
- Class 2 microbiological safety cabinet
- Bunsen burner
- Alcohol for flaming

Method for the spin (centrifugal) freeze-dry method

Cultures are dried from a frozen state by withdrawal of water vapour under vacuum. The dried cultures or spore suspensions are sealed and stored in glass ampoules. Vigorously sporing fungi are best suited to this method. Various recipes for suspending media have been tried for fungi and bacteria.

1. Prepare a spore suspension in a 10% (w/v) skimmed milk and 5% (w/v) inositol mixture.
2. With a Pasteur pipette add 0.2 ml (approx.) of suspension to each sterile ampoule, ensuring that the suspension does not run down the inside of the ampoule.
3. Cover each ampoule with a sterile cap or cover in batches.
4. Load the ampoules into a spin-freeze accessory and place this on the drier.
5. Spin the ampoules for 30 min and cool to −40°C.
6. Evacuate the chamber and continue to spin for 30 min. (The spore suspension will have frozen into a wedge tapering from the base of the ampoule. This gives a greater surface area for evaporation of the liquid.)
7. Leave the ampoules in the chamber and evacuate for a further 3 h (at this point the moisture content of the material will be approximately 5%).
8. Admit air into the freeze-drier chamber and remove the ampoules.
9. Remove the caps and plug the ampoules with sterile cotton wool compressed to 10 mm in depth, 10 mm (approx.) above the top of the freeze-dried material in a microbiological safety cabinet.
10. Constrict the plugged ampoule 10 mm above the cotton plug using the air/gas torch. The bore of the constriction should remain greater than 1 mm, the outer diameter approximately 2.5 mm. (This is the stage where the freeze-dried material is exposed to atmospheric oxygen and moisture and must be kept as short as possible as the exposure of the partially dried material can cause deterioration; Rey, 1977.)
11. Place the constricted ampoules on the secondary-drying accessory of the freeze-drier and evacuate over phosphorus pentoxide desiccant. The ampoules are sealed and removed at the

point of constriction after a 17 h drying process using a glass-sealing torch under a vacuum. At this point, the moisture content should be 1–2% by dry weight. This drying period is selected for the convenience of working practices in the laboratory. However, it is possible to reduce the water content of the samples to a required lower value with a 3–6 h second stage. The presence of sugars in the suspending media reduces the risk of over-drying.
12. Test the sealed tubes with a high-voltage spark tester to ensure that the seal is intact. A purple to blue illumination appearing inside the ampoule indicates low pressure and an intact seal. Any ampoules that fail the test should be discarded.

Materials for the single phase shelf freeze-dry method

- 2-ml flat-bottomed preconstricted glass vials labelled with the culture unique identifier and grooved rubber bungs. These are covered with aluminium foil to prevent aerial contamination and sterilised by autoclaving at 121°C for 15 min, then dried in a drying oven on metal racks for holding 2-ml vials
- Shelf freeze-drier with programmable shelf temperature control
- Class 2 microbiological safety cabinet
- Bunsen burner

Method for single-phase shelf freeze-drying

All culture work should be carried out in the appropriate microbiological safety cabinet.

1. Grow cultures under optimal growth conditions for the species and on suitable media.
2. Prepare a spore suspension in sterile 10% (w/v) skimmed milk and 5% (w/v) inositol mixture.
3. With a Pasteur pipette add 0.2 ml (approx.) of suspension to each sterile vial, ensuring that the suspension does not run down the inside of the neck of the vial.
4. Aseptically insert sterile, grooved rubber bungs into the neck of the vials to the pre-moulded rim so that the groove opening is above the vial lip.
5. Place the vials on the pre-cooled shelf (−35°C) of the freeze-drier.

6. Place a sample temperature probe into a vial containing the skimmed milk and inositol mixture only. When the temperature reaches −20°C, evacuate the chamber; this reduces the temperature of the sample to −45°C as the latent heat of evaporation is removed and this rises again to the shelf temperature.

7. Maintain the shelf temperature at −35°C for 3 h and then raise to 10°C at 0.08°C/min.

8. After 24 h drying, from the time the temperature of the sample reaches −45°C, lower the shelf base to push the rubber bungs into the vials and seal them.

9. Raise the chamber pressure to atmospheric pressure, remove the vials and heat seal them with the air/gas torch, retaining a vacuum of 4×10^{-2} mbar.

10. When cool, test the sealed vials with a high voltage spark tester to ensure that the seal is intact. A purple to blue illumination appearing inside the ampoule indicates low pressure and an intact seal. Any ampoules that fail the test should be discarded.

11. Store the vials in appropriate conditions, in the dark. Storing at a low temperature is thought to give greater longevities: 4°C is favoured by many, and at CABI freeze-dried fungi have survived almost 50 years at temperatures maintained between 15°C and 20°C.

Materials for recovery

- 2-ml flat-bottomed preconstricted glass vials labelled with the culture unique identifier and grooved rubber bungs. These are covered with aluminium foil to prevent aerial contamination and sterilized by autoclaving at 121°C for 15 min, then dried in a drying oven on metal racks for holding 2-ml vials
- Shelf freeze-drier with programmable shelf temperature control
- Class 2 microbiological safety cabinet
- Bunsen burner

Recovery

1. After 4 days storage, test viability of the preserved strain.

2. Score the glass of the ampoules/vials.

3. Snap the ampoules open using a heated glass rod.

4. Aseptically add sterile distilled water and leave to rehydrate for 30 min.

5. Inoculate appropriate medium and incubate under optimum conditions.

Protocol 11.7

Cryopreservation (Ryan and Smith, 2007)

Cryopreservation has been used for the preservation of fungi since the 1960s (Onions, 1971, 1977), although early work involved a very simple procedure. Storage at −196°C in liquid nitrogen or at slightly higher temperatures in the vapour phase is used. Generally, a cooling rate of −1°C min^{-1} with 10% (w/v) glycerol as a cryoprotectant is applied and, to date, over 4000 species belonging to over 700 genera have been successfully frozen at CABI. However, some members of the oomycetes and basidiomycetes survive cryopreservation less well than sporulating fungi; it is anticipated that the use of species-specific cooling rates may provide improved viability.

Cultures, tissue or spore suspensions are treated with a cryoprotectant such as 10% (w/v) glycerol or DMSO before aseptic transfer into sterile ampoules which are frozen to ultra-low temperatures, usually in the vapour phase of liquid nitrogen. The cooling rates are critical, and the best revivals are achieved when the cooling is done slowly. At ultra-low temperatures metabolism is suppressed, and if the organism survives the initial freezing shock viability should be indefinite. The technique requires large and expensive equipment and a reliable source of liquid nitrogen. For best results, experienced staff are necessary to ensure that optimum standards are maintained.

Materials

- Sterile 10% (v/v) glycerol (other cryoprotectants can be used appropriate to the cell being frozen; see Smith and Onions, 1994)
- 2.0 ml sterile graduated cryotubes (ThermoFisher Scientific, Loughborough, UK) labelled with strain number and batch date
- Sterile Pasteur pipettes
- Liquid nitrogen
- Controlled-rate freezer such as the nitrogen-free VIA Freeze™ series (Cytiva Life Sciences) or the nitrogen-based Kryo series (Planer plc, Sunbury-on-Thames, UK)
- Liquid nitrogen freezer with metal drawer rack inventory control system (Statebourne Cryogenics, Sunderland, UK)
- Class 2 microbiological safety cabinet
- Bunsen burner
- Alcohol for flaming

Method

1. Prepare the spore or cell suspensions in sterile 10% (v/v) glycerol or other appropriate cryoprotectant.
2. Pipette 0.5 ml of the cell suspension into 2.0 ml cryotubes. The Pro-Lab Microbank™ system (Pro-Lab Diagnostics, Richmond Hill, Ontario/Neston, UK/Round Rock, Texas) can be used, although the stated method does not involve controlled rate cooling; in this the cryotubes are filled with cryoprotectant and porous glass beads.
3. Cool the cryotubes at a suitable rate in a programmable cooler or Stirling cycle freezer (Ryan *et al.*, 2014). The cooling rate employed depends on the organism. A rate of −1°C min^{-1} will allow most fungi and bacteria to survive but will not give optimum recovery for all. The rate of cooling is controlled over the critical period from +5°C down to −50°C. The initial cooling rate to 5°C is not critical and this is normally at −10°C min^{-1}. Alternative cooling rates (e.g. −0.5°C min^{-1}) can be used for some recalcitrant fungi.
4. Transfer to a metal drawer rack in the liquid nitrogen storage freezer to complete the cooling to the storage temperature.

Recovery

1. Thaw the cryotube rapidly by placing it in a water bath at +37°C until the contents of the ampoule thaw.
2. Where the Pro-Lab Microbank system is used, recovery of the strain can be achieved without defrosting by chipping off a bead and placing it in an appropriate growth medium.

References

Advisory Committee on Dangerous Pathogens (2021) *The Approved List of Biological Agents*, 4th edn. Health and Safety Executive, London.

Bazzigher, G. (1962) Ein vereinfochtes gefriertrochnungs verfahren zur konservierung von pilzkulturen. *Phytopathologie Zeitschrift* 45, 53–16.

Benson, E.E. (2008) Cryopreservation of phytodiversity: a critical appraisal of theory & practice. *Critical Reviews in Plant Sciences* 27, 141–219.

Benson, E.E., Harding, K., Ryan, M., Petrenko, A., Petrenko, Y. and Fuller, B. (2018) Alginate encapsulation to enhance biopreservation scope and success: a multidisciplinary review of current ideas and applications in cryopreservation and non-freezing storage. *Cryoletters* 39, 14–38.

Booth, C. (1971) *The Genus Fusarium*. Commonwealth Mycological Institute, Kew, UK.

Cafa, G., Holden, N., Mauchline, T.H. and Taketani, R. (2021) Cryopreservation of a soil microbiome using a Stirling 1 cycle approach – a genomic assessment. *agriRxiv* DOI:10.31220/agriRxiv.2021.00066.

DGR (2012) *Dangerous Goods Regulations Manual*, 53rd edn. International Air Transport Association (IATA), Montreal, Canada.

Gordon, W.L. (1952) The occurrence of *Fusarium* species in Canada. *Canadian Journal of Botany* 30, 209–251.

Hamad, I., Raoult, D. and Bittar, F. (2016) Repertory of eukaryotes (eukaryome) in the human gastrointestinal tract: taxonomy and detection methods. *Parasite Immunology* 38, 12–36.

Hamad, I., Ranque, S., Azhar, E.I., Yasir, M., Jiman-Fatani, A.A. *et al.* (2017) Culturomics and amplicon-based metagenomic approaches for the study of fungal population in human gut microbiota. *Scientific Reports* 7, 16788.

HSE (2009) *Working with Substances Hazardous to Health: about a Brief Guide to COSHH*, rev. 2012. Health and Safety Executive, Sudbury, UK.

Kurtböke, I., Meyer, W. and Sly, L. (2019) Sustainable use and preservation of biological resources. *Microbiology Australia* 40, 100–102.

Martin, D., Stackebrandt, E. and Smith, D. (2015) MIRRI promoting quality management systems for microbiology. *EC Microbiology* 2.2, 278–287.

Morris, G.J. (1981) *Cryopreservation: An Introduction to Cryopreservation in Culture Collections*. Culture Centre of Algae and Protozoa, Institute of Terrestrial Ecology, Cambridge, UK.

Onions, A.H.S. (1971) Preservation of fungi. In: Booth, C. (ed.) *Methods in Microbiology, Vol. 4*. Academic Press, London, pp. 113–151.

Onions, A.H.S. (1977) Storage of fungi by mineral oil and silica gel for use in collections with limited resources. In: Pestana de Castro, A.F., Da Silva, E.J., Skerman, V.B.D. and Leveritt, W.W. (eds) *Proceedings of the Second International Conference on Culture Collections, University of Queensland, Brisbane*. World Federation for Culture Collections, Brussels, pp. 104–113.

Pertot, E., Puc, A. and Kresmer, M. (1977) Lyophilization of non-sporulating strains of the fungus *Claviceps*. *European Journal of Applied Microbiology* 4, 289–294.

Raper, K.B. and Alexander, D.F. (1945) Preservation of molds by the lyophil process. *Mycologia* 37, 499–525.

Rey, L.R. (1977) Glimpses into the fundamental aspects of freeze drying. In: Cabasso, V.J. and Regamy, R.H. (eds) *Development in Biological Standardisation*. S. Karger, Basel, Switzerland.

Ryan, M.J. and Smith, D. (2007) Cryopreservation and freeze-drying of fungi employing centrifugal and shelf freeze-drying. In: Day, J.G. and Stacey, G.N. (eds) *Cryopreservation and Freeze-Drying Protocols. Methods in Molecular Biology* vol. 368. Humana Press, Totowa, New Jersey.

Ryan, M.J., Smith, D. and Jeffries, P. (2000) A decision-based key to determine the most appropriate protocol for the preservation of fungi. *World Journal of Microbiology and Biotechnology* 16, 183–186.

Ryan, M.J., Kasulyte-Creasey, D., Kermode, A., San, S.P. and Buddie, A.G. (2014) Controlled rate cooling of fungi using a stirling cycle freezer. *Cryoletters* 35, 63–69.

Ryan, M.J., McCluskey, K., Verkleij, G. and Smith, D. (2019) Fungal biological resources to support international development: challenges and opportunities. *World Journal of Microbiology and Biotechnology* 35, 139.

Ryan, M.J., Schloter, M., Berg, G., Kostic, T., Kinkel, L.L. *et al.* (2020) Requirements for the development of biobanks to support European and global microbiome research. *Trends in Microbiology* 29, 89–92.

Sarhan, M.S., Hamza, M.A., Youssef, H.H., Patz, S., Becker, M. *et al.* (2019) Culturomics of the plant prokaryotic microbiome and the dawn of plant-based culture media – a review. *Journal of Advanced Research* 19, 15–27.

Sharma, B. and Smith, D. (1999) Recovery of fungi after storage for over a quarter of a century. *World Journal of Microbiology and Biotechnology* 15, 517–519.

Smith, D. and Onions, A.H.S. (1994) *The Preservation and Maintenance of Living Fungi*, 2nd edn. IMI Technical Handbooks No. 2. CAB International, Wallingford, UK.

Smith, D. and Ryan, M.J. (2012) Implementing best practices and validation of cryopreservation techniques for microorganisms. *The Scientific World Journal 2012*, Article ID 805659.

Smith, D. and Ryan, M.J. (2019) International postal, quarantine and safety regulations. *Microbiology Australia* 40, 117–120.

Smith, D., Ryan, M.J. and Day, J.G. (2001) *The UKNCC Biological Resource: Properties, Maintenance and Management*. UK National Culture Collection Secretariat, Egham, UK.

Smith, D., Fritze, D. and Stackebrandt, E. (2013) Public service collections and biological resource centers of microorganisms. In: Rosenberg, E., DeLong, E.F., Lory, S., Stackebrandt, E. and Thompson, F. (eds) *The Prokaryotes*. Springer, Berlin.

Smith, D., Martin, D. and Novossiolova, T. (2017a) Microorganisms: good or evil, MIRRI provides biosecurity awareness. *Current Microbiology* 74, 299–308.

Smith, D., Silva, M., Jackson, J. and Lyal, C. (2017b) Explanation of the Nagoya Protocol on Access and Benefit Sharing and its implication for microbiology. *Microbiology* 163, 289–296.

Stackebrandt, E., Smith, D., Casaregola, S., Varese, G.C., Verkleij, G. *et al.* (2014) Deposit of microbial strains in public service collections as part of the publication process to underpin good practice in science. *SpringerPlus* 3, 208.

Tan, C.S., Stalpers, J.A. and Van Ingen, C.W. (1991a) Freeze drying of fungal hyphae. *Mycologia* 83, 654–657.

Tan, C.S., Van Ingen, C.W. and Stalpers, J.A. (1991b) Freeze-drying of fungal hyphae and stability of the product. In: Van Griensven, A. (ed.) *Genetics and Breeding of Agarics*. Pudoc, Wageningen, The Netherlands.

Tommerup, I.C. and Kidby, D.K. (1979) Preservation of spores of vesicular-arbuscular endophytes by L-drying. *Applied Environmental Microbiology* 37, 831–835.

UPU (2019) Section V, Prohibitions and customs matters. In: *UPU Convention Manual*. International Bureau of the Universal Postal Union, Berne, Available at: https://www.upu.int/UPU/media/upu/files/UPU/aboutUpu/acts/manualsInThreeVolumes/actInThreeVolumesManualOfConventionMaj1En.pdf (accessed 1 November 2021).

Verkley, G., Perrone, G., Pina, M., Scholz, A., Zuzuarregui, A. *et al.* (2020) New ECCO tools for Core Material Deposit and Transfer Agreements to promote compliance in procedures for deposit and supply of cultures from a public microbial culture collection. *FEMS Microbiology Letters* 375 (5), DOI:10.1093/femsle/fnaa044.

Voyron, S., Roussel, S., Munaut, F., Varese, G.C., Ginepro, M. *et al.* (2009) Vitality and genetic fidelity of white-rot fungi mycelia following different methods of preservation. *Mycological Research* 113, 1027–1038.

Waller, J.M., Ritchie, B.J. and Holderness, M. (1997) *Plant Clinic Handbook*. IMI Technical Handbooks No. 3. CABI International, Wallingford, UK.

12 Biosecurity Procedures for Working with Fungal Plant Pathogens

Rachel Barker[1]* and Belinda Phillipson[2]

[1]Department for Environment, Food and Rural Affairs, Sand Hutton, York, UK; [2]Department for Environment, Food and Rural Affairs, Horizon House, Deanery Road, Bristol, UK

12.1 Introduction

Recent estimates suggest there are believed to be between 2.2 to 3.8 million species of fungi on Earth (Hawksworth and Lücking, 2017). Fungi are the second most species-rich group of organisms behind insects (Purvis and Hector, 2000). It is estimated there are only 120,000 accepted species, with many new fungi yet to be described or studied (Hawksworth and Lücking, 2017).

Most fungi are harmless saprotrophs, although there are approximately 8000 species that have been linked to disease in plants (Fisher *et al.*, 2020). These are viewed as an increasing global threat to food security (Corredor-Moreno and Saunders, 2020) and the intensification of modern agriculture and impacts of climate change make it more important than ever to study them. However, for most countries there is a sub-set of species that are prohibited from being imported due to the significant biosecurity risk that they pose (Fig. 12.1).

To allow vital scientific research on these prohibited fungi to take place, countries adopt a variety of different strategies to control their import. One approach is to determine which species pose the greatest threat and then adopt legislation to control their import and use. In addition, researchers intending to produce and use genetically modified plant pathogens will also have to comply with the regulations for genetically modified organisms.

This chapter focuses on the processes in place that allow the import and work with these prohibited fungi to take place, while ensuring biosecurity is maintained. The specific focus is on fungal plant pathogens, although the same principles will apply to other prohibited plant pests. Some fungal plant pathogens can pose a risk to animal or human health, for example *Fusarium* spp., which can produce mycotoxins, but this falls outside the scope of this chapter.

12.2 Legislation and Process for Importing Prohibited Fungi

The World Trade Organization Agreement on the Application of Sanitary and Phytosanitary measures (SPS Agreement) allows countries to introduce technically justified phytosanitary import controls. Such controls can include a prohibition on the import of specified plants, plant products and harmful organisms. Countries can introduce derogations from import prohibitions for specific purposes, which can include scientific research (Fig. 12.2), trialling, and official testing or diagnosis. To introduce a derogation, a

* E-mail: rachel.barker@defra.gov.uk

DOI: 10.1079/9781800620575.0012

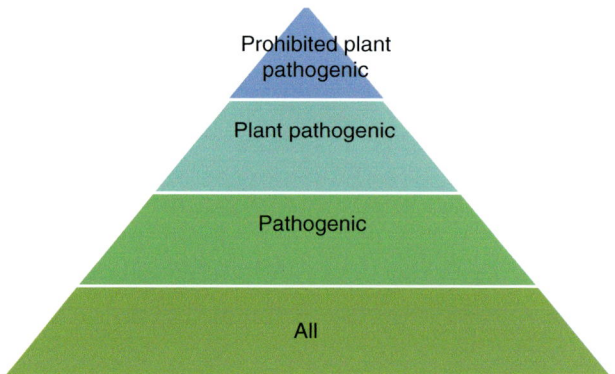

Fig. 12.1. Hierarchy of fungi based on plant health biosecurity risks. Prohibited fungal plant pathogens in the context of this chapter are those which pose an unacceptable risk to the importing country.

risk assessment must have been conducted to justify its implementation (WTO, 2005).

Most countries have a National Plant Protection Organization (NPPO) which is responsible for their phytosanitary import controls. The legislation put in place by an NPPO that allows the import of prohibited species will detail the conditions that must be met, e.g. requirements for import, containment and destruction.

The import of prohibited fungi for specific purposes can be broadly split into the following steps:

- Planning research and application to the NPPO.
- Risk analysis of species and development of special requirements by the NPPO.
- Decision by the NPPO.
- Import of organism if permitted.
- End use of organism.

A general overview of the process that covers all prohibited plant pests can be found in the EPPO standard PM 3/64 (EPPO, 2006).

12.3 Planning Research

Before commencing any work, it is important to think about the biosecurity risks an organism poses and the appropriate level of protection (ALOP) required to protect plant health (WTO, 2005). For prohibited organisms, this will often require approval by the NPPO and use of specialist containment/quarantine facilities. However, even when working with an indigenous organism,

if it has the potential to harm plant health, measures should be put in place to prevent its dispersal.

Protocol 12.1 can be used as a tool to assess the risks posed by the plant pathogen and the planned work. Once fully planned the relevant NPPO needs to be contacted well in advance to ensure full compliance with the relevant legislation. Contact details for the relevant NPPO can be found on the International Phytosanitary Portal (IPPC, 2021).

Laboratory managers and biological safety officers may already be familiar with the application process in a country or have direct contact details for the NPPO.

12.4 Risk Assessment of Application by NPPO

To assist an NPPO undertaking an assessment of an application to import prohibited organisms, the following information will need to be provided:

- **Identity of the organism to be imported:** species, origin, form (e.g. infected plant material, agar plate).
- **Purpose of the import:** description of the work, duration of the work.
- **Importers details:** name, contact details and training or experience.
- **Facilities:** full address including room numbers of where the organism will be held, description of the confinement conditions.

Fig. 12.2. Prohibited fungi can be imported for specific purposes such as research. (© UK Crown copyright – courtesy of Fera.)

- **Procedures:** standard operating procedure (SOP) describing how escape of the organism will be avoided, contingency plan in case of an unforeseen event.

Technical assessment

The NPPO will investigate the biology of the fungi to determine its potential routes of escape: namely through soil, plant material, air, water or via an invertebrate vector. The importer will need to describe and show how they will minimize the risk of escape from the main transmission routes as far as practically possible within their SOP and contingency plan.

This will include the use of suitable laboratories and equipment, as well as specific working practices (e.g. Fig. 12.3). Most containment measures will be applicable to all work with fungi, but some species may require additional measures based on their biology.

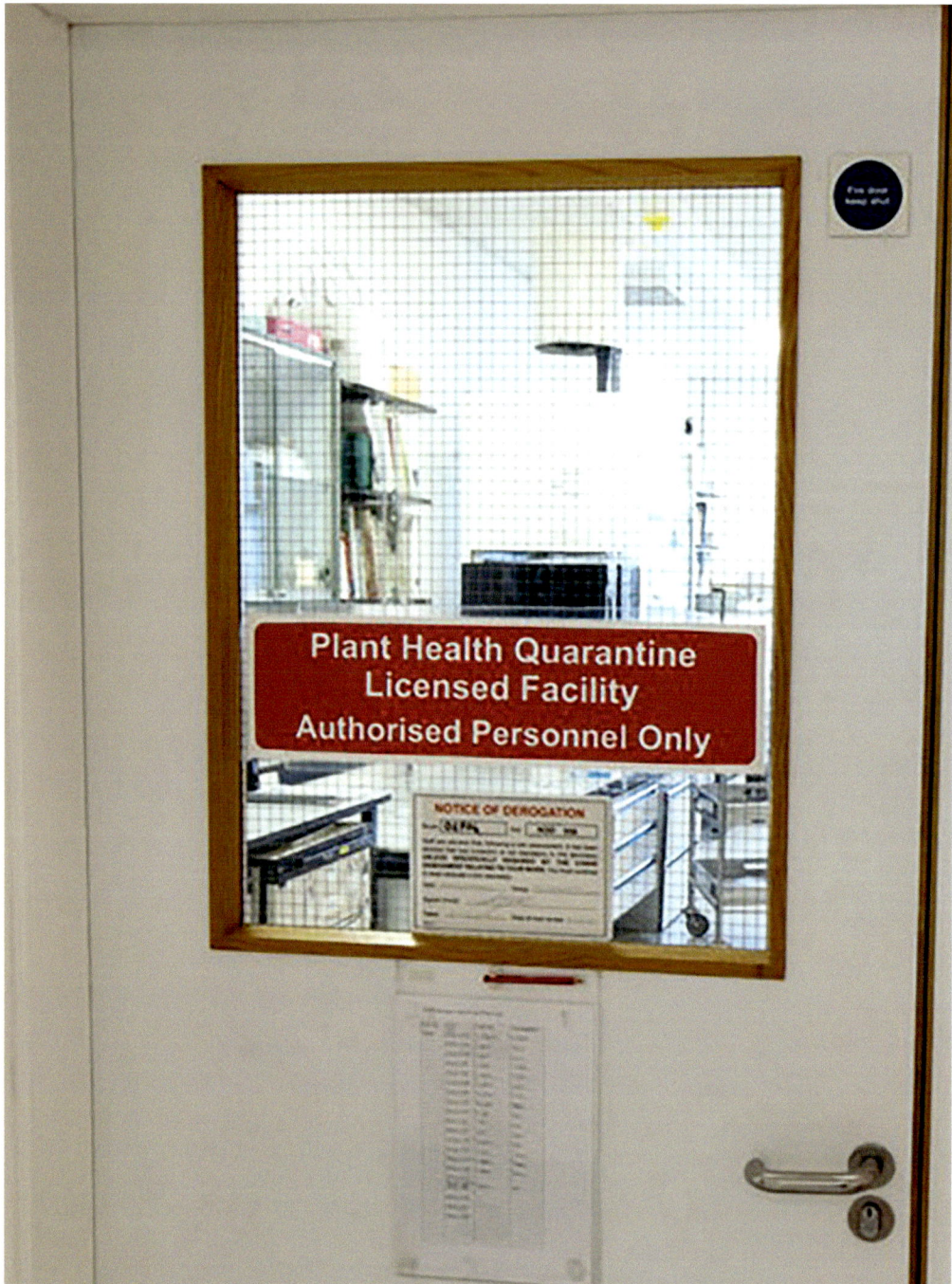

Fig. 12.3. Quarantine facilities should be clearly labelled. (© UK Crown copyright – courtesy of Fera.)

In some countries such as Canada, there are different containment levels for different groups of organisms, with those fungi that have the greatest potential for escape requiring the greatest level of containment. These levels and the requirements for each group of organisms are laid out in the Containment Standards for Facilities Handling Plant Pests (Government of Canada, 2019).

Inspection of containment facilities

Following assessment of the SOP and contingency plan, the containment facilities where the fungi will be held and worked with must be inspected and approved for use at regular intervals.

The type of structure used for containment needs to be assessed. In most cases a laboratory or growth rooms within a laboratory will be most appropriate when working with fungal plant pathogens. Glasshouses may also be suitable if they are secure and suitable to contain the specific species being worked with (Fig. 12.4).

Polytunnels are unlikely to be suitable for working with fungal plant pathogens as they do not provide adequate containment.

When assessing the suitability of the proposed containment facilities, visual inspection will usually be sufficient. However, in some instances, especially for new facilities, further tests may be required to demonstrate containment, for example:

- A smoke test to ensure that there are no cracks and crevices through which material can be released. This test is most frequently used in glasshouses, where panes of glass can fracture or slip.
- Mobile closed-circuit television cameras may be used to assess drains and determine whether they are sealed properly.
- Calibration tests of autoclaves may be carried out to demonstrate that the required pressure and temperature is reached and maintained for sufficient time to inactivate quarantine material.

The measures outlined in Protocol 12.2 cover the main considerations for a containment facility, but those that are most appropriate will vary depending on the biology of the fungus and work being undertaken and the facilities in place. It is ultimately up to the NPPO to decide if the containment facilities are adequate to contain the prohibited organism.

Fig. 12.4. Quarantine containment facility. (© UK Crown copyright – courtesy of Fera.)

12.5   Additional Containment Measures Based on Mode of Transmission

Fungi can re-infect other susceptible host plants via movement of spores in the air, water, soil, infected plant material or invertebrates.

In addition to the measures outlined in Protocol 12.2, the mode of transmission of the species must be considered in relation to its potential escape routes from the containment facility and as a result additional control measures may be necessary.

Spread via water or soil

Organisms such as *Phytophthora* species produce spores which can contaminate drainage water. They also produce long-lived resting spores which can contaminate soil and can survive for years in the absence of their host plant.

Collection and treatment of soil and water that is, or is suspected to be, contaminated by the plant pathogen is vital. Example precautions to take include:

- Installation of soil traps on sinks or label sinks as 'Not for Quarantine Waste Disposal'.
- Ensure existing sterilization methods will be appropriate, e.g. *Synchytrium endobioticum* (potato wart disease) produces thick-walled resting spores with their death point determined to be 90°C for 5 min or 80°C for 15 min. A protocol for heat treatment for 30 min at greater than 80°C therefore provides a safe level of sterilization for waste containing potato wart material.
- Carefully irrigate plants from underneath that are infected with plant pathogens to avoid water splash spread (Fig. 12.5).
- Keep watering to a minimum to prevent accumulation of large volumes of liquid that require treatment.
- A disinfectant (Fig. 12.6), sticky mat or disposable overshoes to prevent egress of any spores that may be present underneath footwear. These must be regularly maintained to ensure continued efficacy.

Fig. 12.5. Trays contain the water used for diseased banana plants. (© UK Crown copyright – courtesy of Fera.)

Spread via airborne spores

A common feature of sporulating fungi is the vast number of spores they produce. For example, a single gall of *Ustilago maydis* (corn smut) is estimated to contain 200 billion spores (Pataky and Snetselaar, 2006). In addition, some fungal spores, such as those produced by *Fusarium graminearum*, can be dispersed long distances, e.g. greater than 100 m (Prussin *et al.*, 2015). Additional precautions must therefore be considered to reduce the risk of airborne fungal spores escaping from confinement facilities, including:

- Entry through two doors separated by a vestibule or growth chamber. Doors should be interlocked so that only one door at a time can be opened (Fig. 12.7).
- Negative pressure:
  - Interlocking of the air supply and exhaust air systems to ensure inward-directional airflow within the heating, ventilation and air-conditioning systems.
    - The maintenance of controlled environment (CE) rooms and glasshouse compartments under negative pressure ensures that air is constantly being drawn into the facilities and that all fungal material remains within the facilities.
  - A back-up electricity supply system should be in place for air systems to maintain negative air pressure gradient and for other essential equipment.

Fig. 12.6. Disinfectant mat outside the confinement facility to sterilize any fungal spores on footwear. (© UK Crown copyright – courtesy of Fera.)

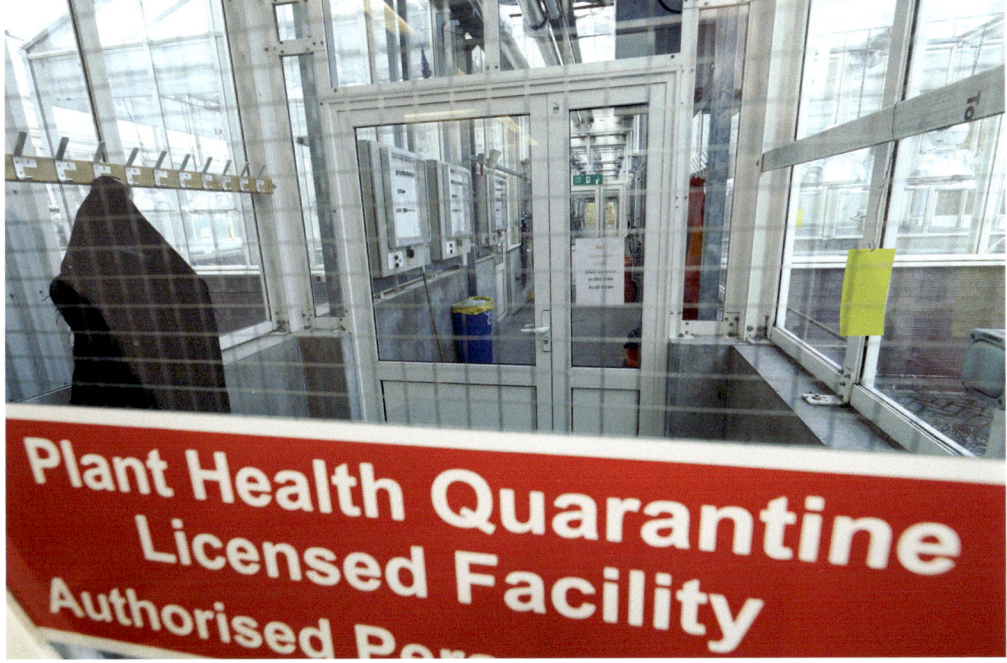

Fig. 12.7. Lobby with interlocking doors before entry to quarantine facilities. (© UK Crown copyright – courtesy of Fera.)

- Consideration of timing:
  - For some fungi, the work could be restricted to times when the fungi are not producing airborne spores.
- Use of a microbiological safety cabinet, class II (MSCII):
  - Sporulating material should be handled in a MSCII which provides protection to the worker, the environment and the material being manipulated by high-efficiency particulate air (HEPA) filtration of the input and extract air.
    - HEPA filters provide the highest level of filtration (standards state that a HEPA air filter must trap 99.97% of all airborne contaminants that are 0.3 μm or larger) but also severely reduces airflow, so are not suitable for all applications.
- Filters:
  - Air drawn into CE rooms or glasshouse compartments should be filtered so that there is no cross-contamination, while extract air should be filtered to prevent fungal spores passing through and being disseminated into the environment. As filters can become contaminated with quarantine material, procedures should be adopted to ensure that they are changed safely without release of material into the environment as outlined in Protocol 12.4.
  - The filter size selected in a containment facility must be small enough to prevent escape of the smallest propagule (Fig. 12.8).
  - Technical information about filter sizes and arrestance efficiencies must be checked with the supplier of the filters. The NPPO may also be able to provide further guidance on the standard filters available in your country and the best to use based on the size of the smallest fungal propagule.
- Secondary containment:
  - Plants infected with fungal suspensions that could potentially sporulate should be grown in specialized CE rooms or glasshouse compartments.
  - The risk of accidental dissemination of the fungal plant pathogen can be reduced further by placing infected plants inside plastic bags in a CE room or growth compartment (Fig. 12.9).

Fungi spread by invertebrate vectors

Some fungal spores (such as *Ophiostoma* spp.) can be dispersed by invertebrates, therefore any facilities working with invertebrate spread fungi must have procedures to prevent the entry and exit of any invertebrate vectors that are there by chance or that are part of the research.

- Consider the mobility of the invertebrate:
  - Life-stage being worked with and its minimum size.
  - How does the life stage(s) being worked with move?
  - How can the life stage(s) be contained?
- Insectary facility:
  - A dedicated facility may be required for highly mobile invertebrate vectors.
  - A light gradient in the corridor is often the most effective means of containment.
  - Appropriate traps for the invertebrate.
- Count numbers:
  - If experiments are focused on vector transmission of fungal spores, the number of invertebrates must be recorded and kept up to date to ensure any escapees are identified.

12.6   Import of Organism

If the importer can meet the requirements of the NPPO, then they may be granted a document that enables the import of the prohibited organism and the work to take place. This document is referred to as an authorization in the EU, a permit in the USA and a dispensation in Canada. A separate document may be attached to the authorization document specifically for import, which must accompany the material.

Any prohibited organism must be transported directly from the place of origin to the approved facilities where the material will be used and stored. In the event of any unsolicited (unexpected) material being sent and not intercepted by customs, the relevant NPPO should immediately be informed to advise on appropriate destruction and decontamination. Packages containing prohibited fungi should only be opened in an environment where dispersal of material can be controlled. When sending or transporting prohibited fungi, three layers of

Fig. 12.8. (a) Maize stem base infected by *Fusarium graminearum* which has an average ascospore size 21 μm × 3.5 μm (Trail *et al.*, 2005), therefore any filtered air exhaust (b) filter size must be smaller than 3.5 μm to contain the ascospores. (© UK Crown copyright – courtesy of Fera.)

containment (tertiary containment) should be used, for example a tube within a sealed bag inside a box (Fig. 12.10).

12.7   End Use of Organism

The aim of plant health controls is to reduce the risk associated with the imported fungal plant pathogens. The work is therefore often time limited, so once work with the organism is completed it should be safely stored, exported or destroyed.

Export

Fungal plant pathogens are often exchanged to share knowledge and provide positive control material for diagnostic tests. It is important that all researchers working with fungal plant pathogens apply containment measures and procedures to reduce these risks. Consequently, before sending any fungal plant pathogens the recipient institute must be able to demonstrate they are compliant with the appropriate regulations. It is important that no material is exported until the recipient institute can demonstrate

Fig. 12.9. Plant infected with a quarantine organism covered with a plastic bag while being incubated in a growth compartment to reduce the risk of spore dispersal. (© UK Crown copyright – courtesy of Fera.)

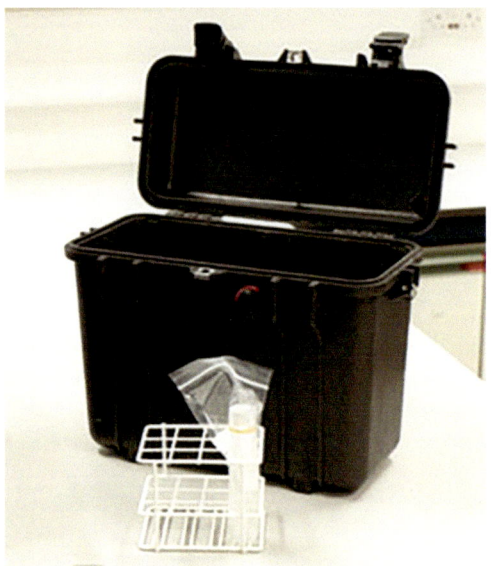

Fig. 12.10. Tertiary containment for sample packaging. Infected material or culture within a sealed tube, within a sealed bag that is, in turn, held in a robust sealed box. (© UK Crown copyright – courtesy of Fera.)

this, usually in the form of official correspondence from their NPPO.

Alternatively, at the end of the work the pathogen will need to be destroyed. General methods for destruction are provided below.

Destruction of solid waste

Waste generated from work with prohibited fungi will pose a risk, as fungal propagules could be released into the environment if not completely inactivated. In addition to the originally imported fungi or infected plant material, any waste generated during the course of research (e.g. consumables, lab coats, packaging materials, personal protective equipment (PPE)) will also potentially be a source of infectious material that could be accidentally released into the environment. All waste from a containment facility must be inactivated regardless of whether it has been in direct contact with the fungi material or infected material is currently present (Fig. 12.11).

Prohibited fungi must be inactivated on site using validated methods that are proven to consistently and effectively inactivate the organisms present in the waste material. Waste should only be collected and transported off site for disposal in exceptional circumstances, because unless strict levels of control are maintained on the waste, there is an increased risk of the material being released into the environment before its inactivation. Permission from an NPPO may have to be requested before waste can be transported off site for disposal.

Currently, the most effective way of inactivating solid waste, including consumables

Fig. 12.11. Material from containment facilities must be collected and treated prior to disposal. (© UK Crown copyright – courtesy of Fera.)

(e.g. pipette tips, gloves, bags, hand wipes), packing materials (e.g. paper, boxes, bags), plants, soil and cultures, is by autoclaving (advice from NPPO should be sought, but typical recommendations are 121°C, 15 psi, for a minimum of 15 min, but 30 min for soil/plant material) or incineration. It is highly unlikely that any fungal plant pathogen could withstand the temperature and pressure conditions used during autoclaving.

A procedure for waste inactivation by autoclaving is given in Protocol 12.5. Autoclaves should be serviced regularly by staff competent in both the technical aspects of the autoclaves and in working in quarantine facilities (Protocol 12.6). The temperatures during the run must be monitored and probes inserted into large volumes of waste to ensure that uniform temperatures are achieved. Ideally, there should be a temperature read-out and a fail-safe mechanism that identifies when the desired temperature and duration of the run has not been reached to indicate that the waste has not been successfully inactivated.

Destruction of liquid waste

The methods currently used for inactivation of liquid waste involve heat or chemical treatment.

Heat treatment involves collecting liquid waste (Fig. 12.12), heating it to a certain temperature and maintaining that temperature for a suitable time. The treatment will depend on the fungal species, other material present and the volume of waste.

Heat treatment has the disadvantage of costly equipment being required to heat and hold liquid waste compared to chemical inactivation. Chemical inactivation involves the addition of high concentrations of disinfectant and incubation for a suitable time. It may require the addition of large volumes of water after inactivation and before release to the main drains to comply with environmental regulations. In the future, the costs and methods of waste disposal are likely to become a greater issue, and other methods such as UV or ozone treatment may have to be considered. Any alternative method of waste inactivation must be validated to demonstrate the effective inactivation of quarantine pathogens before they can be used as standard.

It should also be noted that when carrying out work with plant pathogens, researchers must comply with all the relevant health, safety and environmental regulations that apply and not only those that relate to plant health. The control and inactivation of liquid waste as outlined above and in Protocol 12.5 should inactivate any water-dispersed spores that are likely to be present.

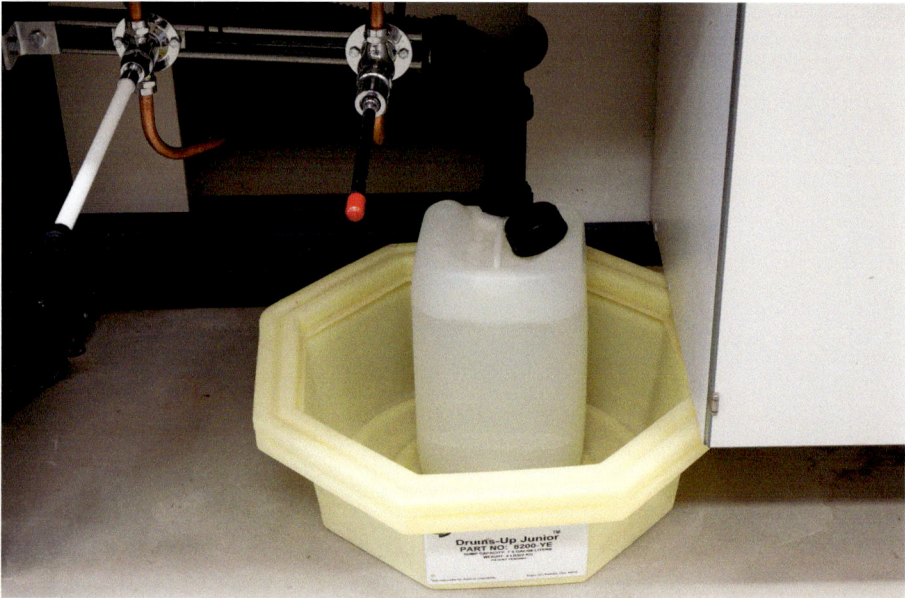

Fig. 12.12. Collection of quarantine wastewater from the sink above in a drum in a suitably sized bund. (© UK Crown copyright – courtesy of Fera.)

12.8   Personal Biosecurity During Fieldwork

It may be necessary for a researcher to collect their own fungi. To do this safely, the same principles of good biosecurity practice applied during work in the laboratory should also be put in place during field work. Key elements to this are good planning and risk assessment to minimize and mitigate the likelihood of introducing or spreading harmful organisms. Good hygiene and sanitation will prevent the introduction, movement around and removal of harmful organisms from any fieldwork site. There are many industry-led good biosecurity practice guides which will help when planning work. One example is the UK Forestry Commission's 'Keep it clean' campaign for good biosecurity practice in arboriculture (Forestry Commission, 2021).

Putting together a simple, relatively cheap, portable biosecurity kit (Fig. 12.13) can help you implement simple measures every day to help limit the introduction and spread of fungal plant pathogens. The following are cheap and easily obtained items to include in your kit:

- bucket (big enough to fit your boot and 5–15 cm of water);
- boot pick;
- brush;
- disinfectant;
- hand sanitizer;
- water container.

Before visiting the site, it is worth assessing the likelihood of encountering a harmful organism that poses a biosecurity risk. Based on the risk, differing levels of good biosecurity practice should be performed.

Areas of lower risk – where a fungal plant pathogen is not suspected:

- Wear footwear and outerwear that can easily be kept clean.
- Clean footwear and outerwear regularly. Ensure they are visually free from soil and organic debris.
- Clean vehicles – do not let mud and organic debris accumulate on tyres, wheels or under wheel arches.

Fig. 12.13. Personal biosecurity kit. (© Copyright Charles Lane.)

- Restrict the equipment taken onto a site – take only what you need for the task.
- Ensure all tools and equipment are clean and free from organic debris.

Areas of higher risk – a fungal plant pathogen has been reported or is suspected:

- Plan to visit highest-risk sites last.
- Clean footwear and outerwear between site visits by removing leaves, soil and other organic material.
- Spray-clean footwear and outerwear with disinfectant until it runs off (boots can be dipped in disinfectant).

- Avoid vehicular access to high-risk sites – park off-site if possible.
- Keep to established hard tracks.
- Remove mud and organic debris from tyres, wheels and wheel arches.
- Clean and disinfect tyres and wheels.
- When taking samples, clean and disinfect cutting tools after each sample.
- Clean and disinfect other tools and equipment before leaving the site.
- Keep any samples in sealed containers with at least three layers of containment.

Protocol 12.1

Risk assessment to conduct prior to contacting NPPO

Rationale
Is the aim of your experiment clear and does it minimize the plant health risks as far as possible?
- What are the aims of the work?
- Could native organisms be used instead?
- Does the work have to take place with live organisms?
- Does the organism pose any risk to human health ?

Biology
Do you know all about the biology of the organism to identify the main ways it could escape containment?
- How can the organism spread?
- Does the organism produce spores?
- What is the lifecycle of the organism?
- How persistent is the organism?

Experimental design
Does the design achieve the aims of your experiment without creating too many samples to deal with?
- How many replicates do you need?
- How many treatments need to be tested?

Facilities
Do you have appropriate facilties to ensure containment of the organism?
- What facilities are available for you to use?
- Are the facilities available suitable for containment of the pest based on its biology?

Practical considerations
e.g. Can any waste produced from the work be dealt with effectively?
- How can waste be disposed of?
- How can the facility/equipment be decontaminated?
- What volume of waste will you have to deal with?
- How will the waste be disposed of/destroyed?
- Is the waste method suitable for the organism?

Protocol 12.2

Measures and considerations for containment facilities handling fungal plant pathogens ('quarantine material') (based on EPPO Protocol 3/64 (EPPO, 2006) and ISPM 34 (ISPM, nd))

Measure/consideration	Explanation and justification
Security and access	
Restrict access	• The facilities should be secure to prevent access by unauthorized people or those who are unfamiliar with the risks posed by such material.
	• The levels of access and supervision given to scientists, maintenance staff and contractors need to be considered. Unaccompanied access should only be granted to trained and competent staff. All other staff (including maintenance staff, visiting workers, cleaners, etc.) should always be supervised.
	• Restriction of access can be achieved by locking facilities and only providing keys, entry codes or swipe cards to those who are competent to carry out work with the quarantine material.
	• For isolated facilities, the use of alarms or other security measures should be considered.
	• The facilities should also be able to withstand the normal climatic conditions of the area and potential deliberate damage/vandalism.
Pest control	• Entry and egress points should be sealed to exclude local pests (e.g. ants, rodents, whiteflies) from entering the contained facility.
	• Appropriate traps can be put in place for known pest problems, e.g. sticky traps.
Entry to facility	• Where possible, entry to the quarantine facilities should be via an entrance lobby with a vestibule or interlocking door arrangement.
Locked and sealed windows and doors	• Windows should be locked and sealed.
	• Door should be self-closing, tight-fitting and have brush seals (also known as brush sweeps) in place around frames.
Infrastructure (physical measures)	
Cordon sanitaire	• Glasshouse only – remove potential host plants around the glasshouse, including weeds, to help to reduce the risk of a fungal pathogen propagating and spreading further in the event of an accidental release.
Accessible, smooth and impervious surfaces (including lab benches and floors)	• The surfaces and floors of the facilities should be easy to clean and disinfect using dedicated cleaning equipment.
	• Floors should be sealed to prevent quarantine material entering the drainage system. The integrity of seals must be checked on a regular basis and any faults rectified immediately.
Clear working area	• Facilities should not be used to house general equipment and chemicals, such as ice machines, reagent stores, fridges and freezers, or large quantities of generic consumables. This will help to reduce footfall and access issues as well as the need to decontaminate or destroy potentially contaminated equipment and consumables.

Continued

Continued.

Measure/consideration	Explanation and justification
Procedural measures	
Protective clothing	• Laboratory coats, overalls, gloves and footwear should be worn in facilities for working with quarantine material and sterilized if being re-used. • Such clothing should be dedicated to a particular facility and should be decontaminated before laundering. • Marking protective clothing, or the use of specifically coloured clothing for specific facilities – for example, green laboratory coats for glasshouse use – should reduce the chances of clothing being removed from facilities and of quarantine material being released.
Signage	• Facilities for working with quarantine material should be clearly labelled so that personnel entering the facilities are aware of the type of material being handled. • Copies of the authorization from the NPPO should be clearly displayed.
Labelling	• Quarantine material should be clearly labelled when not being actively worked with, e.g. in the freezer or fridge, in an incubator or on a laboratory bench.
Storage	• Ideally, a storage facility should only be used for prohibited material and locked. If held in a shared area, then good signage is required to clearly identify quarantine material and it should be stored separately.
Spatial controls	• If possible, shared facilities should be avoided owing to the increased risks of accidental release unless they are carefully managed. • In shared facilities, clearly demarcated areas for quarantine work must be identified. • If non-quarantine material comes into contact with quarantine material, it must be treated as quarantine material, due to the risk of cross-contamination.
Temporal controls	• If facilities are shared, work with quarantine material could be done at a different time to reduce the risk of cross-contamination. • Fungal growth stage: e.g. work could take place before the fungus is sporulating. • Conduct work at a time of year when the host is not susceptible, e.g. winter.
Record keeping	• Keep records of the number of quarantine samples received, the samples being maintained and the stocks that have been destroyed. If records are poorly maintained there is a greater chance of quarantine material being lost and accidentally released. • Records will be key if there is an escape or suspected escape, to establish if it could have come from the facility.
Training and records of training	• All staff must be trained and demonstrate competence in working with quarantine material. • Keep up-to-date copies of the standard operating procedures (SOPs) for working with quarantine material and ensure that all staff working with this material have read, understood and signed the SOPs.
Prohibit plants being grown for ornamental purposes	• No decorative pot plants in quarantine laboratories.
Transport of quarantine material within facilities	• Wherever possible, the facilities in which the material is to be handled (including waste inactivation facilities) should be in close proximity to each other to reduce the likelihood of escape during transfer. • Quarantine material must be transported within three layers of containment.
Waste disposal	

Continued

Continued.

Measure/consideration	Explanation and justification
Equipment for sterilization/destruction of waste	• All waste (e.g. washing-up water, paper towels, boxes, cultures and plants) must be treated as though contaminated.
Drainage	• Drainage water from facilities should be treated to inactivate any quarantine material that enters the drainage system. • If this is not possible, sinks in laboratories (apart from hand/eye-washing stations) and drains in glasshouse facilities/controlled environment rooms should be blocked, or water collected safely in bunded drums for treatment elsewhere. • Procedures should also be adopted to ensure that quarantine material is not disposed of in hand sinks.
Disinfectants for spills and decontamination	• For decontamination, all equipment considered to be potentially contaminated must be treated. • For spills, appropriate chemical disinfectants should be identified and available in sufficient quantities, along with appropriate absorbent material for use in the event of an accidental spill. Further guidance for how to deal with spillages is given in Protocol 12.3. • Chemical disinfectants that are to be used in the event of a spill must be effective against fungi. • The efficacy of disinfectants can be affected by interaction with other material, including soil and plants, so these should be removed prior to disinfection as far as possible. • If in doubt, the efficacy of the disinfectant should be evaluated for the pathogen.
Preparation for emergencies	
Monitoring systems	• Operational processes such as pressure differentials and wastewater treatment must be monitored to prevent failure of essential systems.
Contingency plan	• A plan should be in place so that in an emergency the personnel working in the facility know what to do, e.g. a breach of containment, a flood, a power outage. • Discuss with the NPPO what to do in the event of emergency and who to contact when they visit the facilities.

Protocol 12.3

Dealing with accidental spillage of liquid quarantine waste

In the event of an accidental spillage, the following measures should be taken. It is important to remember that any equipment used to deal with the spillage will require decontamination and that any materials used to absorb the spill will need to be sterilized before disposal. Additionally, any staff entering the contaminated area will need to decontaminate footwear or clothing as appropriate.

The nature of any hazardous liquid needs to be quickly identified as well as whether it poses any health and safety risk or contravenes environmental regulations, in addition to any quarantine concerns. All local risk assessments and standard operating procedures need to be referenced before proceeding and appropriate personal protective equipment worn.

Advice should be sought from the laboratory responsible person and, if appropriate, the relevant authorisation holder, to ascertain the nature of any prohibited material in the facility to help determine the risks involved.

Materials

- Containment and absorbent materials, e.g. specialist synthetic sorbent sheets, booms or pillows, or organic sorbents such as paper, rags, sawdust, clay granules or sand
- Autoclave bags

Method

1. Take actions to stop and contain the spillage – e.g. turn off tap or stopcock or place the leaking item into a large container or bag and use booms or similar material to prevent further spread.

2. Block off any drains in the near vicinity into which any liquid might escape.

3. Absorb any spillage using suitable absorbent materials.

4. Restrict access to the contaminated area.

5. Treat the affected area with a suitable disinfectant (such as bleach).

6. Disinfect any contaminated footwear or equipment.

7. Record details of the amount of spillage, and of the area and measures taken to decontaminate it.

8. Ascertain whether there was any escape to surface drainage and inform authorisation holders and health and safety team immediately, as appropriate.

9. Dispose of all contaminated materials by placing them in a suitable container or autoclave bag for sterilization.

10. Following the incident, implement measures to reduce risk, and review policies and procedures, e.g. in the event of a burst pipe, check all similar pipework and fittings or introduce local easily accessible shut-off valves or, if the accident was due to a glass vessel breaking, look at using a similar item made out of more robust materials.

Protocol 12.4

Changing extract filters in quarantine facilities

Filters may be used to prevent the dispersal of fungal spores or infected debris from containment facilities. They may be disposable and require replacing or may be more durable and require removal and cleaning. Filters may require inspection and cleaning or replacement on a biannual or more frequent basis, depending on the nature of the work and the filters in place. During replacement, the following procedure, in association with the manufacturer's recommendations, should be followed. From a contingency point of view, it is helpful to maintain sufficient stocks of disposable filters or spares of more durable filters.

Materials

- New pre-filters, filters
- Large sealable bag

Methods

1. Before entering or undertaking any work in a quarantine facility, consult with the appropriate responsible person or authorisation holder as appropriate to ensure that local policies and procedures are met and that the work is carried out at a time of the least risk.

2. All equipment required (filters, metal screens, tools, etc.) must be taken into the facility before any work is begun to avoid any unnecessary journeys in and out of the area.

3. Wherever possible, any work should be performed outside normal laboratory or facility working hours to reduce the amount of prohibited material being actively worked on.

4. During the work, no prohibited materials, especially flying insects or organisms that produce airborne propagules, should be worked on.

5. Wherever possible, air-handling units should be turned off before (approx. 15–30 min) the work is commenced to allow any airborne pests or propagules to settle and to prevent any airflow into an unprotected extract system.

6. Some air filtering systems include a damper that allows the exhaust to be sealed off. If this is present, then seal off the exhaust before changing any filter.

7. Carefully remove the filter housing and check for any evidence of damage or breaches in its integrity.

8. Remove any pre-filter and examine for any signs of damage, then promptly place it in a suitable bag and seal.

9. Remove the filter and examine the housing and visible ducting for any signs of damage. Promptly place the used filter in a suitable bag and seal.

10. Immediately replace the filter, check that there are not tears, holes or manufacturing defects and that the filter is correctly seated; also replace any pre-filter and housing as appropriate.

11. The checking of performance (e.g. smoke testing) should be considered, depending on the filter and the risks.

12. In the event of any damage to either the housing or filters, this should be reported immediately to the facility manager or authorisation holder and no further work should be permitted until either repairs have been made or the extract has been sealed off effectively.

13. If the filter is heavily congested, then the use of pre-filters should be considered, and the filter checked on a more frequent basis.

14. All materials must be decontaminated before removal or disposal, as appropriate. This can be done either by chemical or heat sterilization (autoclaving).

Protocol 12.5

Inactivation of quarantine waste by autoclaving

Waste material contaminated with quarantine organisms – such as laboratory consumables, packaging materials, plants and soil, and microbiology culture material, should be sterilized by autoclaving to ensure effective inactivation before disposal. Local policies and procedures with respect to health and safety and environmental regulations must always be followed.

Materials

- Autoclave bags
- Large plastic container with lid
- Bowls or jugs suitable for autoclaving
- Stainless-steel buckets
- Autoclave
- Temperature probes

Methods

1. Place waste material for sterilization into autoclave bags. These bags are frequently labelled with a biohazard sign and 'HAZARDOUS WASTE AUTOCLAVEABLE DISPOSAL BAG'. Guidelines for the correct use of these bags are sometimes also written on the bags.

2. The bottom of the bag should be supported.

3. It is advisable that bags are placed in a robust watertight container, such as a large plastic box, a dustbin with a lid or a stainless-steel bucket. All containers should be clearly labelled as containing quarantine waste. This will help to reduce risk in the event of bags falling over or if they split or tear, including during transport to the autoclave, as any spillage will then be contained.

4. Autoclave bags should be sealed before removal from the quarantine facility. During use within the facility, bags should not be left open but folded over at the top, so reducing the opportunity for release.

5. Bags should not be overfilled and should be taken safely and promptly to the autoclave area for sterilization.

6. Specialist autoclave bags for 'sharps' – such as pipette tips, inoculating loops and spreaders – are available to help reduce the risk of such items piercing normal autoclave bags.

7. Larger pieces of woody plant material or those with thorns or spines should either be wrapped in paper or placed in a paper bag to prevent piercing of the autoclave bag.

8. Dense material, such as soil, wet plant material and potato tubers, should be mixed with bulky materials such as contaminated paper waste to help the sterilization process.

9. Autoclave bags are unsuitable for liquid waste in most circumstances. Liquids should be collected into containers suitable for autoclaving (such as bowls, jugs or metal buckets).

10. Autoclave bags should be loaded carefully into the autoclave and positioned according to the manufacturer's instructions.

11. In general, bags must be pierced carefully once placed in the autoclave (but not before this) to permit steam penetration and successful sterilization. However, on some occasions, when airborne release is highly likely, autoclave bags can remain sealed to prevent the escape of spores. This can reduce the efficiency of autoclaving, so material should be autoclaved twice before disposal.

12. Sterilization programmes for autoclaving are determined by the nature of the waste. Dense waste such as soil can require a longer treatment period to ensure that the entire load reaches the required temperatures. Some fungal spores are extremely thick walled and may survive if not heated sufficiently.

13. Advice from the NPPO should be sought, but typical recommendations are $121\,°C$ (15 psi) such that the centre of the load is maintained at these conditions for at least 15 min for microbiological waste and 30 min for plant and soil waste. Wherever possible, probes should be used to monitor the temperature, and vacuuming or 'free steaming' used to reduce the amount of air in the load. Increasing the temperature and the duration of the process is used by some laboratories as an additional precaution.

14. Ideally, autoclaves should have a temperature read-out and a fail-safe mechanism that clearly identifies when the desired temperature and duration have not been reached. If this occurs, waste has not been successfully inactivated and must be treated again. It is recommended that the autoclave and temperature probes are calibrated on an annual basis or according to the manufacturer's recommendations.

15. If dealing with unusual waste or pathogens, it is advisable to demonstrate the effectiveness of any sterilization process.

16. Once autoclaved, it is advisable to dispose of any culture media immediately and carefully as it may still be able to support microbiological growth.

17. As a precaution, some laboratories choose to dispose of autoclaved waste via clinical waste streams that involve incineration. This is not usually mandatory but can add an additional layer of safety.

Protocol 12.6

General precautions for maintenance work in quarantine facilities

Any equipment/tools/consumables/materials brought into or present in a quarantine facility could potentially be contaminated by a licensed organism. Therefore, it is essential that before anything is taken out of such a facility it is decontaminated.

Materials may either be: (i) *directly in contact* with a fungal pathogen, e.g. filters, pipework, pumps, camera inspection equipment, silt, plant debris, wastewater or (ii) *not in direct contact but potentially contaminated* by a fungal pathogen by presence of or use in a facility, e.g. lamps, toolboxes, tools, ladders.

Materials

- Autoclave bags
- Appropriate chemical disinfectant

Methods

1. Before entering a containment facility, think carefully about what to take in and how to decontaminate and dispose of it.
2. Ensure that the potential risks are understood based on the hazards present and the likelihood of exposure occurring. Advice from the facility manager and authorisation holder should be sought.
3. Wherever possible, use or provide dedicated equipment (e.g. tools, ladders) for the containment facility.
4. Take in only the items required to complete any maintenance work, e.g. take in the tools needed but not the whole toolbox, leave trolleys outside, wherever practical carry out equipment service and maintenance within facilities and not by removing items to a workshop, remove all packaging on consumable items before entering the facility, use disposable plastic sheeting not fabric cloths to cover equipment when decorating.
5. When working in the facility, ensure that you do not contaminate equipment by direct contact, e.g. place tools on clean surfaces, protect from splashing.
6. If equipment or materials come into direct contact with prohibited organisms then the following steps must be taken:
- Clean the contaminated surface to remove any debris.
- Wipe any contaminated surface with an appropriate disinfectant at the correct dilution and duration, seeking guidance from the authorisation or facility responsible person as necessary.
- Dispose of materials for disinfection safely (e.g. by autoclaving).
- Remove equipment immediately once decontaminated.

References

Corredor-Moreno, P. and Saunders, D.G.O. (2020) Expecting the unexpected: factors influencing the emergence of fungal and oomycete plant pathogens. *New Phytologist* 225, 118–125.

EPPO (2006) [EPPO Standard] PM 3/64. Intentional import of organisms that are plant pests or potential plant pests. *Bulletin OEPP/EPPO Bulletin* 36, 191–194.

Fisher, M.C., Gurr, S.J., Cuomo, C.A., Blehert, D.S., Hailing, J. *et al.* (2020) Threats posed by the fungal kingdom to humans, wildlife, and agriculture. *mBio* 11, DOI: 10.1128/mBio.00449-20.

Forestry Commission (2021) 'Keep it clean' under How biosecurity can prevent introduction and spread of tree pests and diseases. Available at: https://www.gov.uk/guidance/prevent-the-introduction-and-spread-of-tree-pests-and-diseases#public (accessed 2 May 2022).

Government of Canada (2019) Biocontainment for facilities handling plant pests. Available at: https://inspection.canada.ca/plant-health/invasive-species/biocontainment/eng/1391707650055/1391707686040 (accessed 2 May 2022).

Hawksworth, D. and Lücking, R. (2017) Fungal diversity revisited: 2.2 to 3.8 million species. *Microbiology Spectrum* 5, 10.

IPPC (2021) List of Countries. Available at: https://www.ippc.int/en/countries/all/list-countries/ (accessed 2 May 2022).

ISPM (nd) 'ISPM 34' under Adopted Standards (ISPMs). Available at: https://www.ippc.int/en/core-activities/standards-setting/ispms/ (accessed 2 May 2022).

Pataky, J.K. and Snetselaar, K.M. 2006. Common smut of corn. The Plant Health Instructor. DOI:10.1094/PHI-I-2006-0927-01.

Prussin, A.J., Marr, L.C., Schmale, D.G., Stoll, R. and Ross, S.D. (2015) Experimental validation of a long-distance transport model for plant pathogens: application to *Fusarium graminearum*. *Agricultural and Forest Meteorology* 203, 118–130.

Purvis, A. and Hector, A. (2000) Getting the measure of biodiversity. *Nature* 405, 212–219.

Trail, F., Gaffoor, I. and Vogel, S. (2005) Ejection mechanics and trajectory of the ascospores of *Gibberella zeae* (anamorph *Fuarium graminearum*). *Fungal Genetics and Biology* 42, 528–533.

WTO (2005) The WTO Agreement on the Application of Sanitary and Phytosanitary Measures (SPS Agreement). Available at: https://www.wto.org/english/tratop_e/sps_e/spsagr_e.htm (accessed 2 May 2022).

13 Quality Assurance and Quality Systems

David Galsworthy*

The Animal and Plant Health Agency, Sand Hutton, York, UK

13.1 Introduction – the Role of Quality Systems in Modern-day Society

There has been an increasing expectation in the world that organizations delivering critical products and services use quality assurance systems to ensure that their quality objectives are achieved consistently. Sectors embracing this philosophy include those for health, food and the environment. The use of quality assurance systems has been driven by the increased reliance of regulators on the demonstration of compliance with various international quality standards.

Advantages demonstrated by the implementation of quality systems include:

- increased efficiency;
- control of risk;
- access to markets;
- promotion of best practices and knowledge transfer; and
- demonstration of due diligence in the event of legal action.

13.2 The Quality Standards Landscape

Increased interest in the introduction of quality systems has resulted in the production of a number of international quality standards that are now universally used as the basis of quality system design and implementation.

The two key standards that potentially have an impact on the work of the plant health laboratory community are ISO 9001:2015 (ISO, 2008) and ISO/IEC 17025:2017 (ISO, 2005). Both standards are maintained by the International Organization for Standardization (ISO) and are administered through accreditation and certification bodies.

ISO 9001:2015 is a generic quality management system standard that can be applied right across an organization. Implementation is demonstrated through:

- a set of procedures covering all the key processes;
- monitoring processes to ensure that they are effective;
- adequate record-keeping;
- monitoring non-conforming work, and taking corrective action when necessary;
- regular review of the individual processes; and
- facilitating continual improvement.

The ISO/IEC 17025:2017 covers laboratory competence and is divided into five main sections that contain the requirements for laboratory accreditation.

* E-mail: david.galsworthy@apha.gov.uk

DOI: 10.1079/9781800620575.0013

ISO 17025 MAIN SECTIONS

Section 4: General Requirements	Section 5: Structural Requirements	Section 6: Resource Requirements	Section 7: Process Requirements	Section 8: Management Systems Requirements

Section 4: General requirements

This section covers impartiality and confidentiality, two requirements that are vital for maintaining the trust and confidence that the users of tests and calibrations place in the laboratories they use. Impartiality implies that the laboratory will not allow commercial, financial or other pressures to compromise the quality of results. Internal issues, personal relationships or other conflicts of interest are addressed and resolved. Confidentiality requires the laboratory to keep all results and information private.

Section 5: Structural requirements

This section defines the basic organizational components of a laboratory, its range of activities and its commitment to an effective management system. It states that an accredited laboratory must be a legal entity or part of a legal entity which is responsible for its testing and calibration activities. Section 5 sets management's responsibilities in an accredited laboratory and their responsibilities to customers, regulatory authorities and organizations that provide recognition. Section 5 also defines the basic requirements for personnel, the authority given to them and the resources needed to carry out their duties.

Section 6: Resource requirements

In this section there are six clauses that address the requirement for the laboratory to have available the personnel, facilities, equipment, systems and support services necessary to perform its laboratory activities.

Section 7: Process requirements

This section covers 11 core processes to improve efficiency. The section begins with the review of requests, tenders and contracts. The selection, verification and validation of methods is one of the most technical and most important parts of the standard. Sampling, the handling of test items and technical record-keeping are covered here. Ensuring the validity of results is the quality monitoring and control function in the laboratory. Several tools for monitoring are listed, and the requirements for proficiency testing are explained.

The standard goes into much detail regarding the reporting of results. Requirements are laid out for dealing with complaints and nonconforming work. A focal point in this electronic age is clause 7.11, on control of data and information management.

Section 8: Management systems requirements

This is where Options A and B come in. Option B applies if the laboratory is part of a larger organization, or if it has its own effective management system in accordance with ISO 9001:2015. Here, the management system requirements specified in clauses 8.2 to 8.9 are covered by the existing quality management system, as long as laboratory activities are included, and the laboratory is capable of demonstrating its fulfilment of ISO/IEC 17025:2017 clauses 4 to 7.

If the laboratory's quality management system is independent of any other management system, Option A applies, and the laboratory must comply with Section 8's requirements.

This section covers eight activities, including quality management system documentation such as policies and objectives, control of documentation and records, addressing risks and opportunities, improvement, and corrective action.

13.3 Accreditation and Certification

Third-party assessment of ISO 9001:2015 compliance is demonstrated through certification by a recognized certification body. Examples of companies that act as certification bodies include the British Standards Institute (BSI), Lloyd's Register of Quality Assurance (LRQA) and Bureau Veritas Certification.

Gaining accreditation from a national accreditation body (such as the United Kingdom Accreditation Service, UKAS) demonstrates compliance with ISO/IEC 17025:2017. Accreditation bodies are nominated nationally in most developed countries.

13.4 Quality Systems for Plant Health Research and Diagnostic Laboratories

The credibility of test results is critical to peer review, successful publication, and technology transfer and uptake. In recent years, there has been a push internationally for diagnostic laboratories to demonstrate this competence through accreditation to ISO/IEC 17025:2017, and this has now become a mandatory requirement for laboratory approval by many plant health regulatory authorities. For example, the European and Mediterranean Plant Protection Organization (EPPO) has developed a useful guidance document, 'Specific requirements for laboratories preparing accreditation for a plant health diagnostic activity' (EPPO, 2010).

The activities of a laboratory covered by the accreditation are defined in a scope, which will list the range of materials analysed, the range of organisms detected (e.g. virus, phytoplasma, fungus) and the techniques used (e.g. PCR, RT(real-time)-PCR or ELISA).

13.5 Core Elements of the Quality System

Quality system design

The overall design of a quality system is critical to its success. Ensuring that the systems involved are intuitive and easy to use will help to reduce the possibility of staff resistance due to a perceived increase in bureaucracy. The design is normally structured in the form of a pyramid (Fig. 13.1). The different layers of the pyramid are structured in terms of policies, procedures and records. Each layer needs to be clearly organized and linked to the others to allow easy movement from one part of the system to the next.

Procedures

Procedures are described as standard operating procedures or protocols and are commonly referred to as 'SOPs'. A suitable format is presented in Protocol 13.1 for an example SOP. Each SOP should provide a brief introduction, a statement of scope that includes what the SOP encompasses but, equally important, what it does not cover (e.g. the assay is suitable for testing plants but not soil), a list of key materials, and then a detailed step-by-step procedure which should be easy to follow and include useful 'tips'. This may refer to other protocols or instruction manuals – if they are easily available – and also to controlled documents.

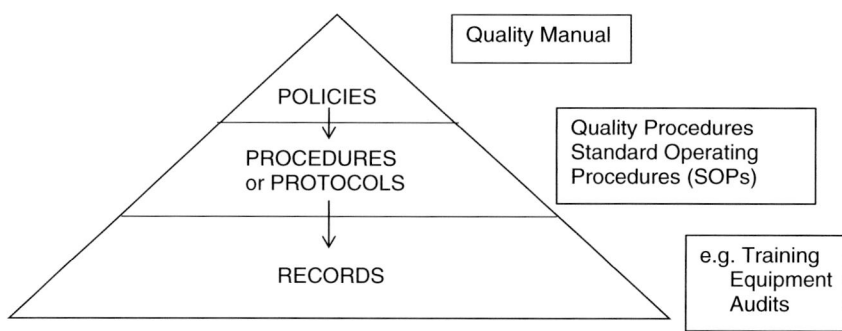

Fig. 13.1. Hierarchical representation of a typical quality system.

Roles and responsibilities

Roles and responsibilities need to be clearly defined throughout the organization. Management responsibilities will include identifying who is responsible for the technical content of all the outputs of the laboratory as well who is responsible for the maintenance of the quality system.

Quality manual

The quality manual defines the laboratory policies and is normally structured to mirror the contents of the quality standards being applied. The manual should be short (no more than 30 pages) and act as a road map to all the other documents that are relevant to the quality system. The use of web-based tools, such as content management systems, to structure the quality manual gives the opportunity to link all the relevant documents electronically as well as giving staff access from any location with Internet access.

Document control

It is necessary to make sure that all staff work to the most up-to-date version of each document. This is ensured by the use of a document-control procedure. Electronic document control solutions are universally applied and reduce the administrative burden that paper-based systems involved.

Purchasing services and supplies

The quality of key reagents is critical to the maintenance of the quality of testing in any plant pathology laboratory. Enquiries should be made about the quality systems being employed by critical suppliers to seek assurance that these reagents are produced in a controlled manner. Reagent stocks should be given sensible shelf lives and containers clearly marked with expiry dates. Confirmation that the reagents are stable for this period will be obtained by showing successful positive controls during their routine use; reagents that are potentially unstable (such as antibodies and PCR primers) are clearly a priority for control in this way.

Complaints

A policy and procedure needs to be produced and used to ensure that if a customer of the laboratory has any concern about the quality of the service provided, then this will be investigated and, if necessary, action taken to address the complaint. An overview and flow chart of these procedures are presented in Protocols 13.2 and 13.3.

Control of non-conforming work

Continual improvement is at the heart of all the quality standards. A policy and procedure is necessary to direct how non-conforming work is identified and how actions are taken to correct the non-conformance, as well as to ensure that non-conformance does not happen again in the future. Flow charts describing these procedures are provided in Protocols 13.2 and 13.3.

Control of records

Records are critical to demonstrate the quality system is under control and in compliance with quality standards. A chain of records will link the various stages of the analysis process from sample receipt to the issue of the analysis report. These records must be capable of recreating the analysis process and are normally checked by 'vertical audits' that trace the analysis process in search of missing links.

Audits

Internal audits are the internal mechanism of monitoring quality system implementation. Internal audit staff should be given adequate training to perform the auditing tasks. Courses are widely available from consultants and other organizations, such as accreditation and certification bodies. An annual audit programme needs to define the specific aspects that will be covered by the audits. An overview of the audit process is given in Protocol 13.4. This would normally include all the different aspects of the quality standards being applied, as well as witnessing specific activities and procedures being carried out by staff. When non-conformances are recorded, these need to be actioned and evidence produced to demonstrate that corrective actions have been carried out.

Review

The review is the annual assessment that the quality system is still fit for purpose and is meeting the requirements of the quality standard and its customers. Inputs to the review include the internal audits, the report from the external assessment body, and details of any complaints and non-conforming work. The output of the review is a set of actions that need to be taken to ensure that the identified changes are made.

13.6. Specific Requirements of ISO/IEC 17025:2017 for Laboratories

Equipment calibration and monitoring

Equipment used in the laboratory needs to be controlled to ensure that the equipment performance tolerances set in the methods and procedures are met. When the absolute value of the equipment settings is important, then calibration traceable through to the internal measurement system should be sought. Examples of this are balance calibration using weights traceable through the internal weights system to the mass used to define the kilogram in Paris since 1889, and critical incubator temperatures using calibrated thermometers or thermocouples. Checks on equipment performance are used for equipment for which the absolute settings are not critical and for monitoring

equipment between calibrations. Examples of these would include balance check weights, thermocycler temperature checks and autopipette usage.

Records need to be maintained of the checks carried out on the equipment and actions taken if the checks show that the equipment goes out of tolerance. An example of such a record log is given in Protocol 13.5.

Validation of methodology

Validation is the production of objective evidence that the methodology being used in tests is fit for its intended purpose. The process of producing validation data for chemical tests is now clearly defined, but this is not so clear in plant health diagnostics. Key to this process is the production of data to demonstrate analytical specificity, analytical sensitivity and repeatability/reproducibility. A good example of how validation data can be used to demonstrate fitness for purpose is taken from the field of immunoassay, as described in Fig. 13.2.

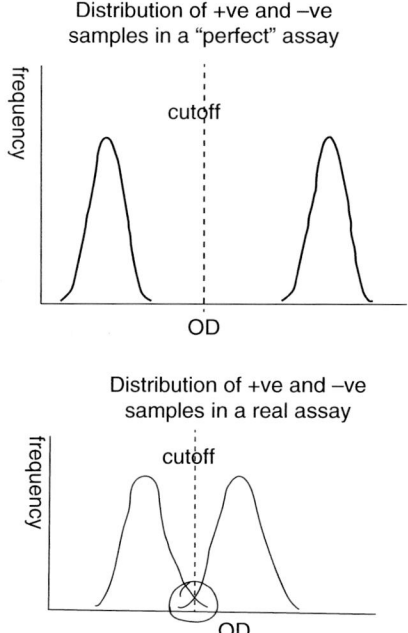

Fig. 13.2. Evaluation of immunoassay data to determine cut-off values.

Validation data should be collated in the form of a report. The report should include:

- Purpose of the analysis.
- Proposed validation plan.
- Validation data.
- Discussion on the data produced.
- Final statement that the method is fit for purpose.

Staff competency

The demonstration of staff competency is critical to demonstrating ISO/IEC 17025:2017 compliance. When introducing the quality system into long-established areas of the laboratory, there is normally a general assumption made that the staff performing the testing are competent to carry out the tests. Ongoing competency of the staff is demonstrated through the day-to-day quality control procedures as well as by external proficiency testing, where available. When new staff are taken on, or when staff are redeployed to areas where they have no experience, then a training procedure will need to be employed and evidence of competence produced.

The training process is normally a four-stage process, as described in Protocols 13.6 and 13.7, and summarized below:

1. Analyst reads the standard operating procedures.
2. The procedures are demonstrated by a trained analyst to the trainee.
3. The trainee carries out the procedure under supervision from a trained analyst.
4. The trainee demonstrates competence by the use of reference materials, previously analysed materials and witnessed analysis, and is then signed off as competent.

Training records need to be produced for all staff. Records of the competency assessment must be included in or referenced from these training records.

Internal quality control

Internal quality control (IQC) procedures are used to check the performance of the methodology in use. Whenever possible, each batch of samples analysed should include some from IQC monitoring to give feedback on both the method and the analyst's performance. Obviously there needs to be a balance between the risk associated with things going wrong with the analysis and the level of the IQC included with each set of samples. The key to an effective IQC regime is the availability of quality control (QC) samples that can be taken through the whole analytical process; this can be problematic for some areas of plant health diagnostics where positive control material is difficult to obtain consistently. Where available, the QC material should be included with each batch of samples, and the results assessed against the determined acceptable range of the material. This range should be established, where possible, by the analysis of 20 samples. Data collected from the QC samples should, again where possible, be graphed to identify any shifts or trends in the results. Where QC samples have failed, the decision-making process of acceptance or rejection of the samples (associated with the QC sample) should be clearly documented, particularly if the decision is made to accept the results and not reanalyse the samples.

Proficiency testing

Proficiency testing is an independent and unbiased assessment of the performance of all aspects of the laboratory, both human and hardware. Proficiency testing should not be confused with collaborative trials, sometime referred to as 'ring-tests', where the method rather than the laboratory is being tested.

Laboratory proficiency testing is a desirable element of laboratory quality assurance. With increasing demands for independent proof of competence from regulatory bodies and customers, proficiency testing has the potential to be relevant to all plant pathology. In proficiency tests, the laboratory is encouraged to use its usual method so that the testing of a routine laboratory sample is simulated as closely as possible. While the outcome of the analysis may be dependent on the choice of method, it could also be affected by the performance of the laboratory equipment or the competence of the analyst.

Each of the participants receives a report which reveals their own laboratory number, allowing them to identify their performance assessment. Anonymous results and assessments are also listed for all other participants in the proficiency test, which allows a laboratory to compare its performance with other laboratories. Reports also contain information on the methods used by participants. Laboratories that do not perform satisfactorily in a proficiency test need to take remedial action to identify the cause of the poor performance.

Proficiency test (PT) schemes covering the detection and/or enumeration of bacterial, fungal, insect, nematode and viral plant pests and pathogens have been developed by several different organizations. These include:

- EU Reference Laboratories, which are obliged to run proficiency schemes for priority pests each year. These are obviously focused on EU member states, but third countries have been allowed to carry out the PT rounds when they have applied to take part.
- Deutsche Sammlung von Mikroorganismen und Zellkulturen (DSMZ), which runs a potato virus scheme from Germany.
- The French Agency for Food, Environmental and Occupational Health & Safety (ANSES),

which runs various PT schemes covering viruses and bacteria.
- The International Seed Testing Association (ISTA), which runs various schemes for pest and diseases in seed materials.

The proficiency test scheme chosen should be run according to the requirements of the international proficiency tests standard, ISO/IEC 17043:2010 (ISO, 2010).

13.7   Benefits of Accreditation

The key to the successful implementation of a quality system is to minimize the impact of those areas associated with staff-perceived problems and to maximize the impact of the 'benefits':

- Improved customer confidence.
- Increased recognition for individual staff and the laboratory.
- Better staff training programme.
- Harmonization of systems across the laboratory.
- Better quality of work.
- Improvements in the accuracy of data.

Protocol 13.1

Standard operating procedure (SOP) pro forma: example

TITLE: **EXTRACTION AND MORPHOLOGICAL IDENTIFICATION OF *TILLETIA INDICA* TELIOSPORES FROM GRAIN OR SEED SAMPLES BY SIZE-SELECTIVE SIEVING WASH TEST**

OWNER:	APPROVER:	ISSUE DATE:
A. Nalayst	M. Anager	January 2022

Introduction

Tilletia indica causes the disease Karnal bunt, or partial bunt, of wheat (*Triticum* spp.). Triticale (× *Triticosecale*) is also naturally infected and rye (*Secale*) is a potential host. *T. indica* reduces grain quality by discolouring and imparting an objectionable odour to the grain and products made from it. It also causes a small reduction in yield. There are other *Tilletia* species that can be confused with *T. indica* which are commonly found in harvested grain or seeds.

A size-selective sieving wash test method is used to detect *T. indica* teliospores in samples of *Triticum* spp. or × *Triticosecale* grain or seed. This method has, on average, an 82% efficiency of recovery and microscopic examinations typically require only a few slides per 50 g sub-sample. When suspect teliospores are found, morphological identification (including comparison with other *Tilletia* species) is carried out. In cases where a molecular confirmatory test is required, a real-time PCR test on *Tilletia* teliospores recovered from microscope slides is used at Fera.

T. indica is a quarantine-listed organism within the EU and is on the EC IAI (EPPO A1) list. Quarantine requirements apply to seed and grain of *Triticum* and × *Triticosecale* from countries where *T. indica* is known to occur.

Operators must have received appropriate training which is documented in their training record before using this method.

! **CRITICAL POINTS** may be highlighted in the text as **crucial** steps in the SOP which will affect the successful outcome of the method.

⚠ **SAFETY POINTS** may be highlighted in the text where any risks associated with performing the method are above and beyond those found in a normal laboratory. Further details of the risk and potential mitigation will be provided in the safety section.

Scope

This SOP is to be used for the detection of *T. indica* using the size-selective sieving method for teliospore extraction, morphological identification and single spore preparation for direct real-time PCR and in conjunction with the EPPO diagnostic protocol for *T. indica* (PM 7/29 (3) *Tilletia indica*) (further referred to as EPPO diagnostic protocol) as necessary. For the real-time PCR method, please refer to SOP PLH/1101 'The identification of *Tilletia indica* by real-time PCR on individual teliospores'.

Safety

This work must be carried out within the Plant Health Quarantine Pathology Facility 04G02-05 and must adhere to the Code of Conduct, SOP PLH/916, and procedures for access, cleaning, waste disposal and decontamination as outlined in SOP PLH/922.

All staff are responsible for their own safety **and** the safety of those working around them. Staff should be familiar with relevant COSHH assessments, emergency response protocols and appropriate waste disposal routes for hazardous chemicals. No staff should perform a method until they have been fully trained and signed off as competent. If any aspect is unclear, seek clarification from your line manager, the laboratory responsible person or a senior scientist (as appropriate) **before** commencing work.

Materials

- 30% Bleach solution
- Sterile distilled water
- Wash water containing 0.01% aqueous Tween-20 (detergent)
- Large weigh boats (approx. 8 × 8 cm)
- Balance
- 250 ml Erlenmeyer glass flasks
- 100 ml measuring cylinder
- Parafilm 'M' or clingfilm
- Laboratory flask shaker (wrist-action)
- Sieve Stack-system: 53 μm* and 20 μm** mesh (nylon sieve mesh from BDH or Spectrum Laboratories)
- Glass bottle that snugly fits the sieve stack-system over its mouth
- Aspirator bottle with distilled water
- Disposable standard micro pastettes
- Pipettes plus disposable pipette tips (filtered)
- New disposable 15 ml polypropylene centrifuge tubes (conical-bottom). ! **CRITICAL POINT** polypropylene tubes must be used as teliospores stick to the sides of polycarbonate tubes, giving false results
- Centrifuge (swing-out action)
- Glass microscope slides, coverslips
- Compound microscope (×100–400 magnification)
- Dissecting microscope
- Glass coverslip pieces (approx. 10 × 10 mm^2), sterilized (autoclaved at 121°C for 15 min or baked at 170°C for 2 h)
- Sterile dissecting needles
- Fine-tip forceps
- Premixed PCR master mix of the amplification of *Tilletia*-specific DNA (from MTU, see SOP PLH/1101)
- 1 μl Tris-EDTA (TE) buffer (from MTU, see SOP PLH/1101)
- Ice

* If 53 μm nylon mesh cannot be sourced, then an alternative mesh size could be used, e.g. 50 or 70 μm mesh.
** ! **CRITICAL POINT** The 20 μm mesh size CANNOT be substituted as trapping of any *T. indica* spores may be compromised as a result.

Protocol-specific safety points

- Protective nitrile gloves must be worn at all times, changing in between samples.
- Safety spectacles must be worn for eye protection when preparing and using bleach solution.

Procedure

1. Size-selective sieving wash test

1.1 According to the EPPO diagnostic protocol, a minimum of three replicate 50 g sub-samples are tested to ensure detection of teliospores if they are present in the sample. Refer to Table 1 below for the number of samples required to detect different numbers of teliospores.

1.2

Table 1. No. of replicate samples required for detection according to level of confidence (%) (from EPPO diagnostic protocol).

Contamination level (No. of spores/ 50 g sample)	No. of replicate samples required for detection according to level of confidence (%)		
	99%	99.9%	99.99%
1	3	5	6
2	2	3	4
5	1	1	1

1.3 Bleach the sieves and glass flasks by immersion for 15 min in a 30% bleach solution. Bleach eliminates the risk of false positives by cross-contamination from previous samples by killing teliospores and making them appear hyaline compared to the normally dark, pigmented viable spores. **!CRITICAL POINT**

Take care not to overexpose the sieve mesh to the bleach as this will soon cause the mesh to tear/disintegrate.

1.4 Rinse off the bleach thoroughly from the equipment with tap water. Allow to drain and dry.

Protocol 13.2

Complaints and non-conforming work follow-up

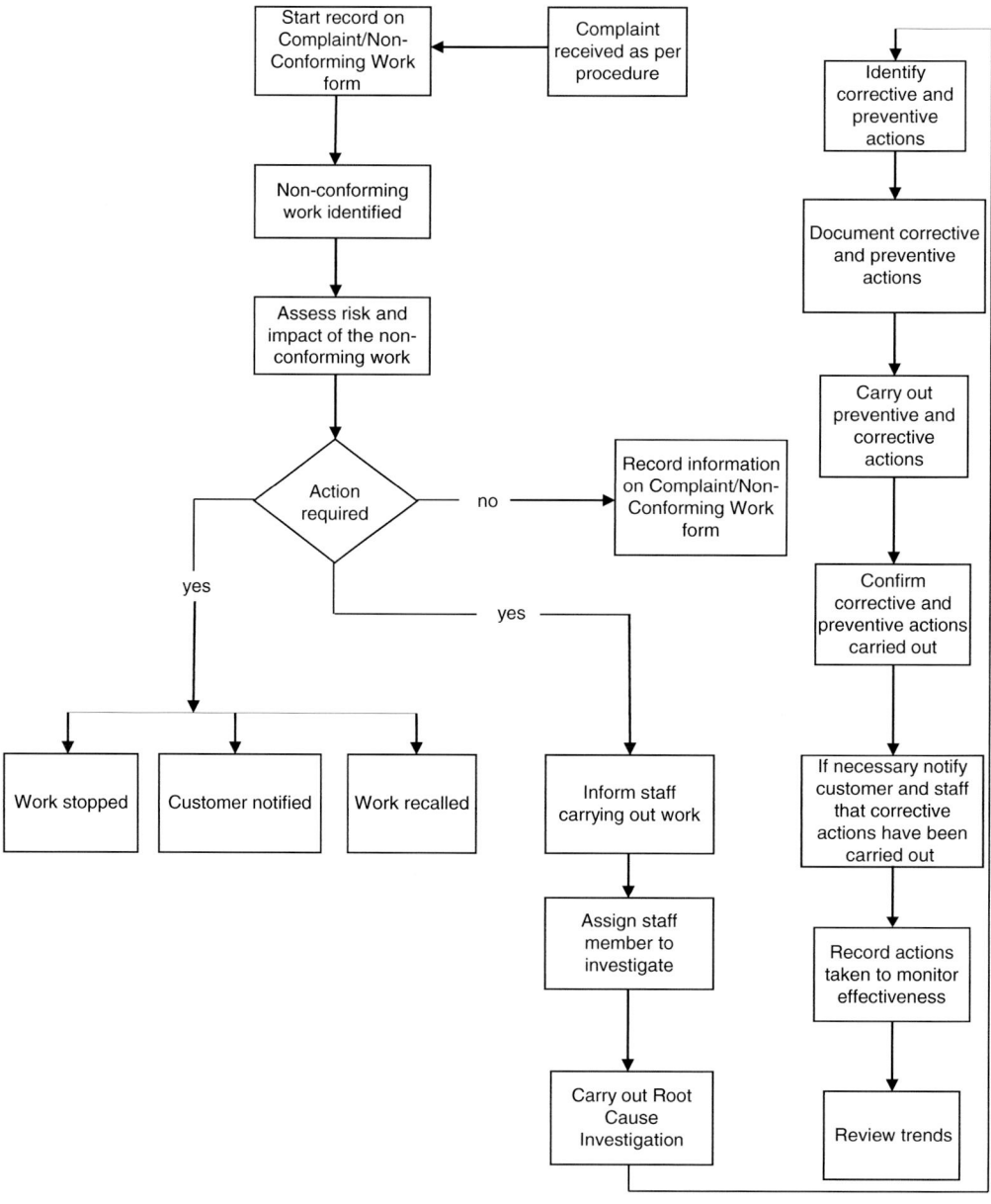

Protocol 13.3

Complaints and non-conforming work report

Complaint/Non-Conforming Work Report		
Complaint initial contact		
Complaint received by:	Date:	
Team/Programme:		
Complainant details (name and address):		
Received by:	FAX ❑ LETTER ❑ EMAIL ❑ OTHER ❑	
Initial referral to:	Date:	
Acknowledgement or reply sent to customer (within 2 working days of receipt):	Date:	

Non-Conforming Work Reporting	
Non-conforming work reported by:	Date: ....
Initial referral to:	Date: ....

Nature of Complaint or Non-Conforming Work		
Project/Sample identity/Area of work:		
Team leader responsible for the area:		
Give a brief description summarizing the details:		
Further Action		
If NO further action is being taken, tick here ❑		
Give reasons why no further action is being taken:		
Agreed by:	Date:	

Root Cause Analysis
Investigate the underlying reason(s) for the complaint/non-conformance. Document your investigation in this section. This evidence will be used to decide on a course of corrective and preventive actions.

Corrective actions identified	Responsibility	Due date: ....
Preventive actions identified	Responsibility	Due date: ....
Actions completed (to be completed by the investigating officer/team leader)	Confirmed by	Date: ....

Protocol 13.4

Audit process

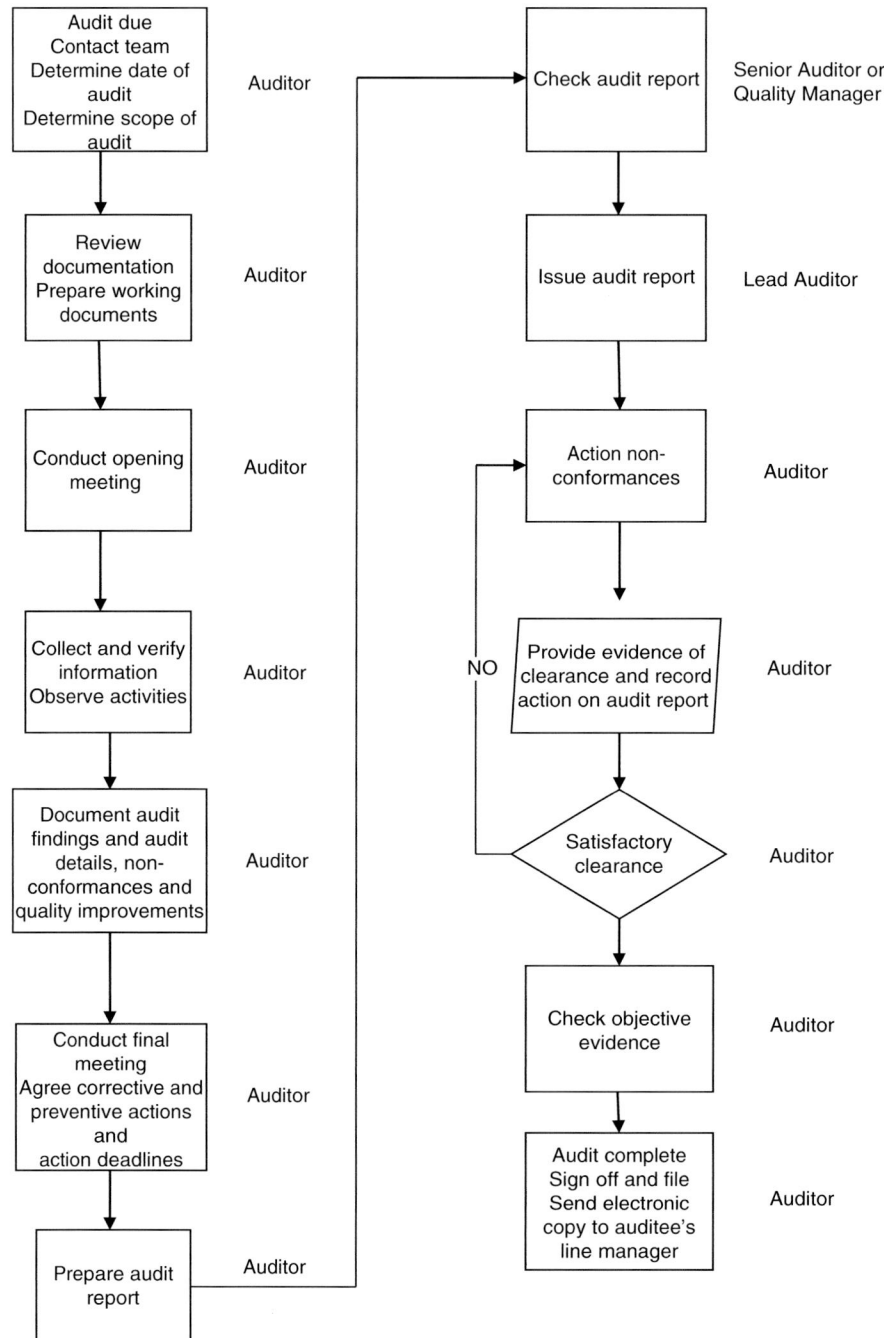

Protocol 13.5

Equipment record and maintenance/calibration log: example

Name of item of equipment: Thermal Cycler Reference number: TC1

Team	Plant health
Name of manufacturer/Supplier	Applied Biosystems
Model no.	9700
Serial no.	A96S8242828
Name of service agent	Applied Biosystems
Address	Lingley House
	Birchwood Boulevard
	Warrington, Cheshire, UK
Service contract ref. (if applicable)	N/a, call out only procurement team
Contact name	Technical Support
Phone no.	01925 825650
Date equipment received	20/01/19
Condition when received	New
Date made operational	20/01/19
Location of operating manual	10GA09
SOP ref. (if applicable)	DNA 8
Calibration/maintenance schedule	Serviced annually

Responsible Person/Deputy Responsible Person

Date	Name	Responsible person (R) or deputy responsible person (D)
20/01/22	M. Anager	R
20/01/22	A. Nalyst	D

Current location of equipment

Date	Room no.	Comments	Initials
20/01/22	10G01		

CALIBRATION/MAINTENANCE/REPAIRS/MODIFICATIONS

Date	Nature of work carried out (attach any relevant documents)	Performed by	Initials

Protocol 13.6

Staff training record: example

Spore trapping

TASK SPECIFIC IN-HOUSE TRAINING RECORD

NAME :   PAGE NUMBER :

Details of task performed including SOP numbers and editions where relevant	Date and Duration of training	Nature of training including records of the tasks undertaken during training Evidence of competence/experience (reference to relevant documents/records)	Training stage	Trainee's initials to confirm training has been undertaken	Trainer's initials to conform training has been carried out successfully	Date authorized	Line manager's initials to confirm competence
-							

* Training stages:
1. Studied task and task documentation   2. Observe procedure
3. Undertake the task under supervision   4. Assessed as competent to undertake the task unsupervised

Protocol 13.7

Staff training process

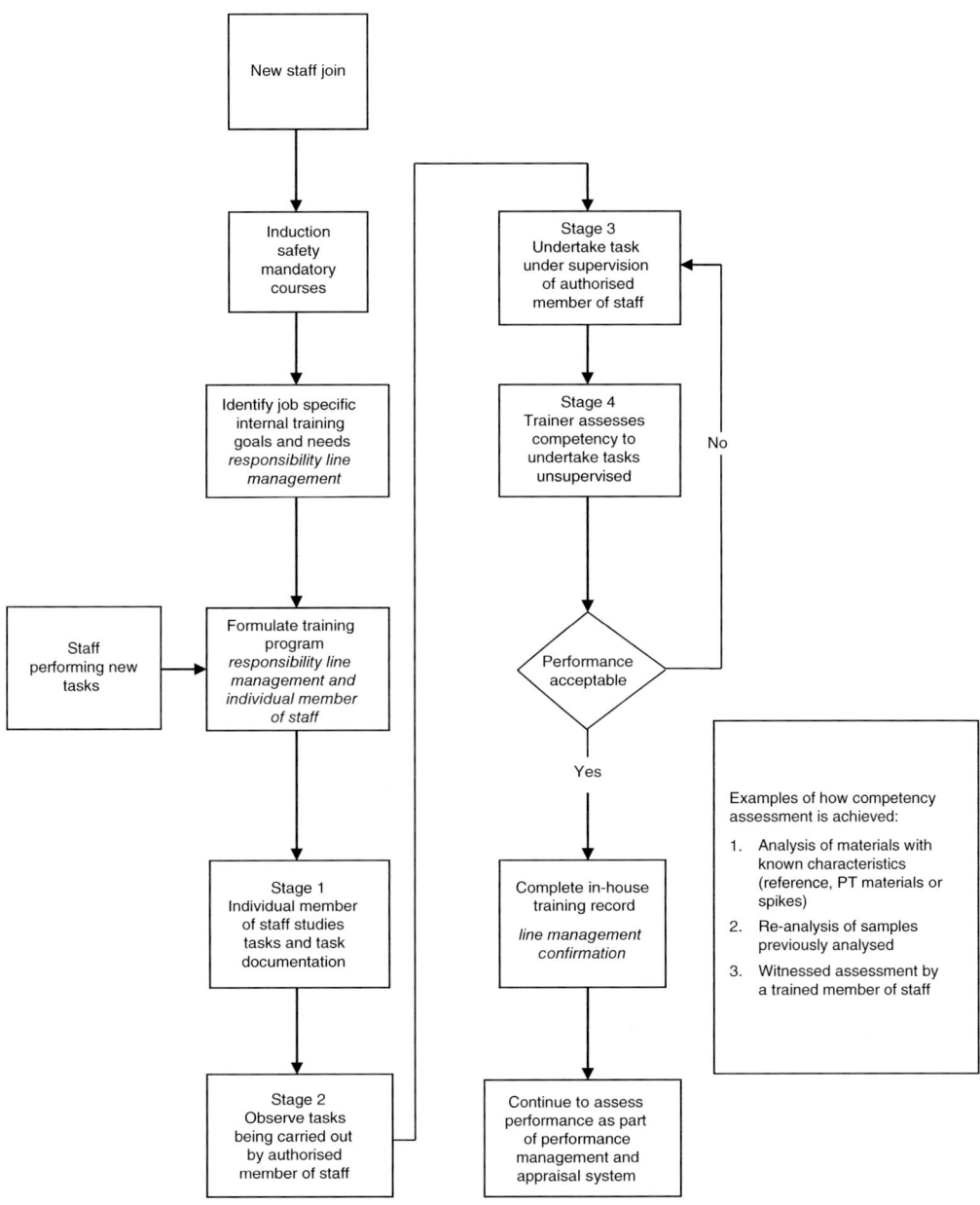

New staff join

Induction safety mandatory courses

Stage 3
Undertake task under supervision of authorised member of staff

Identify job specific internal training goals and needs
responsibility line management

Stage 4
Trainer assesses competency to undertake tasks unsupervised

No

Staff performing new tasks

Formulate training program
responsibility line management and individual member of staff

Performance acceptable

Yes

Examples of how competency assessment is achieved:

1. Analysis of materials with known characteristics (reference, PT materials or spikes)

2. Re-analysis of samples previously analysed

3. Witnessed assessment by a trained member of staff

Stage 1
Individual member of staff studies tasks and task documentation

Complete in-house training record
line management confirmation

Stage 2
Observe tasks being carried out by authorised member of staff

Continue to assess performance as part of performance management and appraisal system

References

EPPO (2010) PM7/98 (1): Specific requirements for laboratories preparing accreditation for a plant pest diagnostic activity. *Bulletin OEPP/EPPO Bulletin* 40, 5–22. Corrigendum (2011) *Bulletin OEPP/EPPO Bulletin* 41, 249–251, 40, 5–22.

ISO (2005) *ISO/IEC 17025:2017: General Requirements for the Competence of Testing and Calibration Laboratories*. International Organization for Standardization, Geneva, Switzerland.

ISO (2008) *ISO 9001:2015: Quality Management Systems – Requirements*. International Organization for Standardization, Geneva, Switzerland.

ISO (2010) *ISO/IEC 17043:2010: Conformity Assessment – General Requirements for Proficiency Testing*. International Organization for Standardization, Geneva, Switzerland.

14 Plant Health Engagement

Lucy Carson-Taylor* and Paul A. Beales
The Animal and Plant Health Agency, Sand Hutton, York, UK

14.1 Introduction

Plant health engagement is a form of public relations if we consider that the strategic role of public relations is to define and manage stakeholder relationships (Oliver, 2001). Effective communication and engagement does not happen by accident. It requires a clear strategic overview, defining purpose and objectives, thinking about individual and business needs, and deciding audiences, as well as deciding on the process and evaluation of impacts of any campaign. It is worth considering the interrelated aspects of the disease tetrahedron (Fig. 14.1) to produce the most effective engagement activities. In a world of highly adaptable pathogens, plant pathologists need to play an integral part in engagement with agricultural practitioners to balance the need of agricultural production with the need for plant protection.

In this chapter, we focus on plant health engagement purpose, people, process and evaluation.

We can learn a lot from historical outbreaks of fungal plant diseases, where, if communication, information and education had been more widely available at the time, introduction or spread of diseases could have been prevented.

For example, the devastating effect of potato late blight (*Phytophthora infestans*) was widely known, since the devastating famine caused by the disease in the mid-1850s. However, during a period of drought in Europe in the mid-1970s, there were shortages of potatoes and large quantities were shipped from Mexico. Exotic strains of the pathogen were introduced and began to spread, including the opposite mating type of *P. infestans* which produced further more aggressive/fungicide-resistant exotic strains. If more awareness had been raised about the impact of exotic strains of pathogens, the importance of quarantine procedures and what to look out for, this could have been prevented.

There is, however, a long history of plant health public engagement. Indeed, the worldwide Colorado potato beetle campaign is one of the best known and could still be considered one of the most successful campaigns. This campaign had a simple, well-illustrated poster, with messaging about the concern, what to look for and what to do if you find the pest (Fig. 14.2).

It helps that the Colorado beetle is a brightly coloured beetle that looks different from common native insects. The campaign was supported by clear messaging and a 'hook' or incentive that

* E-mail: lucy.carson-taylor@apha.gov.uk

DOI: 10.1079/9781800620575.0014

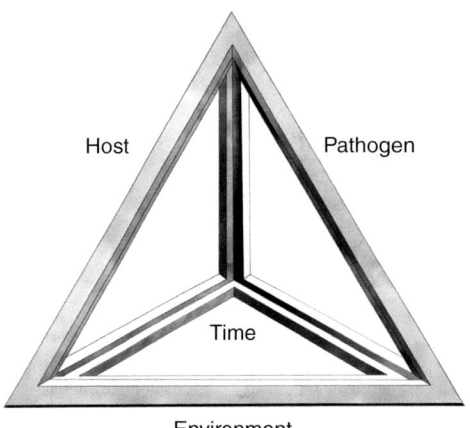

Fig. 14.1. Disease tetrahedron. (© UK Crown copyright.)

14.2   Purpose

The purpose of plant health engagement activity can be quite wide and varied, but commonly falls into categories such as sharing, creating or applying knowledge. Sharing knowledge may be about informing people of your work but may also inspire people to get involved by stimulating their curiosity in the subject. Working collaboratively with stakeholders, both from industry and members of the public, can be helpful in shaping the direction of future work and encouraging participation and engagement with the outcomes of the project. This can play an important part in supplying feedback on projects as they develop and help in applying the knowledge to encourage long-term changes in attitudes and behaviours.

Setting objectives and aims

Before you embark on any engagement activity, it is important to understand what you want to achieve with your audience. Do you want to inform and educate them, to raise their awareness on a basic level; is there a need to respond in some way or to change their behaviour? Therefore, it is essential the campaign is clear about what you want it to achieve and by setting objectives and aims you can help development of the process. Consider what you know about your topic – how well do you understand this, do you need expert guidance to check your thinking and perceptions? What has gone before; did it work, can it be built on? It is important that you carry out research at the planning phase so that you know your subject and its history, and about any controversies and limitations that may impede its progress. Setting objectives that clearly define the measurable outcomes and the aims of what is hoped to be achieved through the activity at the onset of the project will help develop the evaluation criteria and determine whether in-project (formative) feedback is required and will help shape the future of the project. Any objectives and aims should be SMART (Specific, Measurable, Achievable, Relevant and Time-Limited) to ensure successful delivery.

14.3   People

Understanding the people you want to engage with is essential for any successful awareness

the beetle is a threat to food security that has encouraged public engagement and participation in capturing the insect and reporting it. The poster was widely distributed across the UK and, to this day, individuals still remember the campaign through posters that were displayed in, for example, police stations.

Communicating to a broad or non-specialist audience about fungal plant pathogens can be more challenging as often the signs of the pathogen are not obvious (because they are microscopic), although the symptoms of the disease may be dramatic and eye-catching. However, the rise of citizen science projects such as Observatree (https://www.observatree.org.uk/; accessed 22 February 2023) have led to the development of a broad range of resources, including factsheets, images, videos and identification training materials specific to public awareness and engagement. International organizations such as CABI (https://www.cabi.org; accessed 22 February 2023), the European Plant Protection Organization (EPPO) (https://www.eppo.int; accessed 22 February 2023) and the International Plant Protection Convention (IPPC) (https://www.ippc.int/en/; accessed 22 February 2023), as well as professional societies such as the British Society for Plant Pathology (https://www.bspp.org.uk/; accessed 22 February 2023) and American Phytopathological Society (https://www.apsnet.org/; accessed 22 February 2023), have active public engagement and stakeholder-based resources.

Fig. 14.2. Colorado beetle campaign poster. (© Crown copyright.)

campaign. This will help shape the level and the type of communication techniques to be used: at the simplest level, understanding the diversity of languages required would be key. Although it is critical to think about the target audience, it is also important to recognize that there are also partners, funders, government policy makers and delivery teams involved in any campaign. Appreciating the motivations for engagement with the campaign is important. At times, there may be the potential for conflict between different parties, so careful negotiation may be required. Understanding the challenges and rewards of all these groups will ensure a successful campaign.

Audience segmentation and insight

Audience segmentation and mapping allows you to understand and prioritize your campaign targets. A beautifully produced campaign that is poorly targeted will not deliver the outcomes you desire. In a complex campaign targeting multiple audiences, mapping can take a considerable amount of time. Figure 14.3 shows an example of this. This segmentation was developed to look at audiences relevant to a tree health biosecurity campaign; the desired behaviour is based on the evidence that *Phytophthora ramorum* can be spread by people. There are multiple audiences who have been grouped together according to expected behaviour and relationships with woodland. This was based on prior insight and knowledge of these groups. Further insight was commissioned as part of the baseline study to understand more about audiences' attitudes, habits and preferences towards woodland use and biosecurity. This insight is helpful to make sure that the messaging is as relevant, meaningful and effective as possible.

The information that has been captured during the audience segmentation exercise can begin to build audience insight. This insight is valuable as it allows you to build an evidence base of which techniques are known to work best with different audiences and gives the ability to share knowledge about audiences with people or institutions that might be key collaborators. The value of insight goes beyond a single project or campaign; it

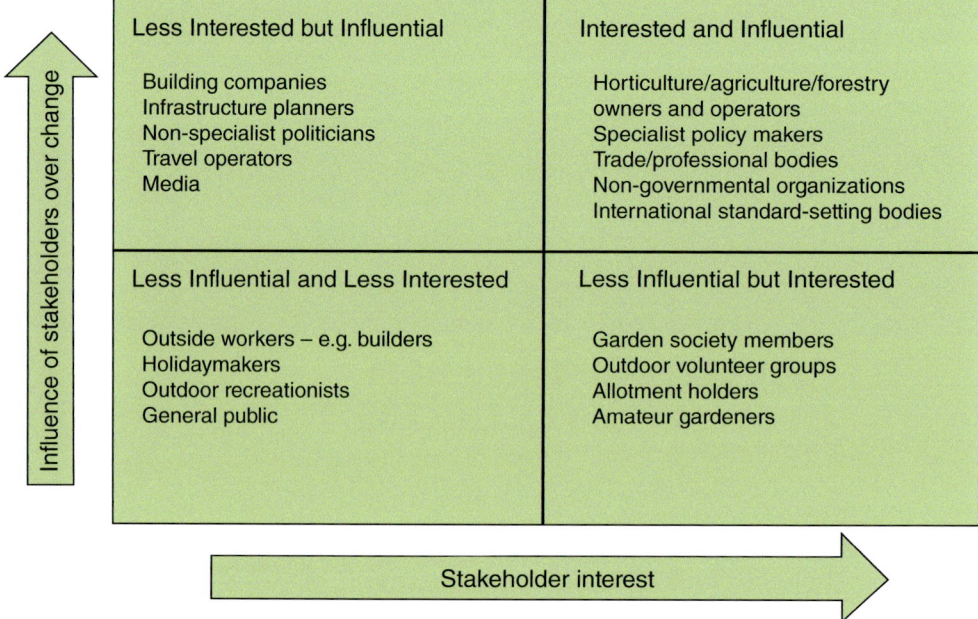

Fig. 14.3. Audience segmentation identifies primary target audiences for engagement campaigns. (© UK Crown copyright.)

allows for the replication of communication approaches that are the most effective for the target audiences across multiple campaigns. It can also be used to understand which other organizations/researchers have active campaigns that target your desired audience, which may allow more effective scheduling of communications with audiences throughout the year. Within your own organization, this could save money by commissioning fewer individual studies but instead a single larger cross-cutting insight study to understand audience habits.

Partnerships

Think about forming partnerships to increase the number of people you engage with ('reach') if you had an objective such as: 'due to recent interceptions, there is a need to raise awareness of the gardening public to the risk of *Phytophthora ramorum* 10% to 30% in the next six months'. In this case, it would be good if you were to partner with a well-known external organization with high membership numbers that had an interest in biosecurity. The organization would be well placed to maximize your chances of meeting your objective by raising awareness in their membership through written articles and information at their shows and events.

Making friends and influencing people

Where there is an opportunity to collaborate it is important to do so, if possible, because a unified message is more powerful than a lone voice and may well be better received from a trusted source. However, building partnerships takes work and trust is a critical element of this. If you are in Government or academia, then working with a membership organization can make your messaging more meaningful to your target audience. It is not just Government or a researcher telling them what to do or giving them information, it is an organization they already know, support and are involved with. The Action Oak Partnership (https://www.actionoak.org/; accessed 22 February 2023) (Fig. 14.4) is an example of multiple sectors working together for a common goal where the partners have different and shared audiences but one voice and joined up messaging.

Losing friends and alienating people

The quickest way to lose friends and alienate people is to ignore what has gone before and what is currently going on. At best the campaign will not be a success with the target audience, at worst you will undermine the successful work of others through message confusion. It is important to accept that there is expertise beyond your own or your organization and to make the most of it. Do some background work to understand who the key contacts are in other organizations because they could be the lever you need to give your campaign a major boost in a sector. As part of your planning, take a moment to understand your stakeholder landscape, who are they, how they interact or compete to raise their profile in a sphere that is competing for members, money, visibility and influence.

14.4 Process

With clarity on the purpose and people, you are well placed to think about the process and the distinct types of intervention that you can make. At the simplest level, this is thinking about whether this is a one-off intervention or a series of events over a set timescale. Broadly speaking, these can be categorized into: written content (e.g. reports, factsheets, leaflets, webpages, posters, postcards, 'tweets', blogs), presentations (conferences, public meetings, vlogs) and events (exhibitions, tradeshows, school and college 'show and tell', public science fairs, etc.) which may be delivered face-to-face or remotely. In the past few years there has been a significant shift to online delivery and access of information through the internet, requiring new skills in the digital age. This has also permitted greater participation and opened new international audiences, leading to a democratization of access to information and engagement because cost is less of a barrier to participation. The move away from printed journals and physical attendance of workshops and meetings has allowed many more people to participate and access information (Fig. 14.5).

Fig. 14.4. Action Oak Partnership event stands at large- and small-scale public shows in the UK. (© UK Crown copyright.)

Fig. 14.5. Improving breadth of communication via live stream of conferences, enabling those unable to attend to participate through video streaming and interactive question-and-answer sessions. (© UK Crown copyright.)

Communication strategy

You have your objectives and understand your audience, so now you need your communications strategy, development of key messages and production of campaign plan (see Protocol 14.1). It is important to identify what information you want the audience to act on. A good example is the 'Don't Risk It!' campaign (Fig. 14.6) (https://www.eppo.int/RESOURCES/eppo_publications/don_t_risk_it; accessed 22 February 2023) (see Section 14.5).

The poster is clear and simple and provides both information and justification for the desired action:

Pests and diseases can hide on plants
Please do not bring home plant, seeds, fruit, vegetables or flowers

Keep your audience at the forefront and challenge your own assumptions. For example, a digital-only strategy will not always be appropriate – think about how your audience accesses and uses information. It may be appropriate to provide both digital and conventional media. Take time to plan: determine the resources you have, what you will need and what can be achieved within your budget.

Your communication strategy will require content or 'messages' that can get complicated very quickly and easily ridden with jargon and acronyms. When thinking about this, use this simplified approach: (1) who, (2) says what, (3) in what way, (4) where, (5) when and (6) to what effect. This highlights the importance of thinking about what the message will look and sound like to the receiver. Is it understandable? Does it make sense? Is there ambiguity that could lead to genuine or wilful misinterpretation?

Make it easy for your audience to understand you, write your message without jargon and use an appropriate vocabulary. If in doubt, keep it as simple as possible. When communicating with the public and stakeholders, it is important remember that scientific terminology will not be widely understood and may result in losing people's attention. Therefore, it is not only the content of your message but how you deliver it that is important – audience insight will aid getting this right and using feedback during the project (formative) will help you refine both content and messaging to make sure it is being understood and people are engaging with it.

For example, the information in Fig. 14.7 is scientifically accurate, but is not displayed in a way that is engaging to any audience. No clear purpose for the campaign is indicated, the text contains jargon and numerous acronyms, there is no clear indication where to look for the disease or how to report if found. None of the images show symptoms on the host plant, and although scientifically accurate morphological structures, they do not represent that which would be seen in the field. The text is difficult to read due to the image/text colour combination.

As part of your communication strategy, you will need to develop the technical content and decide how you will respond ('lines') to take in response to enquiries and questions. By understanding your audience, you will have an appreciation of how the information you are communicating may impact on them and the nature of the questions. It may be helpful to think about providing a simple questions and answers document to help deal with predicted and potential common concerns. The frequency and length of time of a campaign run should also always be considered. Pathogen campaigns that require ongoing engagement should have materials that are fresh and relevant, with considerations of impact on other ongoing or new activities.

As public engagement campaigns frequently are aimed about changing behaviour, it is worth considering mechanisms for encouraging positive change. Behavioural economics and insights or 'nudge theory' are recognized as important, effective and valuable concepts in the world of behavioural change. A useful starting place is the work of the Behavioural Insights Team (https://www.bi.team/; accessed 22 February 2023).

Outputs – products and events

Factsheets, leaflets, posters and postcards

Uptake of information can depend on how it is presented – do you need to give them an essay when a postcard with a compelling image and basic information on the back is more likely to be understood, acted on and kept (Fig. 14.8)? Think about giving people the information that they want or need to know rather than everything

Fig. 14.6. 'Don't Risk It!' campaign poster. (© EPPO Copyright. Used with permission.)

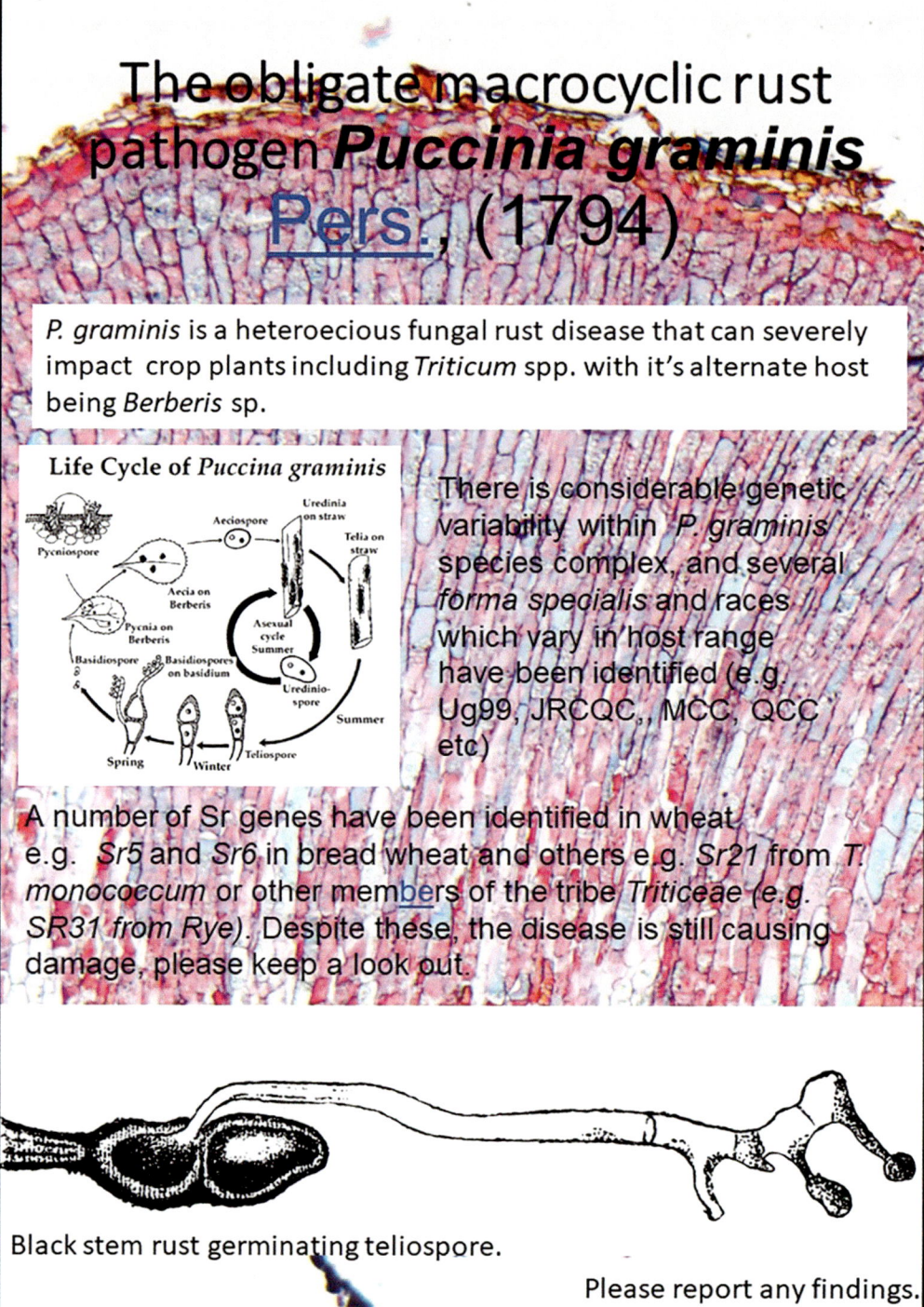

The obligate macrocyclic rust pathogen *Puccinia graminis* Pers., (1794)

P. graminis is a heteroecious fungal rust disease that can severely impact crop plants including *Triticum* spp. with it's alternate host being *Berberis* sp.

Life Cycle of *Puccina graminis*

There is considerable genetic variability within *P. graminis* species complex, and several *forma specialis* and races which vary in host range have been identified (e.g. Ug99, JRCQC, MCC, QCC etc)

A number of Sr genes have been identified in wheat e.g. *Sr5* and *Sr6* in bread wheat and others e.g. *Sr21* from *T. monococcum* or other members of the tribe *Triticeae* (e.g. SR31 from Rye). Despite these, the disease is still causing damage, please keep a look out.

Black stem rust germinating teliospore.

Please report any findings.

Fig. 14.7. An example of a less effective engagement leaflet. (© UK Crown copyright.)

A range of *Phytophthora* spp. have recently been discovered in the UK.

Animal &
Plant Health
Agency

P. *ramorum*: Affects a wide range of shrubby and tree hosts, but is most significant on Larch.
P. *kernoviae*: A range of tree and shrubby hosts are susceptible to this organism, most significant on rhododendron, beech and bilberry
P. *austrocedri*: Affects Juniper trees in the UK.
P. *lateralis*: First found in 2010, principally affects Lawson cypress trees.

Phytophthora spp. Are commonly referred to as water moulds as they require water for part of their life-cycle. Swimming spores can infect leaves, bark and in some species plant roots.

On leaves: symptoms often appear as black lesions with a yellow/water-soaked margin. Wilting or early abscission is also a common symptom.
On stems: black to reddish-brown lesions can appear anywhere on stems, causing stem die-back
On bark: Black sticky "bleeds" appearing anywhere on bark, removal of the outer bark reveals a reddish-brown discolouration.
On roots: Rotting of primarily fine roots, but larger roots can also be killed

Many phytophthoras produce resting spores that survive unfavourable conditions.

Any findings of these or other suspected non-indigenous plant pests should be reported to the Plant Health & Seeds Inspectorate: planthealth.info@apha.gov.uk Telephone: 0300 1000 313

Further information on the recognition and biology of this and many other pests/diseases is available http://fera.co.uk/plantClinic/plantPestDiseaseFactsheets.cfm

Fig. 14.8. Example of an information postcard (front and back). (© UK Crown copyright.)

you want to tell them – you can always provide access to further information online. When producing resources, it is always good to ask for feedback from the target audience before you finalize the material to be published (see Protocol 14.2).

Images

Images will convey more about the pest itself and its symptoms than paragraphs of words. For fungi, where these can be variable (e.g. for fungal pathogens such as the EPPO A1 listed Japanese pear rust (*Gymnosporangium asiaticum*) – the aecial stage on pear is significantly different from the telial stage observed on juniper), clear photographs or drawings showing the range of symptoms on each host, including early-stage symptoms, can be helpful, particularly where you are asking for people to report suspect findings. The Colorado beetle (discussed in Section 14.1) insect was distinct from other indigenous pests, enabling production of clear and effective campaign engagement materials. However, fungal plant pathogens identification will often be more challenging because signs of the pathogen are not always visible. Reliance on host symptoms is therefore required, showing a range of symptoms on common hosts (Fig. 14.9a, b). Pictures of healthy plants, and of 'lookalikes', are useful for comparison, differences that seem obvious to

you may be easily overlooked by someone who is unfamiliar with the pathogen. Drawings of life cycles are a safe way to explain the biology and improve understanding; try to make these as uncluttered as possible to avoid people being put off by 'visual noise' from too much information that may not be relevant.

When developing resources to support your campaign, it is not always necessary to commission new photographs. For example, web pages and databases dedicated to plant pests from reliable sources such as EPPO (https://www.eppo.int/; accessed 22 February 2023) and Bugwood (https://images.bugwood.org/; accessed 22 February 2023) should be helpful to aid your campaign or communication.

Videos

Videos can be a useful resource to raise awareness of a pathogen or a desired behaviour of your target audience. Using an appropriate title and tags will ensure that videos can be found through internet searches. If videos are to be displayed at events, they need to be short (30 s to 2 min) and in a repeated loop. In such a case, consideration should be given to whether audio is needed or not. For example, at a busy show it either will not be heard or will need to be played at a volume that will be inconsiderate to your fellow exhibitors.

Fig. 14.9. (a) Pear rust showing obvious signs of the pathogen emerging from host plant. (b) Indistinct leaf chlorosis and necrosis, which is less effective for a fungal disease awareness campaign because the symptoms are common and could be the result of various pathogens, pests or even abiotic. (From Shutterstock.)

Videos for social media need careful consideration because it is challenging to hold people's attention; the videos should rarely be longer than a minute, be of suitable format and size, and should start powerfully to grab the attention of those passing by. When producing videos, plan carefully using storyboards. Consider location, lighting and sound quality when filming as well as post-editing production time, which can be at least three times the length of the filming process.

Shows and events

For gathering intelligence, have deeper conversations with, and immersion in the attitudes of, your target audience; there is huge value to face-to-face conversations and direct engagement. Although conversations may appear relaxed, you should have your campaign's objective in mind.

Creating an exhibit that is visually appealing is important when engaging people and attracting attention (Figs 14.10–14.12), so consider adding further sensory engagement. For example, using a scent such as mushroom can bring forward thoughts about decay, or woodland sounds add further sensory elements, making your exhibit more engaging. Think about what will make you stand out: striking images of pathogens can be extraordinarily beautiful and

Fig. 14.10. Oak powdery mildew model. (© UK Crown copyright.)

Fig. 14.11. Badges are a good engagement tool that can be handed out at events. (© UK Crown copyright.)

great imagery can unlock your subject or re-search for visitors. Do not be frightened of trying something different. For example, the 'Don't Move Firewood' team in the USA (https://www.dontmovefirewood.org/; accessed 22 February 2023) have dressed up as giant beetles at state fairs: it has been fun and attention-grabbing! Similarly, the Maltese Government used the car-toon character 'Xylellu' to explain Xylella risks to travellers.

When you are exhibiting at a show, you are only as strong as the people telling the story. Your team need to be well briefed. Use a standard document that can be adapted for each event, covering issues such as: event information, health and safety, appropriate dress (e.g. branded clothing), behaviour (e.g. don't stand in a group chatting and ignoring visitors), campaign mes-sages, how to follow up enquiries and any evalu-ation you want to be done on the day (see Protocol 14.3). It is also helpful to have some background information notes for those working the stand to cover wider information about the plant pest and diseases, and lines to take if asked specific ques-tions. Where possible, have a lead person to take ultimate responsibility for delivering your goals; if you have someone experienced then they can mentor and support less experienced but keen members of staff to develop their engagement skills. It can be helpful to have a pre-briefing of all the team working on the stand prior to the event to explain all the elements of the stand and pro-vide an opportunity for questions and answers. After the event, remember to thank all those who participated in the activity.

14.5 Evaluation

It is essential that evaluating the impact of any campaign is part of the initial planning pro-cess and not something done as an extra at the end of the project once the information has been gathered. Evaluation may occur during

Fig. 14.12. Giant wicker model of an Asian long horned beetle. (Courtesy of Fera.)

the project (formative) to help develop how the project evolves (Fig. 14.13) or may occur at the end of the project (summative) to measure success against initial outcomes. Data collection may be qualitative or qualitative and requires both analysis and reporting in a style(s) suitable for the audience. This may require both a detailed technical report and a stakeholder-facing report.

Measuring success

It can be challenging to measure and demonstrate changes in knowledge, awareness or behaviour, and impossible without knowing what is happening before you begin your campaign. Setting a baseline will help you to measure the success of an activity (Protocol 14.4). A baseline is your measure of the current state of play and

Fig. 14.13. Promise jars at an engagement event can give a rough evaluation on actions people may be willing to take and can inform future campaigns. (© UK Crown copyright.)

from there you can measure any changes in audience behaviour following interventions. Setting a baseline can also give justification for your campaign by proving a knowledge gap or need to encourage a behaviour that may be absent. A simple example would be if you do not know how many people are signed up to your newsletter, how can you prove that a campaign has led to an increased sign-up? Or if 99% of your target audience are already signed up to your newsletter, is there any point spending time and money on a campaign to increase sign-ups?

For more complex cases involving knowledge gain, awareness-raising or behavioural change, there are two potential ways to gain a baseline. If you have a budget, then commissioning qualitative and quantitative research with questions targeted at the specifics you want to know is ideal. Using your own commissioned research allows you to design a baseline that takes into account the desired metrics for evaluation further down

the line. This is particularly useful for complex campaigns around behavioural change.

If you have a limited budget, then it is possible to make use of existing data from other related sources (e.g. publicly available data, desk research using articles from peer reviewed journals and trade data) but this limited and may not achieve the degree of specificity you require. A baseline could be designed to measure current reach, revenue or reputation of your business or institution. It could be something straightforward, such as current numbers and sectors registered on a scheme, or as complex as the public's attitudes and behaviours around biosecurity when using woodlands, for example.

Using data gathered from social media activity, website activity, information downloads and registrations can give you the justification that your campaign activity has created the uplift in audience engagement required. Looking at the analytics behind the online activity will tell you where the interactions are coming from. This can be important. For example, authors of a webpage were delighted to find a sudden uplift in webpage interactions for a campaign they were running and assumed it must be due to their latest Twitter activity. Following analysis of the data, it was found to be due to a webpage link sent out in an email by a garden centre to their customer database.

Consider inputs, outputs and outcomes. This could be looked at in terms of input (e.g. your time into developing guidance), output (e.g. leaflet, exhibit or film) and outcome (what has changed because of your output). It is quite easy to become focused on outputs: they are simple to measure and evaluate (e.g. 1000 views of a film on YouTube, 100 retweets of biosecurity advice, 200 downloads of guidance). These figures are useful for demonstrating reach, campaign visibility and awareness and sometimes are enough. If the campaign is based on behavioural change, then demonstrating this requires the third step of achieving and measuring an outcome.

For example, if you are looking for improvement in a behaviour, measuring this requires numerical data: quantify how many more people have taken the positive action because of your activity. It is tempting to have a vague objective that might state 'increase sign-up to the new tree notification scheme', since even one new sign-up would demonstrate that the objective has been met and you have been successful. But if you are questioned about numbers, this would not stand up to outside scrutiny. If you have a target of an increase in sign-up to the tree notification scheme by 10% or by 250 people within six months, then you have something defined to measure your campaign against. If it is the first time a campaign has run and there are no data available to form a baseline, then take the opportunity to use the campaign as a benchmark and gather data that will be useful to inform future campaigns.

It is crucial that audience needs are understood. A researcher may want to communicate every detail about the fascinating pathogen they are working on to a non-specialist audience, but this is a quick way to lose the attention of that audience. Taking an outcome-focused approach with a good understanding of an audience will help to focus the communications to a greater effect.

Example campaign

The 'Don't Risk It!' campaign mentioned earlier was developed centrally by EPPO, the European and Mediterranean Plant Protection Organization. This is an example of a campaign with one clear message to one audience that has been popular with member countries of EPPO, helped by the translations that have been provided on materials that can easily be downloaded for use. The team at EPPO have been flexible in allowing different countries to put their own interpretation on the artwork to suit their cultural differences. The objective of this campaign is to raise public awareness about the risks of moving plants and their associated pests during international travel and to encourage responsible behaviour. It is primarily intended that the materials should be displayed in airports or any other sites where international travellers will see it (e.g. seaports, railway stations, travel agencies, embassies).

An airport or event is a busy place, and it is important that the messaging does not get lost among the other visual stimuli. Using colourful artwork with a commercial look in the UK helped it stand out (Fig. 14.14). The single simple message asks for a default action; the audience do not have to go to any extra effort to comply with the campaign. The campaign is supported in the UK by multiple partners, including one who shared the original baseline data to demonstrate that travellers who were keen gardeners were bringing plants into the UK from overseas trips. The airport location of the posters was originally in arrivals, as that is where another campaign partner could place them at no cost. The campaign team gathered evidence and were concerned that by then it was already too late – the plants were in the country. Using continuous feedback and evaluation, the focus of the campaign was switched to target people before they travelled, with signage at departures, at horticultural shows and events (Fig. 14.15) and on television gardening programmes, all supported by a social media campaign promoted by celebrity gardeners and many others. Other resources, such as leaflets and luggage tags, were also made available to support this campaign (Fig. 14.16).

Fig. 14.14. 'Don't Risk It!' campaign raising awareness at RHS Chelsea flower show event 2022. (© UK Crown copyright.)

Fig. 14.14. Continued.

Fig. 14.15. An example of the UK artwork developed for airports. (© UK Crown copyright.)

Fig. 14.16. 'Don't Risk It!' Luggage tags and downloadable leaflets. (Available at: https://www.eppo.int/RESOURCES/eppo_publications/don_t_risk_it (accessed 22 February 2023) © EPPO copyright. Used with permission.)

Protocol 14.1

Annual event planner

Fill in the table as fully as possible
Year:
Objectives:
Aims:

Activity/Event	Jan	Feb	Mar	Apr	May	Jun	Jul	Aug	Sept	Oct	Nov	Dec
Trade												
Public												
Government												
Academia												
Conferences												
Workshops												
Talks												
School/children												
Other												

Protocol 14.2

Feedback form

Event		Venue	
Date		Completed by	

Number of visitors to the stand per day		

Type of visitors (please tick all that apply and indicate % of visitors if possible):			
Type	Y/N	% of visitors or actual number	
Retailers/Nursery owners or staff			
Professional gardeners/landscapers			
Importers or their agents			
General public			
Youth group leaders			
Teachers/lecturers			
Members of gardening associations/clubs			
Other (please list):			

Summarize the main topics of discussion

Any other points to note?

Protocol 14.3

Exhibitors' briefing checklist

	Completed
Address of event, including map	
Travel arrangements (e.g. public transport, bus routes, nearest train station)	
Local accommodation	
Event passes/tickets	
Contact details for event leads	
Stand location within event	
Event dates, set-up and show times	
Staff rota	
Roles and responsibilities: Before event, during and after event (each day)	
Exhibit images – design plan, showing key elements such as storage, electrics, etc.	
Exhibit description – featured pests and diseases, plants	
Key objectives, aims and messages	
Frequently asked questions and answers sheets	
Photographs/videos	
Approaches to media questions	
Evaluation	
Specific organizers' requirements	
Dress code	
Exhibit set-up and breakdown	
Refreshments and on-site facilities	
Health and safety requirements, risk assessments	
Emergency procedures	

Protocol 14.4

Baseline evaluation survey

This survey aims to give an understanding of participants' basic knowledge of selected quarantine plant pests and diseases.
Fill in as much of the form as possible:
General information:

Name	
Age group	
Gender	
Education level	
Work area	

What knowledge do you have of the following quarantine plant pests/diseases?

Plant pest/disease	No knowledge	Limited knowledge	Significant knowledge

Reference

Oliver, S. (2001) *Public Relations Strategy*. Kogan Page, London.

Appendices

Appendix 1

Commonly Used Growth Media

A diverse range of solid and liquid media has been developed to study fungi. These may be simple basal media to encourage a wide range of fungi or highly specialized media to select for specific groups or even genera. The addition of antibiotics may be used to reduce bacterial growth or fungicides to restrict growth of undesirable fungi. A comprehensive review of media is provided by Shurtleff and Averre (1997). However, in practice, the following media described below will be sufficient on most occasions for routine work. Once poured, media should be dated and stored in appropriate conditions (usually at 4°C in the dark) for a defined period (typically up to a week if antibiotics are present or even several months) to ensure active ingredients do not degrade.

General principles for production of agar are as follows:

- Ensure all ingredients are well mixed and dissolved.
- Autoclave in suitable vessels – usually at 121°C for 15 min. Overheating during sterilization will cause darkening, precipitation, poor gel strength and pH drift.
- Allow the agar to cool until it reaches a temperature of 45–50°C (agar sets at 38°C and melts at 84°C). Do not keep molten media at 50°C or more for more than 3 h as overheating effects will occur.
- Sterile antibiotics (e.g. chloramphenicol, penicillin, streptomycin) or fungicides may be added to cooled agar. Thermally unstable additives may be sterilized by microfiltration.
- Once autoclaved, media may be stored for up to 6 months at low ambient temperatures in the dark in screw cap bottles or dispensed for immediate use.
- Dispense the agar aseptically avoiding air bubbles – typically 10–20 ml per 9 cm Petri dish.
- Once the agar has solidified it may be labelled, the batch dated and stored in the dark preferably at 4°C.

Several manufacturers (e.g. Oxoid) produce proprietary dehydrated agars.

Media Recipes

Distilled or tap water agar (TWA) – a low nutrient agar to encourage sporulation but sparse mycelial growth

Dissolve 15 g of agar in 1 l of either tap water or distilled water, as appropriate.

DOI: 10.1079/9781800620575.appx

Petri's mineral solution – to encourage sporulation of *Pythium* and *Phytophthora*

$CaNO_3$	0.4 g
$MgSO_4.7H_2O$	0.15 g
KH_2PO_4	0.15 g
KCl	0.06 g
Distilled water	1 l

Potato dextrose agar (PDA) – a general purpose medium, good for both mycelial growth and sporulation, typically varies from 1% to 4%

Scrub clean 200–500 g of potato tubers (do not peel them) and cut into small pieces (*c.*15 mm cubes), boil until cooked in 1 l of water. Strain through a cheesecloth or fine mesh and adjust volume of the filtrate to 1 l. Add 10–20 g of dextrose and 12–17 g of agar, mix thoroughly and autoclave. Quantities may be varied depending on requirements.

Malt agar (MA) – a rich medium that encourages mycelial growth

Heat 20 g of malt extract in 1 l of water until dissolved. Add the agar (15–20 g) and boil until the agar is well dissolved.

Carrot piece or carrot agar (CPA or CA) – frequently used to encourage sporulation of *Phytophthora*

Carrot piece agar (Werres *et al.*, 2001): agar 22 g, carrot pieces 50 g, distilled water 1000 ml.
Carrot juice agar: agar 5–22 g, carrot juice 50 ml, distilled water 950 ml.

SNA – a specialist media for *Fusarium*

Speziellar Nährstoffarmer agar (SNA): KH_2PO_4 1 g, KNO_3 1 g, MgSO4.7H_2O 0.5 g, KCl 0.5 g, glucose 0.2 g, sucrose 0.2 g, agar 15 g, distilled water 1000 ml.

Cherry decoction agar (CHA) – a general purpose medium for primary isolations

Agar 60 g, distilled water 3600 ml, cherry juice 400 ml. Filter the cherry juice and adjust the pH to 4.4 with KOH. Dissolve the agar thoroughly first, then add cherry juice.

Oatmeal agar (OMA) – a rich medium that may help with fungal fruiting body production such as *Phoma*

Blend 60 g of oatmeal (rolled oats) in 600 mg water and heat to 45–55°C. Add 12 g of agar dissolved in 400 ml of water.

V8 agar – frequently used for *Phytophthora*

Vegetable juice agar (V8): vegetable juice 250 ml, $CaCO_3$ 5 g, agar 15 g, distilled water 1000 ml. Add $CaCO_3$ to the vegetable juice and stir firmly during 15 min. Centrifuge the mixture for 20 min at 5000 rpm and pour off the supernatant. Make up the resultant to 1 l with distilled water and autoclave.

PARP + PARP-H – semi-selective media for *Pythium* and *Phytophthora* respectively

PARP-H (Jeffers and Martin, 1986): cornmeal agar 17 g, distilled water 1000 ml. Autoclave, then cool to 50°C in a water bath. Then prepare pimaricin 5 mg, ampicillin (Na salt) 250 mg, rifampicin (dissolved in 1 ml 95% ethanol) 10 mg, PCNB 100 mg, hymexazol 22.5 mg and dissolve all in 10 ml sterile distilled water. Add to cooled media, pour, store at 4°C in the dark and use within 5 days.
PARP: if hymexazol is unavailable, then PARP (without it) is still very useful.

References

Jeffers, S.N. and Martin, S.B. (1986) Comparison of two media selective for *Phytophthora* and *Pythium* species. *Plant Disease* 70, 1038–1043.

Shurtleff, M.C. and Averre, C.W. (1997) *The Plant Disease Clinic and Field Diagnosis of Abiotic Diseases*. The American Phytopathological Society, St Paul, Minnesota.

Werres, S., Marwitz, R., Man in't Veld, W.A., De Cock, A.W.A.M., Bonants, P.J.M. *et al.* (2001) *Phytophthora ramorum* sp. nov., a new pathogen on *Rhododendron* and *Viburnum*. *Mycological Research* 105, 1155–1165.

Appendix 2

Commonly Used Mounting Agents, Stains and Cements for Sealing Slides

A wide range of mounting agents and stains have been developed by mycologists over the years. However, in practice, only a few are used routinely.

Mounting Agents

Water

A very safe mounting agent, but it will dry out quickly.

Lactic acid

A general mounting agent which may be used alone, or with the addition of cotton blue or trypan blue.

Lactoglycerol

A general mounting agent which can be used in preference to water as it does not dry out over time or during warming. Lactoglycerol is a safer alternative to lactophenol.

Glycerol	100 ml
Lactic acid	50 ml
Distilled water	50 ml

Proportions may be varied depending on the viscosity of the mounting agent required.

Indian ink

Produces an opaque background useful when examining spore ornamentation, e.g. smuts, bunts and rusts.

Stains

Cotton blue or trypan blue in lactoglycerol

A general-purpose reagent that preferentially (although not exclusively) stains fungal structures within plant material.

Lactoglycerol	67 ml
Distilled water	33 ml
Cotton blue or trypan blue	0.1 g

This gives a 0.1% stain, but the intensity of the stain may be varied as required.

Melzer's reagent

An iodine-based stain useful for examining fungal structures, especially ascomycetes.

Chloral hydrate	100 g
Potassium iodide	5 g
Iodine	1.5 g
Distilled water	100 ml

Cement for Sealing Slides

For best results, apply to a dry and clean slide. Remove all excess mountant with a dry absorbent tissue (any excess mountant may soften the sealing agent and result in an imperfect seal). Apply at least two coats (preferably three), allowing each coat to dry thoroughly before reapplication.

Fingernail polish

A general-purpose sealant (available in a range of colours).

Appendix 3

Commonly Used
Sterilizing Agents

70% ethanol

Dilute industrial methylated spirits (IMS) with water to desired concentration. Neat IMS is less effective than the diluted product.

Bleach

Thin household domestic bleach (sodium hypochlorite) should be diluted with water to achieve the desired concentration of a 0.12% sodium hypochlorite solution. This is commonly achieved by a 10% dilution of household bleach with an active ingredient of 10%. However, some domestic bleach is only 5% a.i., so dilution will have to be altered to achieve about 0.12%. Thick bleach and perfumed products should not be used. Diluted bleach solutions should be dated and fresh supplies made on a regular basis (daily). Avoid diluted solution being kept in bright light to prevent degradation.

Bleach should not be used to disinfect plant material if trying to detect *Phytophthora* or *Pythium*.

Index

Note: Page numbers in **bold** type refer to figures
Page numbers in *italic* type refer to tables

CABI – who we are and what we do

This book is published by **CABI**, an international not-for-profit organisation that improves people's lives worldwide by providing information and applying scientific expertise to solve problems in agriculture and the environment.

CABI is also a global publisher producing key scientific publications, including world renowned databases, as well as compendia, books, ebooks and full text electronic resources. We publish content in a wide range of subject areas including: agriculture and crop science / animal and veterinary sciences / ecology and conservation / environmental science / horticulture and plant sciences / human health, food science and nutrition / international development / leisure and tourism.

The profits from CABI's publishing activities enable us to work with farming communities around the world, supporting them as they battle with poor soil, invasive species and pests and diseases, to improve their livelihoods and help provide food for an ever growing population.

CABI is an international intergovernmental organisation, and we gratefully acknowledge the core financial support from our member countries (and lead agencies) including:

    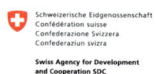

Discover more

To read more about CABI's work, please visit: **www.cabi.org**

Browse our books at: **www.cabi.org/bookshop**, or explore our online products at: **www.cabi.org/publishing-products**

Interested in writing for CABI? Find our author guidelines here: **www.cabi.org/publishing-products/information-for-authors/**